T0321105

WHAT EVERY ENGINEER SHOULD KNOW ABOUT

DEVELOPING REAL-TIME EMBEDDED PRODUCTS

WHAT EVERY ENGINEER SHOULD KNOW
A Series

Series Editor*

Phillip A. Laplante
Pennsylvania State University

*Founding Series Editor: **William H. Middendorf**

WHAT EVERY ENGINEER SHOULD KNOW ABOUT
DEVELOPING REAL-TIME EMBEDDED PRODUCTS

Kim R. Fowler

CRC Press
Taylor & Francis Group
Boca Raton   London   New York

CRC Press is an imprint of the
Taylor & Francis Group, an **informa** business

CRC Press
Taylor & Francis Group
6000 Broken Sound Parkway NW, Suite 300
Boca Raton, FL 33487-2742

First issued in hardback 2018

ISBN 13: 978-1-138-41344-3 (hbk)
ISBN 13: 978-0-8493-7959-8 (pbk)

Library of Congress Cataloging-in-Publication Data

Fowler, Kim.
   What every engineer should know about developing real-time embedded products / Kim R. Fowler.
       p. cm.
   Includes bibliographical references and index.
   ISBN-13: 978-0-8493-7959-8 (alk. paper)
   ISBN-10: 0-8493-7959-8 (alk. paper)
   1. Embedded computer systems--Design and construction. 2. Real-time data processing. 3. Electronic apparatus and appliances--Design and construction. I. Title.

TK7895.E42F69 2007
004'.33--dc22   2007013382

Visit the Taylor & Francis Web site at
http://www.taylorandfrancis.com

and the CRC Press Web site at
http://www.crcpress.com

Contents

What Every Engineer Should Know: Series Statement

What every engineer should know amounts to a bewildering array of knowledge. Regardless of the areas of expertise, engineering intersects with all the fields that constitute modern enterprises. The engineer discovers soon after graduation that the range of subjects covered in the engineering curriculum omits many of the most important problems encountered in daily practice—problems concerning new technology, business, law, and related technical fields.

With this series of concise, easy-to-understand volumes, every engineer now has within reach a compact set of primers on important subjects such as patents, contracts, software, business communication, management science, and risk analysis, as well as more specific topics such as embedded systems design. To understand the topics covered in these books requires only a lay knowledge, and no engineer can afford to remain uninformed of the fields involved.

Preface

Purpose

This book focuses on the processes and trade-offs used to develop real-time embedded products. It uses case studies and examples that allow you to compare and contrast design decisions made for different projects in different markets. My hope is that a consistent format will serve as a guideline when you develop new products.

Another goal, admittedly a distant one, is to encourage change and improvement in the business of developing embedded systems by helping you to see some of the relationships between various disciplines.

Scope

The book covers smaller, self-contained devices and subsystems, ranging from handheld devices to appliances to racks of equipment. While the processes described in the book are important steps, they are not necessarily all the steps needed to guarantee success. Furthermore, these processes do not sufficiently encompass large systems, such as transportation vehicles or systems of systems.

Audience

The book is primarily for design engineers (electronic hardware and software engineers and industrial designers), their managers, and people who should have an overall perspective on product development. I hope anyone wanting to better understand how other people and disciplines interact in engineering and product development will benefit. Some technical background, two years or more of technical school or university, is needed to understand the material.

How to read

This book is a curious hybrid between a textbook and a reference. Reading all the way through probably will be incredibly boring for the vast majority of you. Most likely, reading the first chapter or two and then several case studies should efficiently satisfy the knowledge uptake for your market or field of interest.

The first three chapters lay the basic ground work for good processes, followed by eleven case studies. The final three chapters contain some selected observations for specific products and markets. (I will not inflict you with a long summary here.) The case studies do follow a standard format so that you may find it easier to compare them. The 15 topics within each case study are as follows:

- Concept and market
- People and disciplines
- Architecting and architecture
- Phases
- Scheduling
- Documentation
- Requirements and standards
- Analyses
- Design trade-offs
- Tests
- Integration
- Manufacturing
- Support (training, logistics, maintenance, and repair)
- Disposal
- Liability

Author

Kim Fowler is a consultant and developer of new products; he lectures internationally on developing real-time embedded systems. Kim has over 25 years of experience in designing, developing, and managing medical, military, and of satellite equipment projects. He cofounded Stimsoft, a medical products company, which he sold in 2003.

He has written the textbook, *Electronic Instrument Design: Architecting for the Life Cycle*, published by Oxford University Press in 1996. He is the editor-in-chief for the *IEEE Instrumentation & Measurement Magazine* and a columnist. He has published widely in biomedical and engineering journals, has three patents, four filed, and eleven invention disclosures.

Acknowledgments

I could not have written this book without help from a number of folks, particularly the people who graciously allowed me to interview them. I acknowledge each person at the end of the case study for which he or she gave me an interview. Most acted as peer reviewers, as well, for which I am very grateful. I thank Rudy Marshall for reading and editing the entire book.

I also thank the folks at CRC Press, a Taylor & Francis Company, who made the production of the book possible, particularly Allison Shatkin, editor for computer engineering, design automation, and system-on-chip, Marsha Pronin, project coordinator, and James Miller for preparing the front cover. I particularly thank Jennifer Genetti for her delightful cooperation in composing and correcting the typeset copy.

Finally, to my friend, David Paul, who encouraged me to work on the book and do the most important things first—thank you, David.

Kim Fowler

List of Abbreviations

ADC	analog-to-digital converter
AFD	arc fault detector
ANSI	American National Standards Institute
APT	automated parametric test
ASIC	application-specific integrated circuit
ATE	automatic test equipment
ATP	acceptance test procedure
BDM	background debug mode
BIT	built-in-test
BITE	built-in-test equipment
CAD	computer-aided design
CAFE	corporate average fuel economy
CARB	California Air Research Board
CB	certified body
CCSDS	Consultative Committee for Space Data Systems
CDR	critical design review
CDRL	contractor data requirements list
CE	Conformite Europeene
CMD	command
CMMI	capability maturity model integration
CoDR	conceptual design review
COGS	cost of goods sold
COTS	commercial-off-the-shelf
CPIN	computer program identification number
CPU	central processing unit
CRR	controlled release review
CSA	Canadian Standards Administration
D-Level	depot level
DAC	digital-to-analog converter

DC	direct current
DFi	design for improvement
DFA	design for assembly
DFM	design for manufacture
DFT	design for test
DFt	design for transfer
DFx	design for x
DHF	design history file
DID	data item description
DMA	direct memory access
DoD	Department of Defense
DOS	disk operating system
DSL	digital subscriber line
DSP	digital signal processor
DUT	device under test
ECM	engine control module
ECP	engineering change proposal
ECU	engine control unit
EDU	electronic data unit
EGSE	electronic ground support equipment
EMC	electromagnetic compatibility
EMI	electromagnetic interference
EPA	Environmental Protection Agency
ESCM	energy storage control module
ESD	electrostatic discharge
ETA	event tree analysis
FAT	first article test or factory acceptance test
FCC	Federal Communications Commission
FDA	Food and Drug Administration
FET	field-effect transistor
FFT	Fast Fourier Transform
FMEA	failure modes effects analysis
FPGA	field-programmable gate array
FTA	fault-tree analysis

FW	firmware
GPS	global positioning system
GSE	ground support equipment
GUI	graphical users interface
HALT	highly accelerated life test
HASS	highly accelerated stress screen
HAST	highly accelerated stress test
I-Level	intermediate level
IC	integrated circuit
ICD	interface control document
ICT	in-circuit test
IEC	International Electrotechnical Commission
IEEE	Institute of Electrical and Electronic Engineers
I/O	input/output
IP	intellectual property
IR	infrared
ISO	International Organization for Standardization
JHU/APL	The Johns Hopkins University Applied Physics Laboratory
JTAG	Joint Test Association Group
LAN	local area network
LCD	liquid crystal display
LED	light-emitting diode
LEV	low emissions vehicle
LOC	lines of code
LRU	line-replaceable unit
LUT	look-up table
MEMS	microelectromechanical systems
Mil-Std	military standard
MOU	memo of understanding
MRD	marketing requirements document
MTBF	mean time between failures
NASA	National Aeronautics and Space Administration
NHTSA	National Highway Transportation Safety Administration
NIST	National Institute of Standards and Technology

NRE	non-recurring engineering
OEM	original equipment manufacturer
O-Level	organizational level
OSHA	Occupational Safety and Health Administration
OSV	on-site verification
PC	personal computer
PCB	printed circuit board
PDA	personal digital assistant
PDR	preliminary design review
PER	pre-environmental review
PHO	production handoff
PID	proportional integral differential
POST	power-on-self test
PRB	product review board
PROM	programmable read-only memory
PSR	pre-ship review
PWB	printed wiring board
PWM	pulse width modulation
PWR	power
PZEV	partial zero emissions vehicle
QA	quality assurance
RF	radio frequency
RFP	requests for proposal
RoHS	Restriction of use of certain Hazardous Substances
RPM	revolutions per minute
RTG	radioisotope thermionic generator
RTL	register transfer level
RTN	return
RTOS	real-time operating system
SAE	Society of Aerospace Engineers
SBC	single board computer
SCT	system compatibility test
SEB	single-event burnout
SEE	single-event effect

SEI	Software Engineering Institute
SEL	single-event latch-up
SERD	support equipment requirements document
SET	space environment testbeds
SEU	single-event upset
SIL	systems integration laboratory
SME	subject matter experts
SOW	statement of work
SRA	shop replaceable assembly
SRU	shop replaceable unit
SULEV	super-ultra-low emissions vehicle
SW	software
TPU	timer processor unit
TLM	telemetry
TSP	test script processor
UART	universal asynchronous receiver/transmitter
UI	user interface
UL	Underwriters Laboratory
ULEV	ultra-low emissions vehicle
UML	Unified Modeling Language
UUT	unit under test
VAR	value-added reseller
VOIP	voice-over-internet-protocol
WEEE	waste from electrical and electronic equipment
WRA	weapons replaceable assembly
ZEV	zero-emissions vehicle

1

Development Processes

1.1 Introduction

Real-time embedded devices, products, and systems touch every part of our lives. Generally they are unseen, "buried" inside things (Figure 1.1). In spite of their invisibility, people still expect those products to function—for example, microwave ovens, automobiles, or aircraft with hundreds of microcontrollers and embedded systems.

This book introduces various development processes for real-time embedded products; actually, it is more like a "keyhole" view of how some products come to market. It will focus on development processes through examples and case studies. This will hopefully give you a deeper appreciation and understanding how you might go about designing and developing real-time embedded products.

1.1.1 Basic Definitions

So, what is an embedded, real-time system? What are the common elements of such a system? What is the "language" used in developing them? Here are some definitions that will provide you a foundation to compare and contrast ideas and products.

An embedded system reacts to stimuli and generates actions or output displays (Figure 1.2). An embedded system generally has three basic building blocks: input, data processor, and output. The input can include signal receivers, sensors, signal-conditioning circuitry, and analog-to-digital converters. The data processor can do a variety of things: fusing data, sophisticated filtering, and making complex decisions. The output converts the processed data into displays, usable signals, and physical or mechanical operations. Often the embedded system is a self-contained module with these three building blocks; sometimes it is a group of modules.

Ganssle and Barr define an embedded system as "a combination of computer hardware and software, and perhaps additional mechanical or other parts, designed to perform a dedicated function" ([1], pp. 90–91).

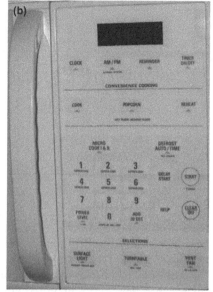

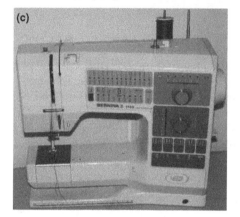

FIGURE 1.1

Examples of products containing real-time embedded systems. (a) The engine and cabin controls have embedded processors. (b) The microwave oven controls have an embedded microcontroller. (c) The sewing machine has an embedded processor to control many different types of stitching. (d) The toy robotic system has an embedded processor and programming system. (e) The lock on a hotel room has embedded electronics and processing. (© 2006 by Kim Fowler, used with permission. All rights reserved.)

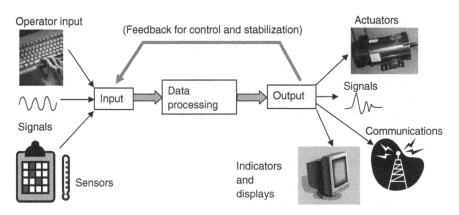

FIGURE 1.2
Diagram of a real-time system. (© 2006 by Kim Fowler, used with permission. All rights reserved.)

Ganssle and Barr further define real-time as "having timeliness requirements, typically in the form of deadlines that can't be missed" ([1], p. 228). *Real-time means completing tasks within specified deadlines; it is not defined or limited by a specific execution speed.* Just because a system is real-time it does not mean that the processor is necessarily fast—it just meets the deadline requirements. There are many things that affect performance including the execution speed of the processor, the rate of data transfer (bandwidth), the latency (or response time of the system to a stimulus), and memory size.

For more information about real-time performance, consult References 2–5.

1.1.2 Purpose

This book focuses on the processes to develop a successful product and the trade-offs that go into designing that product. It does not focus on the performance of a product.

The book compares and contrasts decisions made between different projects. This means that I need not present all possibilities for the architecture of a system—other books do a better job at that.

My goal is to provide a basis for comparing designs in different markets. Hopefully that basis will have a fairly consistent format that can serve as a guideline—or even a checklist—when you develop a new product.

Another goal, albeit a "stretch goal," is to foment change and improvement by encouraging you to see some of the interrelationships between various disciplines. I hope that these changes go in all directions—up, down, and sideways—which means to management, support staff, customers, and clients.

So, plan for change by planning for excellence, thoroughness, and consistency. This book will provide a framework for some of the needed thoroughness and consistency. The excellence? Well, that is left to you!

1.1.3   Scope

The processes and case studies presented in this book are for smaller, self-contained subsystems, such as a motor controller, a data acquisition system, or a handheld chemical sensor. The book's scope does not encompass larger systems, such as automobiles, aircrafts, ships, and command-and-control centers, and it does not address system-of-system concerns.

1.1.4   For Whom Is this Written?

This book is written for design engineers (primarily electronic hardware and software engineers), their managers, and people who should have an overall perspective on product development. Also benefiting from this book will be those wishing to better understand how other people and disciplines live in the world of engineering and product development.

1.1.5   Outline of Efforts—Basis of Comparison

The basis for comparison between the different case studies are 15 topics encountered during the life cycle of most projects. These topics are as follows:

- Concept and market
- People and disciplines
- Architecting and architecture
- Phases
- Scheduling
- Documentation
- Requirements and standards
- Analyses
- Design trade-offs
- Tests
- Integration
- Manufacturing
- Support (training, logistics, maintenance, and repair)
- Disposal
- Liability

Each one of these topics is an important phase or activity. The remainder of this chapter will expand on them. Each case study will then discuss the development trade-offs on the basis of these 15 topics.

1.2 Concept and Market

1.2.1 What, Who, Why, Where, When, and How

First things first.

- What is the product?
- Who is going to use it?
- Why will people use it?
- Where will they use?
- When will they use it?
- And finally, how will they use it?

These are very basic questions that must be answered before you exert any further effort. Otherwise, two things probably will happen. One, you could seriously stretch development time each time you "respin" the product design because new features are added. Two, you could end up in a fruitless search to fit an inappropriate or ill-advised design into a useful niche.

Understanding customers and their needs is often considered the sole domain of marketing. I think that is far too restrictive; engineers and designers need to see how people actually perceive, buy, use, abuse, repair, and dispose products. Once you know these things, you will be better able to speak "marketing" and work with other functions within your company.

1.2.2 Revolution, Disruption, or Evolution

New products develop in different ways. They can develop through revolutionary paradigm shifts, disruptive technology, and evolutionary change. Each type of change has a unique development process.

Technical revolution is a true breakthrough, or paradigm shift, brought about by a visionary or a very small group of visionaries. Several characteristics mark every technological revolution. (1) The revolution always involves simultaneous major improvements in several areas, such as capability, performance, and utility. (2) The change begins when someone questions a basic assumption, which most people accept as unchangeable, but which the visionary eventually proves wrong. (3) The revolution has four to six separate, simultaneous, and quite significant developments that cross technical disciplines. (4) Revolutionary change eventually leads to changes in the infrastructure of society. You can see these characteristics in the telephone, automobile, the airplane, the television, and the computer.

"Disruptive technologies bring to a market a very different value proposition than had been available previously. Generally, disruptive technologies under perform established products in mainstream markets. But they have other features that a few fringe (and generally new) customers value. Products based on disruptive technologies are typically cheaper, simpler, smaller, and, frequently, more convenient to use" [6]. Often these new technologies do not even come close to the performance of current products. They also have trouble gaining a foothold in the market before taking off. Finally, most established companies do not take on a disruptive technology; a risk-taking visionary typically must bring it to market.

Evolution is the refinement of current technology. Going from one model of music system to another is an example. The features tend to multiply—with more buttons than you care to look at. Generally, evolutionary change follows some of these priorities and trends:

1. Performance increases
2. Features and convenience multiply
3. The product becomes smaller or more dense
4. Power density increases
5. Finally, efficiency increases

This book will focus on the evolution of new products and how you might improve those processes.

1.2.3 Economics

Economics drive most product development. If the product is not economically viable, it should not go to market.

Many components play into the economics of product development. Cost figures as a major part of the economics of a product. Many things comprise the cost: purchasing components and materials, manufacturing and assembly, distribution, maintenance, and disposal. The cost associated with nonrecurring engineering (NRE) must be amortized over the life of the product. NRE includes all design, test, and qualification performed during the initial development of the product. Then there is the "cost of goods sold" (COGS) that refers to the ongoing expenses and recurring costs that go into the production of each unit. COGS includes buying components and the labor to assemble and build each unit.

Another concern is the margin between cost and price. Specialized low-volume products generally have high margins, that is, a large ratio (or difference) between the price of the final product and the cost to develop and manufacture it. High-volume products generally have low margins, that is, a small difference between the cost and the final price.

1.3 People and Disciplines

1.3.1 Focus

Every project involves many people. In this book, I will focus primarily on those making design decisions during development. These include several different disciplines for the design engineer: electrical, software, mechanical, and manufacturing.

Smaller and simpler projects have smaller teams—in some cases, it may be only two or three people. Bigger, more involved projects, such as satellite instruments or higher volume systems, have bigger teams that may include hundreds of people.

1.3.2 Team

Here is a listing of many of the types of people involved in the development process:

- Engineers—electrical, software, mechanical, and manufacturing
- Manufacturing personnel
- Marketing and sales
- Management and administration
- Purchasing and procurement
- Technical support
- Legal department
- Customers

Electrical engineers are involved in specifying, building, fabricating, testing, and fielding systems, circuit boards, components, cabling, connectors, and interfaces. Those interfaces include human operations, I/O with other systems, and mechatronics.

Software engineers are involved in specifying, building, fabricating, testing, and fielding systems, algorithms, the user interface, and I/O.

Mechanical engineers are involved in specifying, building, fabricating, testing, and fielding systems, physical components, the user interface, enclosure, materials, environmental testing, and mechanical I/O—mechatronics.

Manufacturing engineers are involved in specifying, building, fabricating, testing, and fielding systems, design for assembly, design for disassembly, and manufacturing tests.

For bigger projects, design teams will include industrial designers, human factors specialists, trainers, and educators. Many other people are involved besides the design team; they are outside the purpose and scope of this book.

1.3.3   Teamwork

Every person brings a different skill set and personality to the project. We need to accommodate the inconsequential differences and yet be able to confront and question problems. This can only be done through integrity, communication, openness, and trust. Teamwork encompasses these virtues.

One obstacle to teamwork is the "us vs. them" attitude—you see it in engineers vs. sales and marketing, workers vs. management, or hardware designers vs. software developers. Some people claim that competition is good within a team; I disagree. Human nature is such that it often degenerates and polarizes people into opposing camps. Until you have been on both sides of an issue, you cannot imagine the challenges and pressures each side faces.

Get rid of the "us vs. them" attitude! A particular concern is engineers vs. managers—popular cartooning may be descriptive of some situations, but lampooning does not ultimately improve the situation—it only serves to divide. Cartoons and humor can be good, but they should only serve to warn of problems in relationships. Both sides should actively work to avoid these pitfalls.

Never trivialize. A friend of mine often points out that, if something is not in someone's field of expertise, then they can tend to consider it easy to do. Please do not tell a software developer, "It's only a few lines of code to add." The implication is, "Hey, it's easy, particularly for software." It demonstrates a lack of understanding in the process—configuration control, regression testing, documentation changes, more reviews, and so on.

1.3.4   Leadership and Managing the Project

Many books have been written on leadership and management. I am only going to outline a summary of what I have seen and what I believe is effective.

Leadership personalities vary. Some are visionary; others are administrative. Both types are needed to develop the concept and then manage the details. Few people can do both equally well.

Leadership styles also vary. Some are more formal and tend to mandate directions; others are more informal and tend to cheerlead. The formal style works for certain teams and projects—sometimes it works for a government project. The informal style uses influence, agreement, and consensus to move the team forward.

Recognizing strengths and weaknesses in leadership personality and style will help compensate and balance out the team. Combinations of personalities and styles, when alloyed in teamwork, give the best strength and creative genius for developing projects.

1.4   Architecting and Architecture

Architecting is the process of defining the architecture of your product. Rechtin and Maier define systems architecting as both art and science.

The art of systems architecting bases judgments "on qualitative heuristic principles and techniques, that is, on lessons learned, value judgments, and unmeasurables" [7]. The science of systems architecting bases judgments "on quantitative analytic techniques, that is, on mathematics and science and measurables" [7].

Architecting matches, balances, and compromises between form and function in a product. It identifies, specifies, and trades off components, subsystems, their configurations, and their interactions. Architecting does the following:

1. Selects the process
2. Identifies parameters
3. Defines the interfaces
4. Performs feasibility analyses
5. Recommends architecture
6. Manages features
7. Certifies completion and ready for use

All of these activities flow into and through the generation of the requirements. A change in requirements can force a change in architecture. Architecting is that repetitive process that hones a system solution and sets the requirements.

Architecting is a necessary foundation stone to developing a product. Without it, development schedule and effort spin out of control. A team of management, designers, and engineers should be involved in architecting. Often a systems engineer or product leader will organize the effort.

1.4.1 Process

Architecting, the act of establishing a design's architecture, is all about process. There are several different models for process development: waterfall, prototyping, and spiral (Figure 1.3) ([2], Chapter 2). Waterfall development came out of military programs in the 1950s and 1960s; it still is used for one-off projects, such as satellites, that have to be right the first time. Some medical devices and mission-critical systems use a modified waterfall scheme called the V-model (Figure 1.4). Spiral development received definition during the 1980s and 1990s; it fits well into large-volume manufacturing of products and provides for steady evolution of designs. Prototyping is a more informal type of development that works well for smaller, more focused applications that do not have tight certifications and regulatory requirements.

One activity within architecting is to partition the design. This involves all aspects of development: hardware, software, testing, integration, and installation. Partitioning considers the features that characterize a product; interactions (and complexity) expand in a factorial fashion as features increase (Figure 1.5). The goal in architecting is to reduce the interactions

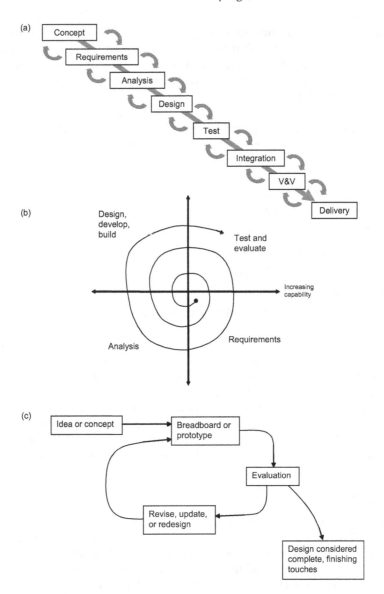

FIGURE 1.3
Three different process models for development: (a) the traditional waterfall process model, (b) spiral development model, (c) a more ad hoc process model using prototyping. (© 2007 by Kim Fowler, used with permission. All rights reserved.)

to a manageable set of interfaces between modules, which should constrain complexity.

Another goal in architecting is to balance design complexity with resources. Reference 8 gives some dimensions of design complexity and outlines the availability of resources to address those design complexities. Table 1.1 outlines the components of design complexity. Table 1.2 outlines

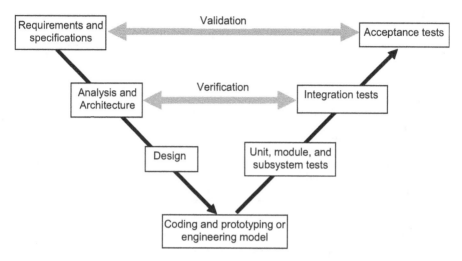

FIGURE 1.4
V-model for process development in mission-critical systems or medical devices.
(© 2006 by Kim Fowler, used with permission. All rights reserved.)

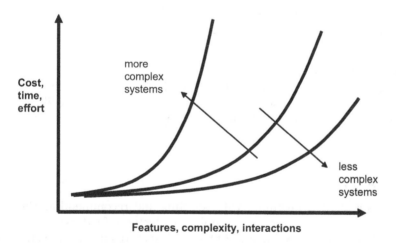

FIGURE 1.5
Complexity exponentially increases cost, time, and effort to design and build systems. (© 2006 by Kim Fowler, used with permission. All rights reserved.)

the resources to address design complexity. This simple analysis can give you a good idea of just how feasible your project is.

1.4.2 Parameters

All products have defining parameters: cost, size, weight, performance, power, or dependability. Each of these has multiple levels of parameters within themselves. Performance, for instance, may have concerns with speed, throughput, loading, and memory size. Power may have input filtering issues or operational cycling to conserve energy.

TABLE 1.1

Components of Design Complexity, Following Reference 8

Metric	Range
Design type—from redesign to innovation to revolutionary breakthrough	0–15
Knowledge complexity—from common to expert specialist knowledge	0–10
Number of steps to complete the design	0–10
Quality implementation effort	0–10
Process design	0–5
Aggressive goals for selling price	0–5
Total design difficulty score	0–55

TABLE 1.2

Components of Resources to Address Design Complexity, Following Reference 8

Metric	Range
Cost—NRE to develop product	0–15
Time—from definition of project to delivery of the first unit	0–10
Infrastructure—tools, administration, company capabilities to perform	0–10
Total resources effort required	0–35

Many products, particularly simpler ones, have a critical parameter that drives design. You should first optimize the architecture for that critical parameter. (More sophisticated systems, however, are less likely to have a single critical parameter driving the design. Optimization is much more difficult.)

1.4.3   Analysis

Analysis is tightly integrated with selecting and recommending the architecture. You analyze various possible architectures for feasibility in your application before selecting one. Often analysis turns up concerns that need to be addressed by a revision in the architecture. You iterate between architecture and analysis until you converge on a solution that is "close enough."

1.4.4   Architecture

Architecture is the structural plan that allows you to accommodate the design intent; the requirements then attach to or "hang from" this structure. Architecture defines both form and function. Some of the potential structures and structural trade-offs that you might consider are distributed vs.

central design, modular vs. custom design, loose vs. tight coupling, types of processing, testability, manufacturability, and the human interface ([2], Chapter 15).

Distributed vs. centralized: A distributed architecture has functionally defined modules that are sometimes physically separated from each other and often connected by a network. A centralized architecture, on the other hand, tends to lump functionality into a single "mainframe" unit. A PC with peripheral cards and a central power supply is an example of a centralized architecture.

Distributed architectures tend to be more robust, easier to test, and easier to upgrade, but they are larger and heavier than centralized architectures for small products (Figure 1.6). Distributed architectures fit larger systems better. Centralized architectures can optimize to perform one task or one type of tasks quite well. Large, tightly coupled, centralized systems can be very difficult to test because of the many interactions between components.

Modular vs. custom monolithic: A modular architecture generally reduces the complexity of system interactions. It can also allow parallel effort in design, test, and fabrication, which speeds development. By definition, distributed architectures are modular, but modular design is not precluded by centralized architectures whose components and subsystems can be modular.

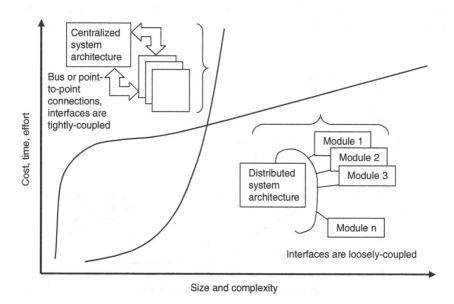

FIGURE 1.6

Comparison of distributed and centralized architectures. (© 2006 by Kim Fowler, used with permission. All rights reserved.)

A custom, monolithic design can optimize a single function (such as performance, weight, or size) for a particular application. Generally, monolithic designs are better for single-purpose devices. Modular designs, on the other hand, generally add size and weight to a design over a custom, monolithic design, and potentially can be more expensive in COGS.

Loose vs. tight coupling: Loose coupling reduces dependence between modules and subsystems, which means that communication between modules reduces, as well. Distributed architectures can use loose coupling to advantage because they are more tolerant of change, rework, and test.

Tight coupling makes modules highly dependent on each other. Tight coupling can optimize performance around a single parameter but not for multiple parameters. Tight coupling also makes testing of the product much more difficult.

Distributed architectures tend to be loosely coupled, which means that the interfaces are clearly defined and the communications between units are minimal. Centralized architectures tend to be tightly coupled, which means that the communications between modules can be voluminous.

Types of processors: Processors take a variety of forms (Figure 1.7). They range from specialized circuits with dedicated hardware to powerful, general-purpose processors that offer a plethora of services. One major trade-off in selecting a processor is between computational speed and sophistication of operation. Dedicated hardware and field-programmable gate arrays (FPGAs) can be extremely fast but have very limited forms of processing. Digital signal processors (DSPs) are great at number crunching,

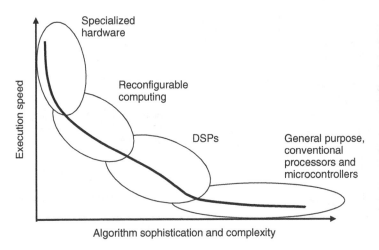

FIGURE 1.7
Comparison of different computing architectures. (© 2006 by Kim Fowler, used with permission. All rights reserved.)

as in filters and data compressors, but do not generally integrate a wide variety of peripherals. Microcontrollers, on the other hand, do integrate a wide variety of peripherals but are not nearly as fast as DSPs. Microprocessors generally have the best tool-development suites and the widest the variety of capabilities but tend to be big, power-hungry, and expensive.

Testability: Making a product amenable to testing has mechanical, electrical, and software implications. First, specify the level of repair that will be allowed. Next, determine the types of test equipment necessary for your product—whether commercially available or custom-built. Then consider the ease of disassembly and connection to test equipment. Finally, determine how much effort will need to go into the software to support testing.

Manufacturability: Making a product easy to manufacture is a big concern for selling in high volumes. Manufacturability primarily has mechanical and electrical concerns. Labor is one of the biggest components of cost in manufacturing; making a product easy to assemble and package is the most direct way to reduce cost.

Human interface: All embedded systems eventually produce actions that humans initiate and perceive. The goal should be to make the operations as obvious and intuitive as possible. Buttons should have obvious functions; there should not be any awkward or unusual placement or sequences of their use. Displays should be clear and not display cryptic symbols or abbreviations or jargon. Software operations should have a clear flow; simple wizards are a good example of simple and understandable flow.

1.4.5 Interfaces

A good interface design hides complexity by capturing, organizing, and communicating information within the project. This is true for the human interface, as well as for the cables, connectors, and communication channels. A good interface design reduces the burden of testing and verifying system performance because it allows separate testing of individual functions.

1.4.6 Features

All features interact with each other and with the system functionality. New features affect other parts of the design in unforeseen and unpredictable ways—one goal of a suitable architecture is to limit the untoward effects of new features. Please understand that *most people prefer control over performance. Control means that people want to understand the system and find it easy to use.*

Adding features is not a simple linear process. It is iterative and expands in a nearly factorial manner. You need to regression test the entire system after you add each new feature.

A good team leader and team will handle the features carefully. They will allot priorities to what is required, what is desired, and what is requested.

1.5 Phases of a Project

Let us move into some of the "nuts and bolts" of project development. Every project goes through different phases. This section will outline some phases of development often encountered on many projects. Table 1.3 summarizes the phases.

Many parameters can alter the actual phases of development, depending on size of the project, complexity, regulations, certification, and development cost. A simple, small project with a small team of three or four people, for instance, may combine the first two phases into a feasibility study, have a single design stage, and then combine production and maintenance and disposal into a third phase. Here I present a basic set of phases; the case studies will comment how these might vary for specific situations.

1.5.1 Concept

The concept phase explores the feasibility of an idea or new project. The team uses various means to analyze, simulate, and determine feasibility. Any obstacles that could prevent realization should be uncovered and trigger a decision to stop the project. This phase defines the requirements and goals for the project. It also proposes a development approach to meet those requirements.

At the end of the concept phase is a review, often called the concept design review or CoDR. The review presents the mission goals, objectives,

TABLE 1.3

A Summary of One Possible Set of Phases That Might Occur During Project Development

Phase	Name
1	Concept
2	Preliminary design
3	Critical design
4	Production (manufacturing)
5	Operation, maintenance, disposal

and constraints. The team should understand and present the requirements for the project and the approach to meet those requirements.

Summary: Concept Phase

What—Explore possibilities and determine feasibility.

Who—Marketing and engineers; bigger projects will include more disciplines such as manufacturing, logistics, and support.

Why—Feasibility and any no-go decisions from "show stoppers."

How—Studies, focus groups, calculations, and simulations.

When—Complete the phase when either schedule or mutual consent determines enough information is available to make the decision (this is not always possible).

1.5.2 Preliminary Design

The preliminary design phase begins the basic approach to a project. The team completes nearly all of its analyses. Usually, breadboards are built and lab tests are completed. This phase prepares the basic design with block diagrams, signal flow diagrams, initial schematics, packaging plans, configuration and layout sketches, and early test results.

At the end of the preliminary design phase is a preliminary design review or PDR. A PDR should present the basic system in terms of the software, mechanical, power, thermal, and electronic designs with load, stress, margins, and reliability assessments. It should also present the software requirements, structure, computational loading, design language, and development tools to be used in development. Finally, it should present the preliminary estimates of weight, power, and volume.

Summary: Preliminary Design Phase

What—Flesh out the design approach.

Who—Marketing, engineers, and manufacturing; bigger projects will include more disciplines such as industrial designers, logistics, and support.

Why—Confirm the feasibility and make a go or no-go decision.

How—Calculations and simulations, design tools, and early prototype tests.

When—Complete the phase when either schedule or the design approach is clear.

1.5.3 Critical Design

The critical design phase completes the final design of the project. The product is ready to go to production after critical design.

At the end of the critical design phase is a review, called the critical design review or CDR. A CDR presents the final designs—schematics, software code, and test results. It also presents the engineering evaluation of the breadboard model of the project. A CDR must be held prior to the design freeze and signed off before any significant production begins. The design at CDR should be complete and comprehensive.

Summary: Critical Design Phase

What—Finalize the design.

Who—Marketing, engineers, and manufacturing; bigger projects will include more people such as test team, industrial designers, logistics, and support.

Why—Make the final go or no-go decision on the product.

How—Reviews, evaluation tests, and field tests.

When—Complete the phase when the design is finished, appropriately tested, and manufacturable.

1.5.4 Production Handoff

Sometimes an additional phase, called the production handoff (PHO) phase, includes integration, test, and delivery; it might be the only fabrication or production that occurs in the project. The product is installed after this phase. Its purpose is to assure that the design of the product has been validated and that all deviations, waivers, and open items have been satisfactorily closed. It also confirms that the project, along with all the required support equipment, documentation, and operating procedures, is ready for production.

This particular phase generally applies to larger custom projects that deliver a single unit. It has both elements of the critical design phase and the production phase in those projects. In large-volume manufacturing situations, PHO focuses solely on preparation for manufacturing.

Summary: Production Handoff Phase

What—Validate the design and close all open issues.

Who—Marketing, engineers, and manufacturing; bigger projects will include more people such as test team, industrial designers, logistics, and support.

Why—Final step in the quality assurance of the design.
How—Tests for integration, environmental survival, and qualification.
Commissioning and training.
When—The customer signs off.

1.5.5 Commercial or Support

This is an open-ended period that completes the life cycle of the product—hopefully, it is a long one for the sake of your company and profitability. This phase is where manufacturing, logistics, maintenance, and disposal are the primary concerns. You and your team should monitor the product and its use periodically to prepare feedback for future development efforts.

1.6 Scheduling

1.6.1 General Philosophy

Planning should be an evolving process; we should continually learn about what works and what does not and then apply these lessons to new projects. Alas, this seldom happens.

A colleague said to me, "Everyone wants to do the right thing, but schedule always wins" [9]. For many markets this is true. Most of us wrestle with the big three—schedule vs. cost vs. quality. Schedule is usually fixed, which leaves us with the difficult choice of either reducing cost or improving quality. The holy grail of much engineering is to improve all three; NASA during the 1990s had the goal of "better, faster, cheaper."

I think one more factor is assumed but not as clearly defined—requirements (or features). While we grapple with the schedule, cost, and quality within a project, most of us just assume a full-feature set (full requirements). Making features, and therefore requirements, a variable along with schedule, cost, and quality will help make the project management and development tractable—see Sections 1.8.4 and 1.10.1 for more details.

Bad schedules often arise from failure to plan properly. Many reasons exist for the failure to plan and to learn planning. This is a shared problem—no one person has total responsibility for planning and scheduling. Only a committed group of people will plan well, work the plan, stick to it, and learn from their mistakes.

You can schedule a project many ways; I will focus on two methods: top-down planning and bottom-up planning. Top-down planning sets delivery dates and milestones first and then finds out what you can fit into the periods between dates. Bottom-up planning starts with the basic details of everything that needs to be done and builds up to the dates when components and stages can be finished. Top-down planning can be easily abused when it over-constrains a project—requiring too much to be done in too short a time. Bottom-up planning can build inefficiencies into the process and extend the schedule too far.

I use both methods in combination to attempt an appropriate schedule. I start with a top-down plan and then see where a bottom-up plan meets it. This kind of exercise helps me see and remove the pitfalls of either method.

1.6.2   Covering the Bases

The first thing to do when planning and scheduling is to understand all the activities that make up a project development. These include meetings, desk work, lab work, communications, documentation, travel, debugging, testing, fabrication, support, installation, and training—but this is not an exhaustive list.

Did I mention meetings, communications, and documentation? Most of us do not have a clue how much time we spend on meetings, communications, and documentation. Firing off an e-mail to explain a procedure or clarify a point often takes a lot longer than the 3 min we thought it would take. Merely walking to and from a 1 h meeting can take up to 15 min each way with distance or interruptions to chat. That is 50% over the allotted meeting time!

Simple telephone calls can fill a day. I have found that calling a parts distributor and asking for price and delivery on a single, generic, run-of-the-mill, stocked item will average out to 8 min; that means I can only inquire about six (6) or seven (7) components in the space of 1 h.

Finally, you need to build in some margin for "surprises." These contingencies are critical to good planning; they should not be used to cover activities that you already know about—that is being dishonest. The best way to include contingencies in your schedules is to study where problems have occurred in previous projects and try to discern a correlation between the activities and the results. Once you better understand the trouble spots, you can apportion margin to those activities.

1.6.3   Software Tools

A simple spreadsheet can provide a reasonable "bottom-up" estimate of time, effort, and cost for your projects. Other software tools, such as Microsoft Project®, can give an estimate of "top-down" plans; if sufficiently exercised it can give a detailed picture of a "bottom-up" effort, too. Chapter 3 gives several examples of tools that you might use to schedule activities.

1.6.4 Problems—Fates, Constraints, and Mandates

Business cannot make headway without initiative and risk. Consequently, planning and scheduling are unavoidable, but traps abound. We need to bring sanity to the process of planning and scheduling. Here are some of the problems that you should understand and avoid:

- Mandating deadlines and requiring a fixed feature set is doomed to failure. Invariably it overconstrains the development. Either fix the milestone dates and let the features evolve, or fix the features of the product and settle for the schedule to be flexible.
- Scheduling more than 50% of someone's time to a task will probably overload them [10]. This means that even in a full-time effort, most people lose about half of their time doing things not directly job-related. They may be moving between meetings or responding to non-essential e-mail or surfing the Internet or chatting.
- Not allowing enough time for testing and review will either stretch the schedule or cut short important qualification steps. I do not know anyone who gets everything right the first time.
- Not accounting for essential work not immediately recognized as project-related (for example, meetings, telephone conversations, and e-mail messages) will result in overloading personnel in a shortened schedule.
- Not recognizing that documentation is time consuming. Crafting a simple memo can occupy hours.
- Not recognizing unexpected delays in procurement when components are out of stock or a subsystem must be modified to suit your requirements will stretch out the schedule.
- Not allowing schedule margin (contingencies) for these problematic areas always results in conflict, delayed deliveries, bugs, and reduced feature sets.

1.7 Documentation

1.7.1 Purposes

Over the years, I have seen three fundamental truths about documentation ([2], Chapter 4):

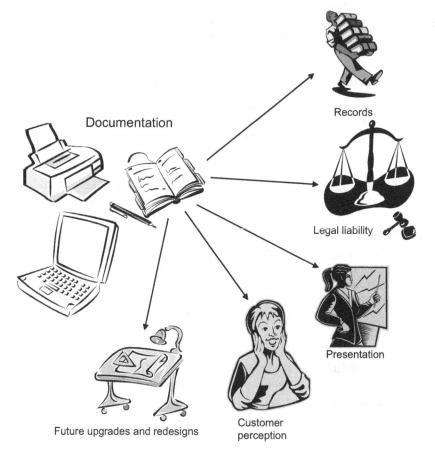

Documentation

Records

Legal liability

Presentation

Future upgrades and redesigns

Customer perception

FIGURE 1.8
Documentation pervades every aspect of a system. (© 2007 by Kim Fowler, used with permission. All rights reserved.)

- *Documentation is integral to every product* (Figure 1.8)
- *Good documentation cannot help a poor product*
- *Poor and inadequate documentation can destroy a good product*

Documentation generally serves three purposes: to record the specifics of development (the "who, how, and why"), to account for progress (the "what, when, and where"), and to instruct the extent of functionality (and thus establish the liability) of your product ([2], Chapter 4).

Records include your engineering notebook, software source listings, schematics, and test reports. This is not an exhaustive listing of all records; each case study that follows will have specific and unique types of documentation. These records will help when modifications, upgrades, fixes, and recalls occur.

Documentation also accounts for the progress toward satisfying requirements and provides an audit trail of the development. These types of documents include memos, meeting notes and minutes, review action items, and project plans. An appropriate plan for documentation can also support rigorous testing, validation, and verification. Again, this is not an exhaustive listing of all plans and archives—you will see specifics in the case studies.

Good documentation also instructs users and owners as to the function and extent of operation for the product. It can limit product liability. Instructions include user manuals, DVDs with instructions, labels and warnings, presenting concise instructions, and listing necessary details about the use of your product. This, too, is not an exhaustive listing of all instructions.

One heuristic that has always worked for me is, "There are good products with poor documentation, but I have NEVER found a poor product with good documentation" ([2], Chapter 4). All this to say, documentation is a necessary but not a sufficient condition for building a quality product.

1.7.2 Types

Many types of documents exist. Each set is project-type specific. They might include notebooks, letters and e-mail messages, memos, project documents, manuals, brochures, and presentations.

The set of documents for a project should answer the "who, what, when, where, why, and how" of product development. Here is a general outline of the types of documents that are part of most products:

- Plans (who, what, where, when, why)
- Design documents (how, why)—source listings, schematics, and engineering notebooks
- Reviews and reports (who, what, where, when, why)
- Instructions (what)—user manuals, brochures, training materials, and maintenance and repair guides

The case studies that follow will cover sets of documents specific for their markets.

1.7.3 General Formats for Documents

Regardless of the document, basic principles apply. Every document should have these attributes in the following order of priority:

1. Correct
2. Complete
3. Consistent
4. Clear
5. Concise

First, you must understand the audience—who will read the document and how they will use it. Tailor your documents to the readers [11]. They should have an appropriate level of detail and reading comprehension (simple, short, and concise is always best). The layout of text and graphics should have an intuitive format or instructional flow. The necessary warnings and updates should also be present.

A users' manual can be a good example of how documents might be organized. Good manuals are simple, clear, and easy to use; these characteristics are achieved through attention to the following points:

- Basic content
  - What: prominently note the model number and version
  - Why: clearly explain the need or theory
  - How: detail the operation
- Clear organization
- Modular format
- Clearly illustrated figures and tables
- Detailed schematics
- References to source listings and other technical records
- Table of contents, index, and cross-references

1.8 Requirements and Standards

1.8.1 Markets

I see six basic markets for real-time embedded products and systems: high-volume consumer appliances, mission-critical consumer (e.g. automobiles), industrial, military, aerospace, and medical. I will touch on each in the case studies. I will tend to focus on the military, space, and medical because of my own experience and because these tend to need more thorough processes.

1.8.2 Standards and Government Regulations

Each market needs to adhere to specific standards and government regulations. If it is a consumer appliance, it must have Underwriters Laboratory (UL) and Conformite Europeene (CE) markings. If it is avionic or military, it has some narrowly defined standards and regulations—such as DO-168B and DO-178B. Certain industrial processes follow Institute of Electrical and Electronic Engineers (IEEE) and International Electrotechnical Commission (IEC) standards. Finally, if it is medical, it not only follows UL and IEC standards but also follows the Food and Drug Administration (FDA) guidelines in the United States, which often apply to other countries, as well. The case studies will give specific instances of these standards and regulations.

For commercial markets, the IEEE and the American National Standards Institute (ANSI) both provide standards for individual components, devices, and systems. In Europe and worldwide, you will have to deal with IEC regulations and the CE mark [12–15].

Beyond commercial concerns, government regulatory concerns are becoming a bigger part of product design. Restriction of use of certain Hazardous Substances (RoHS) regulations are forcing manufacturers to develop recycling programs to take back products at the end of market life and recover the hazardous materials such as lead and heavy metals.

Standards and regulations feed directly into the design requirements for each project. Know your market and its particular set of standards and regulations.

1.8.3 Preparing Requirements

Once you have some grasp on the market and the necessary standards and regulations, then you can begin setting down requirements. Requirements are the codification of intent.

Robert Oshana writes, "Both functional and financial success is affected by the quality of requirements. So what is a requirement? It may range from a high-level abstract statement of a service or of a system constraint to a detailed mathematical functional specification.

Requirements are needed for several reasons:

- Specify external system behavior
- Specify implementation constraints
- Serve as reference tool for maintenance
- Record forethought about the life cycle of the system, that is, predict changes
- Characterize responses to unexpected events" ([4], p. 507)

Good requirements define the metrics of the design. Measurement can then feedback the design output to the requirements so that you can compare and determine progress.

Intent relates to the desire that a product have appropriate performance. When someone says, "It has to be the fastest"—that is intent. The requirements will answer how fast and what "fast" means.

Who defines the intent? How do you encode intent? Those are tough questions. Customers ultimately express the intent. Marketing and engineering then interpret that intent and translate it to definable metrics in the requirements. There are several ways to understand intent: customer visits, focus groups, marketing discussions, and the "bright idea."

For most products, you need to visit customers and ask them about what they really want; this can be an art unto itself, and marketing usually handles much of this. Really tricky issues, though, need attention from the designers and face-to-face discussions. You may not like it, but visiting customers is very important. I will bet that, as you do it, you will come to appreciate and may be even enjoy visiting customers.

Makers of consumer and medical devices often use focus groups, which provide a carefully controlled environment with trained facilitators. These can illicit ideas and directions from potential customers that may not have occurred to you.

Finally, the occasional "bright idea" supplies the intent that customers may not even foresee. This certainly occurs with revolutionary changes. Here a lot of customer training and instruction is necessary. In fact, the customer usually becomes an integral part of the development team. Regardless, you must still submit your "bright idea" to customer review.

Ultimately, intent flows into requirements, which define and set measurable quantities for subsequent testing. Verification means that the design is tested against the requirements. Another important concern is validation; it means that the product is tested to demonstrate that it meets the desired intent.

Note: Some people differentiate between requirements and specifications. They call the natural-language description of the desires as "requirements" and the crafted metrics as "specifications." It can be somewhat artificial in some situations but useful in others. I like the distinction for complex systems.

You need to know what people are thinking. You need to know their desires. Insist on involvement with the customer. Understand them and how they do things. All this will help you define more accurate requirements (or specifications).

Oshana writes, "A good set of requirements has the following characteristics:

- *Correct:* Meets the need
- *Unambiguous:* Only one possible interpretation
- *Complete:* Covers all the requirements
- *Consistent:* No conflicts between requirements
- *Ranked for importance*
- *Verifiable:* A test case can be written
- *Traceable:* Referring to requirements [is] easy
- *Modifiable:* Easy to add new requirements" ([4], p. 508).

The process of developing requirements is still inexact science. Use cases in unified modeling language (UML) have proven useful to reveal some interactions between the system and external inputs. Oshana gives a good example of sequence enumeration that is a way to specify inputs and responses in an embedded system. It uses prior stimuli to the inputs and current ones to cover all possible inputs; these then map to a state machine implementation. Its strength is that it forces you to consider obscure sequences. I refer you to Oshana's book and Appendix D for more information ([4], pp. 508–509).

1.8.4 Managing Specifications

Potentially half or more of all information technology projects are afflicted by shifting, poor, or missing requirements, and end up exceeding the budget and schedule. Thirty-seven percent of all projects suffer from one or more of the following: lack of user input (12.8% of all projects), incomplete requirements and specifications (12.3% of all projects), and changing requirements and specifications (11.8% of all projects) [16]. Table 1.4

TABLE 1.4

Summary of Some of the "10 Requirements Traps to Avoid" [16]

Trap in the Specification Process	Symptoms	Solutions
Confusion over definition of requirements	• Different perspectives (from vision to detailed design to solutions) but no qualifying adjectives • Miscommunication between users and developers that tends to focus only on functionality	Recognize different types of requirements: • Business • User • Functional • Quality
Inadequate customer or user involvement	• Users too busy to refine the requirements • User surrogates (user managers, marketing staff, or software developers) supply all of the input • Developers make requirements decisions without confirmation	• Identify the different user classes • Identify a customer representative for each user class • Build a collaborative relationship between customer representatives and the development team
Vague, ambiguous, or unused requirements	• Varying interpretations • Cannot think of test cases for specifications • Guessing at specifications	• Avoid subjective and ambiguous words • Trace every functional requirement back to its origin

(Continued)

TABLE 1.4

Continued

Trap in the Specification Process	Symptoms	Solutions
	• Glitzy user interface vs. necessary functionality • Developer gold plating	• Inspect the requirements document • Write test cases early and derive requirements from them • Develop prototypes
Unprioritized requirements and scope creep	• Declaring all requirements, or 90%, to be equally critical • Poor definition of initial scope • Changes that sneak in through the back door	• Align use cases with business requirements • Promote consistent classification and common expectations • Plan for changes in requirements and manage those changes
Inadequate change process and version control	• No defined process for changes in requirements • Bypassing process • Working on changes before approval • Poor understanding of implications • Keep finding more implications with further development • Can't distinguish different versions of the requirements documents	• Set up control board to monitor changes • Use a problem- or issue-tracking tool to collect, track, and communicate changes • Systematically analyze the impact of each proposed change • Use a version control tool

provides a summary of symptoms and solutions for generating and managing specifications.

Many companies, including automotive companies and suppliers, are using or moving toward model-based specifications for subsystems to capture requirements. Model-based specifications use a model of the device or system with mathematical constructs, such as sets and sequences, and define system operations by how they modify the system state. Use cases and UML are an example of model-based approaches.

I am not a proponent of the dictum, "Get the requirements right and you'll get the design right." Requirements and specifications are a *necessary but not a sufficient* component of good development process. Historically, civil engineers and architects have shown that few people, if anyone, can ever specify all the details of a building. Changes do occur during construction. We embedded developers and system engineers must develop processes to do likewise.

1.8.5 Speed-Up Schedule and Requirements

Many problems arise because the requirements are either too sparse or too complex. Either situation leads to unforeseen problems and unpredictable interactions later in product development when change is much more difficult. Obviously, sparse requirements leave out important concerns, which lead to unexpected problems. On the other hand, complex requirements can "overspecify" a product, which leads to unforeseen interactions that cause problems.

Reinertsen reports that many projects choke on specifications that are too complete, too stable, and too accurate [17]. Specifications that are overdone tend to shift attention from the few, critical specifications to the many, unimportant ones. One reason for overdoing specifications is because writers elaborate what they know; yet the amount of detail does not relate to what the customer wants. According to Reinertsen, successful products do not necessarily start with complete specifications—the Apple Powerbook was based on a one-page specification. More recently, one successful model of cell phone derived from a group of designers building a prototype with features that they liked; it was not directed by marketing. Even history shows that short specifications can lead to a successful product: a short letter with single page of specifications from Trans World Airlines to Douglas Aircraft Corporation in 1932 lead to the development of the DC-2 followed by the highly successful DC-3 [18].

There are ways to improve scheduling while still establishing useful requirements:

- Write the catalog page for the product before anything else. Years ago Hewlett-Packard instituted this practice. If a feature warrants mention, then it is important. Anything else that is left out is deemed unimportant [17].

- "Brief specifications almost always force higher levels of customer contact ([17], p. 29)." More customer contact leads to a deeper understanding of customer needs, quicker resolution of interpretations, and faster reaction to market shifts. Please note: good quality makes for concise specifications; the converse is not true—short specifications do not necessarily lead to good specifications.

- Allow flexibility in the requirements and make development a "closed-loop system." Reinertsen claims that the initial accuracy is not as important as assuming that the specifications have inaccuracies and then using feedback to detect and correct them [17].

- Pazemenas argues for shortening "the fuzzy front end" [19]. Identify new product opportunities and improvements in technology often—quarterly evaluations may be necessary—so that the fuzzy front end does not drag out from a long learning curve.

- Pazemenas also promotes the effort to define the right product—the one with greatest value to the customer. To do this, you need to get close to the customer; use prototypes to clarify interactions and operations [19].

Note that these will also speed development when properly implemented. Section 1.10.1 on design trade-offs has a more complete presentation.

1.9   Analysis

1.9.1   Feasibility

Analysis predicts the performance or suitability of the product. (That's predicts, not guarantees!) It is faster and cheaper than building prototypes and testing—usually. The concern is, "Where do you stop analysis and begin building?" Let us look at analysis first.

Analysis can have a variety of forms, which typically occur early in the project. Analysis has four general stages: rules of thumb, analytical approximations, numerical simulation, and test. Each stage is important and serves a particular need.

I included testing because it confirms analysis. Testing consumes a whole section in this chapter and in each case study because of its importance in not only establishing the product's specifications but also in confirming analysis.

1.9.2   Heuristics

Heuristics are rules of thumb that generally indicate an appropriate direction. They help constrain a problem during the early concept phase of development. Heuristics provide boundaries or limits to a situation and serve as intuitive checks to avoid gross errors. They are quick and cheap.

Experience is a collection of heuristics that fit together—sometimes ambiguously. They are the things that just work correctly and do not generally contradict each other. Engineers and designers should continually build their experience—stockpile heuristics that outline a meaningful, overall perspective to development.

Tables 1.1 and 1.2 are good examples of heuristics that arise from careful research.

1.9.3   Calculations

Calculations are definitions of physical phenomena and approximations [20]. Most of our formal technical education focuses on calculations as a basis for explaining or describing physics. We calculate everything from the power dissipation of transistors to nonlinear efficiency curves for power converters to spreadsheets of business models.

While more involved than heuristics, calculations can be more refined than heuristics. Like heuristics, calculations are still fairly fast and cheap. Most of us use spreadsheets because they are fast and easy to set up and use.

1.9.4 Numerical Simulations

Numerical simulations are a good next step up from calculations. Simulations operate from known, basic assumptions to project results from complicated situations. You can simulate many different things: circuit operation, electric fields or electromagnetic interactions, software operation and performance, kinematic performance, to name just a few different concerns.

Simulations provide a more in-depth view of potential operations and circumstances than do calculations alone. Their primary forte is to make sense out of complicated interactions. Simulations tend to be much more expensive to purchase than the simple tools that do calculations, but they generally are much cheaper and faster than building a prototype and testing (though not always).

Simulations are usually part of most designers' and engineers' toolboxes. They often provide a reasonable basis for doing "what-if" scenarios in feasibility analysis.

Beware! Simulation is not proof that something will work. You still need to test it.

1.9.5 Testing

Testing confirms analysis through verification and validation. The question is, "When is testing sufficient?" A full suite of tests to cover all possible circumstances is impossible in most cases—if not in all.

Testing covers the mechanics, electronic circuits, software, and usability. It can also help establish dependability (which includes reliability, availability, maintainability, and fault tolerance).

Certainly designers and engineers are involved, at the very least, in specifying the requirements that will be tested. Most of us also take part in portions or all of the testing.

1.10   Design Trade-Offs

1.10.1   Processes to Speed Development

You have several different avenues through which you can speed development; they deal with the "big four": scheduling, establishing requirements, reducing cost, and improving quality. Most of the techniques affect the requirements, which have a direct effect on the other three concerns of schedule, cost, and quality.

The most important thing to realize is that these four elements are interrelated and interdependent. As you add a feature or requirement, you are adding schedule dependencies, cost impact, and quality interactions. You need to strike an appropriate balance between them.

Here are techniques to speed development:

- Write the catalog page first [17]. This exercise will point out the important features and therefore the important requirements. It will help you set priorities for the requirements. A corollary is to write the users' manual next. It will help you uncover and fill in the intermediate requirements, as well as clarify operations and set expectations.
- Write concise specifications. Beware of specifications that are too complete, too stable, and too accurate [17].
- Shorten the fuzzy front end [19].
  - Doing everything you can to identify the need. Identify new technologies and product opportunities often—may be even quarterly evaluations—so that the fuzzy front end does not drag out from a long learning curve.
  - Define the right product—the one with greatest value to the customer; use prototypes with only a subset of functions to clarify operations and expectations.
- Make development of requirements a "closed-loop system" [17]. Keep requirements flexible to allow trade-offs [19].
- Get the right architecture—use modular design for concurrent development and partition the design for functionality. Modularity can contain uncertainty and isolate problems, thereby allowing change and flexibility [17,19].
- Minimize the required work, reuse designs and modules as appropriate. But as soon as you start modifying modules, then redesign them! Shortening development time relies on proven modules that remain unmodified.
- Get the right team and organize for speed [19].
  - Remove long justification/approval phases, elaborate documentation, limited use of outside resources
  - Get enough people to do the job adequately (but not too many!), give responsibility to do the job (e.g. do not remove responsibility with things like long signature approval lists)
  - In some cases, a good contract engineering firm can really help out with needed expertise and resources
- Increase your parallel effort; do concurrent development, which can only happen with flexible requirements and the right types of people to do job (see getting the right team above) [19].
- Management should provide clear objectives and priorities without overconstraining the team. Lack of appropriate objectives and priorities leads to spinning wheels and wasted time [19].

- Manage risk—business, technical, and regulatory. Use prototypes to confirm requirements; test and validate designs as soon as possible. This follows the same reasoning as setting clear objectives and priorities [19].

1.10.2 Intent and Requirements

Change is constant in product design. As it occurs, refer back to the intent—has it changed, too? If so, then completely rework the requirements. If the intent does not change, then review the requirements and determine which items within the requirements need to be updated to reflect the changes. (Remember that requirements should be a "closed-loop feedback system.")

Maintain a system perspective. When change is suggested, ask if it is really needed. Seek to understand the potential interactions.

The whole team needs to be aware of the intent and requirements. The whole team will contribute in various ways. The engineers and designers will spend the most time considering and analyzing them. Usually, the project lead or system engineer is responsible for maintaining the requirements.

Eventually, you will have to validate the product design against the intent and verify the design against the metrics in the requirements. Either the designers or test engineers handle the majority of validation and test.

1.10.3 Hardware

The hardware of a design has many potential trade-offs: type of processing, candidate processors, peripherals, memory, and fabrication. The fabrication of circuits and mechanics has great potential for many different configurations, including printed circuit boards (PCBs), cables, connectors, displays, and input devices. These trade-offs have many consequences, from manufacturing and assembly ease to testing access to cost to customer acceptance.

Remove cables and connectors from the design if possible. They are the source of many failures. They do provide access to the circuitry, however, and may be important in certain systems for testing or upgrades.

Analog circuitry vs. digital circuitry is one area of trade-offs that electrical engineers will consider when designing circuits. Is it better to build an analog filter in the front end, or is it more appropriate to provide digital signal processing to do the filtering? Can an analog amplifier increase the signal magnitude sufficiently to reduce the amount of processing later? Is reprogramming and reconfiguring the processing important enough to put functions in software and remove the analog circuits?

All of these trade-offs aim at reducing cost, unnecessary functions, development work, and circuitry. Some also provide the capability to maintain, repair, and upgrade a product once in the field.

Several different people make trade-offs in hardware design. Product designers, such as industrial designers, constrain the external package, particularly its form and the user interface (buttons, keyboards, and displays). Sometimes management steps in and mandates specific form or package. All these constraints will drive some requirements on the circuitry. Generally, though, engineers make most of the hardware trade-offs.

1.10.4 Cooling

Bigger processors and larger power-handling circuits consume and dissipate more power, though they sometimes are necessary to accomplish the design task. Greater power dissipation generates large thermal gradients in the circuit boards that can stress them and eventually lead to failure. Cooling of these components and circuits allows higher energy density and reduces thermal gradients. Unfortunately, cooling increases cost, adds greater complexity, and lowers reliability of the system. Cooling adds to the cost of the system both in up-front component costs and downstream in greater maintenance needs.

Cooling of electronics can take many forms: radiant, convection, forced convection, liquid cooling, and refrigeration. Each successive type of cooling, from radiant through refrigeration, increases the capacity of heat flow between 5 and 100 times [21]. But, each successive type of cooling also increases the COGS and the maintenance costs by a similar amount.

1.10.5 Power

One aspect of hardware design and trade-offs is the source and conversion of power. Often it is line power from the mains—at various voltages and frequencies. Sometimes, a more portable or remote source of power is needed, possibly batteries, fuel cells, or more exotic forms such as solar cells or mechanical energy.

The product's application will often suggest the choice of power. Portability is probably the biggest factor in determining the power source. Sometimes, the customer or industrial designer or management will request, possibly require, a particular form or source of power.

Once the source of power is determined, you or your engineering team needs to deal with converting the power into a usable form, typically low-voltage DC power. Several forms of power conversion exist: primarily switching DC-DC converters and linear regulators.

1.10.6 Software

Religious wars are waged over software. What language? What development platform and tools? What development process? Should code reviews be implemented? (Yes!!)

Software should be carefully planned, developed, reviewed, tested, and integrated with the system. As with hardware, the product application, standards, and regulatory environment will narrow the choices and trade-offs for software.

The engineer or engineering team often determines the software issues. Sometimes standards, particularly in medical devices, will significantly drive software development and processes. Unfortunately, and too often, company legacy will drive software issues too. As I said, religious wars are fought over software.

1.10.7 Hardware vs. Software

Hardware vs. software is a classic trade-off. There is no final answer for the best trade-off. Hardware wears out and fails under stress; software does not. Hardware and physical devices can have quite predictable limits and failure modes. Software easily hides complex interactions. Hardware usually is very cheap, while software development is very expensive.

The development effort for software can be thousands or millions of times more expensive than the component hardware costs; it usually is much more expensive than the hardware development too. You may specify a 40¢ microcontroller and spend several months to design the circuitry; software development, on the other hand, may require several years of engineering effort to program, test, and certify, particularly if it is in a medical product.

Because software consumes so much effort in developing most products, it should drive the selection of hardware in most cases. Hardware and software engineers generally make this trade-off. Unfortunately, company tradition or management unwillingness to buy new tool support can really hamstring your efforts and end up extending development time and costing much more than it should.

1.10.8 Buy vs. Build

A classic trade-off is to decide between buying components and building them. Components and subsystems can be hardware or software or both; examples include motherboards, standard interface modules, or commercial operating systems.

Purchasing components or subsystems can reduce the time to develop the product. Building, however, allows you to optimize for a narrow set of requirements. Furthermore, if production quantities are low (less than hundreds per year) and the market demands a short delivery time then you should buy commercial-off-the-shelf (COTS) components.

To make the decision between buying and building, first define the parameters of your decision. Then list their priority. Reference 2 (Chapter 15) explains some of the relationships and heuristics that go into the decision. Seven important parameters (Figure 1.9) are cost, quantity, time to market,

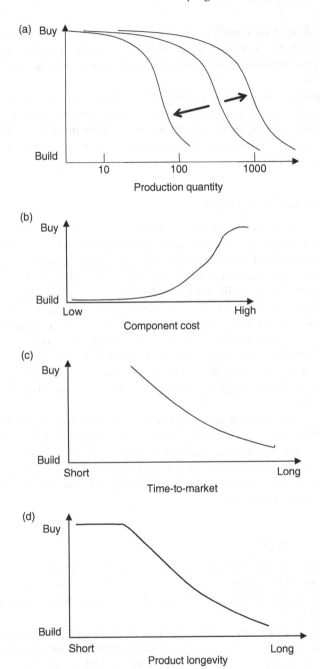

FIGURE 1.9
Seven parameters that factor into the buy vs. build decision: (a) production quantity, (b) component and subsystem costs, (c) time-to-market, (d) product longevity, (e) specification complexity, (f) resources, and (g) technical support. (© 2006 by Kim Fowler, used with permission. All rights reserved.)

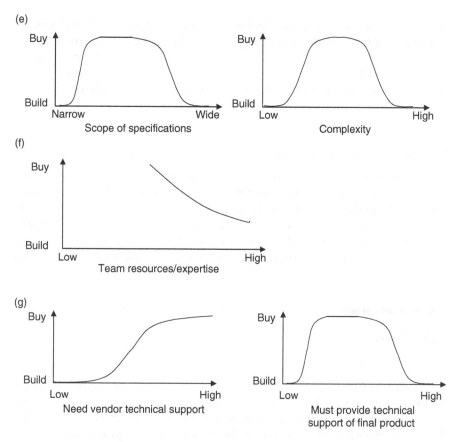

FIGURE 1.9
Continued

longevity on market, specifications, company resources, and necessary technical support you must receive and provide.

Generally, engineers in conjunction with management make the trade-off to buy or to build.

1.10.9 Manufacturing, Assembly, Disassembly, and Disposal

Mechanics, packaging, and circuitry all affect manufacturing, assembly, disassembly, and disposal. Manufacturing and assembly ease can mean one-piece castings and press fits rather than multiple pieces screwed together. Disassembly means that a piece of equipment can be taken apart for repair or upgrade or disposal.

The trade-off is between design time and ease of manufacturing (or assembly or disassembly or disposal). Products sold in large volume need a lot of effort in design to make manufacturing efficient and as inexpensive as possible. A product sold in small volumes will tend toward more labor-intensive manufacturing in trade for faster development.

Inputs from both the engineering team and management go into the decision to trade-off design time with ease of manufacturing (or assembly, or disassembly, or disposal). The authority to make the trade-off depends on the company, the industry, and the application.

1.10.10   Test and Maintenance

Test and maintenance intertwine with manufacturing, assembly, disassembly, and disposal. Whatever makes assembly or disassembly easier, will usually make test and maintenance easier. Test and maintenance does require two further stages. One is access to the functioning internals: circuitry, mechanisms, and replaceable substances (e.g. lubricants). The other is specialized servicing equipment to test the product or replenish its consumables.

1.11   Tests

1.11.1   Types of Tests

Tests come in a variety of shapes, sizes, and purposes. Some confirm aspects of simulation and analysis. Some help debug problems. Some assess manufacturing quality or screen components and subsystems. Some can validate the design intent. Many verify that the design meets requirements.

Tests can be formal or informal. Their results become part of the audit trail of the project design and typically should be properly archived (bet that does not happen in most situations—but it should!).

Most of the testing that I will focus on in this book is that done by engineers. Some, such as design reviews, can involve the entire team or company.

1.11.2   Laboratory Tests

These tests serve one primary purpose—to reveal basic underlying causes. They might be truly scientific tests where you hypothesize, test, interpret the results, and repeat to understand the fundamental physical phenomenon. Or they might be simple debugging to understand the anomalous behavior of a piece of software. The more *ad hoc* tests are also called bench tests.

These laboratory tests are strictly the purview of engineers and software developers. They can perform these tests just about any time during development. Usually, the tests constrain the environment to reduce interactions and clarify a single mechanism or problem.

1.11.3   Inspection

Inspection has two purposes—to find problems and to assure quality. Debugging a problem often involves inspection; it is looking at the circuit or the source code to find a misalignment or incongruity. Inspection also is used in manufacturing to give a visual assessment of quality—checking the workmanship for defects in components.

Engineers and technicians typically use inspection during debugging in the lab. I have had one technician tell me that more than 60% of problems he found were through inspection. Manufacturing personnel use inspection as one criterion for accepting components or subsystems.

1.11.4   Peer Review

Peer review is another form of test. It can be either informal or formal.

Informal peer review is primarily used to help debug. How many of you have asked a colleague to review a piece of code or circuit? (Is it not funny how just stating the problem sometimes will suggest a solution?) Often this informal peer review is among the engineers and designers.

Formal peer review should be a part of the development process. Code inspections are a type of peer review. Usually, this form of peer review only requires people immediately involved in the product design.

Giving design reviews, as covered in Section 1.5, is another type of peer review. As mentioned, formal design reviews can involve many more people—designers, engineers, technicians, marketing, manufacturing, and customers. These can occur throughout the development cycle.

1.11.5   Subsystem Tests—Hardware

The subsystem tests for hardware are formal exercises to confirm that the module performs according to requirements; these are primarily a verification of operation. Examples of hardware subsystem tests might include the following (this is not an exhaustive list):

- Cable orientation and connector mating
- Weight and volume
- Power consumption
- Signal levels
- Communication protocol and bandwidth
- Electromagnetic capability (EMC)
- Display function and brightness
- Keyboard operation

Generally, the design engineer or design team specifies the necessary tests; they may also be the ones to run the tests. For larger projects or mission-critical systems or medical devices, a dedicated tester or engineering team might be responsible for running the tests; in this situation, selecting tests is a collaborative effort between the design team and the test team.

These tests are performed before integration either in the laboratory or in a test facility.

1.11.6   Subsystem Tests—Software

The subsystem tests for software are formal exercises to confirm that each software module performs according to requirements; often these are called unit tests and are primarily a verification of operation. Examples of software subsystem tests might include the following (this is not an exhaustive list):

- Human interface—input and display
- Proper function and calculation of parameters
- Interface protocols
- Rejection of out-of-bounds values
- Fault tolerance
- Completion of tasks within deadlines

Tests of software take two different forms, static and dynamic. Static tests are sometimes called "glass box" or "white box" tests because you can view the structure of the code. Dynamic tests are sometimes called "black box" or "closed box", tests because you do not see the structure of the code, rather the tests exercise the operating software.

Static tests examine and critique the structure of the code through either manual or automated means. Static manual tests include code inspections and walk-throughs, which can be very effective in finding problems with the design and the translation of the original intent. Static automated tests use software tools to check for statement coverage, branch coverage, and path coverage [22].

Dynamic tests are behavioral tests that exercise the inter-module coordination in the operating software. These tests rely on understanding the requirements of the design and testing the software to see if it meets each requirement [22].

Just as in the previous section, generally the design engineer or design team specifies the necessary tests; they may also be the ones to run the tests. For larger projects or mission-critical or medical devices, a dedicated tester or engineering team might be responsible for running the tests; in this situation, selecting tests is a collaborative effort between the design team and the test team.

These tests are performed before integration either in the laboratory or in a test facility.

1.11.7 Environmental

Environmental tests confirm hardware operation to appropriate extremes of temperature, humidity, pressure, and vibration. Again, the requirements specify the parameters exercised in these tests; they are another form of verification of operation.

The specific application drives the environmental tests. Consumer appliances usually do not need to operate in an ambient environment below 0 °C or above 50 °C. They may have to withstand occasional liquid splashes or high humidity. Military devices and avionics do need to operate over wide temperature ranges, often specified between −40 °C and +80 °C; sometimes even wider. Military devices may also have to endure wide humidity ranges to 100% condensing (i.e., meaning water is soaking the circuitry). Medical devices may not need wide temperature tolerances but implanted devices certainly need to withstand the corrosive fluids within the human body. Finally, spacecraft have many environmental parameters to consider: radiation tolerance, pressure, outgassing, launch vibration, and huge temperature swings.

Generally, the design engineer or design team specifies the necessary environmental tests; they may also be the ones to run the tests. For larger projects or mission-critical or medical devices, a dedicated tester or engineering team might be responsible for running the tests. For such a situation, selecting tests is a collaborative effort between the design team and the test team.

These tests may be performed separately. Usually, they are reserved for final tests following integration in a specialized test facility.

1.11.8 Manufacturing

Manufacturing tests assure quality of the fabricated item. Examples of manufacturing tests might include the following (this is not an exhaustive list):

- Fit and finish
- Visual inspection of assembly
- Weight and volume
- Power consumption within limits
- Signal levels within limits
- Display function and brightness within limits
- Mechanism operation within limits

Please note: manufacturing tests are not verification of the design! Manufacturing tests only assure quality of construction. They do not test whether the device is a suitable solution of the requirements, and they are not a thorough exercise of all modules. Verification of design does that and it is done separately.

Generally, the design engineer or design team specifies the necessary tests in collaboration with manufacturing. Generally, dedicated technician or test team run the manufacturing tests.

These tests are performed in the manufacturing facility.

1.11.9   BIT, BITE, and Simulators

Mission-critical equipment sometimes incorporates build-in-test (BIT) and built-in-test equipment (BITE). This is a more specific concern when high availability and reduced maintenance time is imperative. BIT becomes an integral portion of the system.

For larger systems and projects, you may need various simulators that provide a subset of system functions and interfaces. Simulators can substitute for various subsystem modules and allow preliminary tests of the core system modules; they can improve the parallel effort often needed to reduce time. Simulators range from a simple box that provides a rudimentary interface function to a complex cockpit simulator for testing avionic instruments. Simulators are a specialized form of test equipment. Sometimes, they migrate over to become BITE or support equipment for fielded systems. Engineers design and use simulators to develop the design of the target system. Technicians might run the more complex simulators when doing later test and integration.

Generally, the design engineer or design team specifies the BIT and BITE. For larger projects or mission-critical, a dedicated engineering team might be responsible for specifying both BIT and BITE; such specification of test equipment is a collaborative effort between the system design team and the system test team.

These tests are typically performed by trained technicians in the field.

1.12   Integration

1.12.1   Difference Between Integration and Test

Integration is concerned with the multiple interactions between modules and subsystems. Integration includes the building of the hardware modules and software system in a controlled fashion. Integration verifies the total system operation according to specification. Integration also moves toward validation of the intent of the design.

Tests tend to focus on single-point results, while integration focuses on subsystem interactions. Integration also helps qualify the design, so it occurs only toward the end of development. Afterwards, manufacturing tests take over from integration. Integration assures the quality of design while manufacturing test assures the quality of production.

1.12.2   Hardware

Integration of hardware includes fit checks, cable routing, connector mating, power transients, power consumption, compatible signal levels, compatible communication protocols, mechanism operation, and optical alignments. The idea is that things should physically fit and function together.

Generally, the same groups of people who design and perform the hardware testing also do the hardware integration. Integration usually takes place either in the laboratory or in a dedicated facility. Sometimes it is on the factory floor.

1.12.3   Software

Integration of software includes assuring correct interaction between modules, proper and expected input/output responses to the system, and fault tolerance. The goal is that software performs as specified and handles unexpected circumstances in an appropriate and predictable manner.

Generally, the same groups of people who design and perform the software testing also do the integration of the software. Integration usually takes place either in the laboratory or in a dedicated facility.

1.12.4   System

The system integration brings all the elements together—hardware, software, mechanics, optics, and so forth. Integration is a systems concept; hardware and software integration are somewhat meaningless outside the context of the system function.

System integration addresses both verification and validation. Integration includes the following examples (this is not an exhaustive list):

- Correct interaction between all modules
- Proper and expected input/output responses to the system
- Human interface—input and display
- Fault tolerance
- Confirming the boundary of operations (the "envelope")

Generally, the same groups of people who designate and perform the testing also do the integration of the system.

Integration usually takes place either in the laboratory or in a dedicated facility. Sometimes it is on the factory floor. In the vast majority of cases, integration should take place before a field test. Even one-shot deals, like a missile booster, still go through rigorous testing and integration before launch.

1.12.5   Environmental

Environmental tests are often completed during integration. Usually, the interactions within a system need to endure environmental extremes as part of the qualification of the design. Typically, a dedicated team of technicians performs environmental testing during integration; they work in concert with test or design engineers. Environmental testing almost always requires some sort of dedicated facility to control parameters such as temperature, humidity, and pressure.

1.12.6   Field Tests

Field testing can be a form of integration because it exercises the system in day-to-day circumstances with real users. It occurs either with prototypes to refine a design or after rigorous integration. Depending on the size of the project or its phase of development, anyone of the following might conduct the field tests: design engineer, technician, test personnel, or marketing.

1.13   Manufacturing

1.13.1   Electrical and Electronic

Manufacturing of electronic circuits is a complex business. There are integrated circuit (ICs, which constitute an entire industry unto itself, and I will not discuss them in this book), PCB fabrication (another subculture within manufacturing), component assembly and soldering, cable and connector assembly, and module placement and connection.

1.13.2   Mechanical

Even though this is primarily a book on electrical and software issues, mechanical issues and manufacturing never go away; they are always a consideration. Products that are primarily electronics and software still have mechanical concerns—mechanisms (motors, gears, linkages, doors, etc.), circuit-board attachment, and cabling tie down.

There is always a chassis that supports the circuit boards and power supply. If the application is avionics, then the chassis might be one piece, which requires casting or machining. Otherwise, a chassis might be a COTS item built up from plastic or sheet metal, fasteners, and a back-plane PCB. Any type of enclosure requires fabrication and assembly.

1.13.3 Assembly

Assembly is the putting together of components into a subsystem. It can also be the putting together of subsystems into a final product. Manufacturing assembly design focuses on ease, reducing mistakes, and reducing the cost of labor.

Many companies outsource the manufacturing and assembly of the PCBs, which are labor- and capital-intensive activities. In essence, the company becomes an integrator of the final product. For high-volume manufacturing, companies might have a dedicated facility or they might send the entire production to low-cost business locations around the world.

1.13.4 Tests

Each stage within manufacturing usually includes some form of test, even if it is only a visual inspection. Besides inspection, manufacturing tests might include automated tests, functional tests of circuits, mechanical alignment tests, and power-on tests. The case studies will outline some of these manufacturing tests.

1.14 Support

1.14.1 Installation and Commissioning

Larger projects with many subsystems that are very large or very expensive require a well developed installation and commissioning stage. For all the products in these case studies, that is not the situation. I will leave installation and commissioning to the process industries and large industrial applications.

1.14.2 Training

Training varies widely from product to product. Generally, the higher the volume, the more intuitive the product's operation should be. I really do not want to spend more than a few moments learning how to use my new coffeemaker. On the other hand, the more specialized the product, the greater is the need for training.

Training requires that you understand the customer, the user, and their needs. You must also understand the product's purpose and operation. The user's manual is the first line of defense in training the user. It must be well written and easy to use. It should almost entice the user to read it—but do not despair when they do not. Users will read it when they cannot get the product to work!

If the product has high-volume sales and is fairly intuitive, then little training is required. For most training in less-specialized equipment, a technician or an instructor can give the training. If it has few sales and is highly specialized equipment, then an engineer or even the designer might train users. In this case, training is often factored into the cost of sale. Otherwise, training is a separate expense for the customer.

Training can be either on a customer's site, at a classroom facility, or in the factory (company's location).

1.14.3 Logistics

Logistics vary widely from product to product and depend solely on the application. Portable devices need battery recharging or replacement. Various expendables in other applications need to be replenished. Sometimes, it is new lubricants or fuels or cooling liquids. Logistics and supply chain is also a field of operation unto itself, and falls outside the scope of this book.

Non-professional personnel or even the customer can be trained to handle logistics for some of the devices described in these case studies. Highly trained technicians and even engineers might handle logistics for certain mission-critical and safety-critical devices.

1.14.4 Maintenance

Maintenance first depends on the philosophy of the application and the product. Smaller, lower-cost consumer appliances are typically "throw-away." Larger, more expensive consumer appliances tend toward subsystem replacement (e.g. replacing the entire circuit board instead of expending additional labor costs to diagnose and repair the defective component on the circuit board). Economically, it is nonsensical to repair smaller, cheaper appliances. Much more expensive devices and equipment will usually have a defined maintenance philosophy. Here are few of those philosophies:

- Replace subsystems
- Repair subsystems
- Periodic diagnosis and preventive maintenance
- Condition-based monitoring
- BIT and BITE

If you adopt a philosophy of repair rather than one of replacement, be aware that it might take down the equipment and operation for a significant amount of time, reducing its availability. Repair also requires trained labor and dedicated equipment. Labor ranges from highly trained technicians and engineers to depot-level skilled labor to minimally trained personnel. The equipment ranges from simple multimeters to specialized automated instruments.

1.14.5 Technical Support

Technical support is the superset that includes maintenance and repair. It also includes a variety of customer support functions e.g., answering basic questions about operation, handling warranty claims, or training of sophisticated and complex functions. A call center staffed by trained personnel with capabilities of a technician can handle numerous basic questions. Training can range from simple marketing concerns staffed by junior staff to highly sophisticated operations handled by expert engineers. Training can be on the factory site or at the customer's location, depending on the size and cost of the product. A large, expensive product may include on-site training in its initial cost. Smaller, less expensive products may warrant fee-based training at the factory.

1.15 Disposal

1.15.1 Recycling

Recycling of old and discarded electronic products is an international concern. Salvaging and recycling in rising economies return some materials for minor income. Unfortunately, the economic gain offsets the health effects in some of the poorest places. Improperly disposed heavy metals are leaching into the ground water and soils. Trash litters these regions, which increases the health risks from disease and infection.

International pressures are mounting to recycle the entire product: enclosures, batteries, lead, precious metals, lubricants, and plastics. The RoHS regulations, instituted in Europe and in the United States (as of July 2006), require that manufacturers take back and recycle disposed products. This concern is greater for appliance and consumer manufacturers than for those who build specialized, custom equipment.

As engineers and designers, we must be aware of these regulations and incorporate them into the requirements during design. Regulatory and even marketing personnel will need to be involved in specifying these regulations.

1.15.2 RoHS and WEEE

The RoHS Directive places a new burden on manufacturers. RoHS requires the phasing out of lead, mercury, cadmium, chromium, halogen, bromide, and some fire retardants from all products in an effort to limit the environmental damage on disposal. This directive particularly affects electronic products and devices because lead is a significant component of traditional solder, and cadmium is used in nickel–cadmium rechargeable batteries. RoHS became effective in the United States on July 1, 2006.

Until 2006, about 90% of all electronic devices and equipment ended up in landfills when disposed. There the heavy metals—lead, cadmium, mercury, and chromium—leached into the soil. The waste from electrical and electronic equipment (WEEE) Directive 2002/96/EC requires the treatment, recovery, and recycling of electric and electronic waste after August 13, 2006 in the European markets. Complying products must have a sticker that shows a wheeled disposal barrel, called the "Wheelie Bin."

1.16   Liability

1.16.1   Safety

Every product has some level of safety defined by the market and by disparate sets of government regulations. The simplest products, such as hobby products, have essentially no regulations, if they do not connect to line power and do not supply batteries in the product. The most complex products, such as satellite subsystems, may have to conform to international treaties in some situations (e.g., containing a thermoelectric nuclear power system). Otherwise, they only need to adhere to excellent design practices and processes. Medical products arguably have the highest level of safety concerns, particularly those that sustain life or can threaten life if they fail.

Regulations range from UL to IEC (in Europe and the United States and spreading all over the world) to FDA approval for medical devices in the United States. Each case study will outline safety regulations needed.

Everyone in a product team or company should have some awareness of safety. Engineers and designers must define the level of safety in the requirements early in development. For medical devices, regulatory personnel made up of either company staff or hired consultants will also be involved in specification of safety and the requirements.

Once specified during development, each safety requirement will have to be tested by your test team to show conformity to safety standards, as well as to be a satisfactory solution to the operational requirements.

1.16.2   Legalities

There are legal liabilities other than safety.

We engineers and designers have to be aware of the intellectual property rights of others and not copy competing designs [23,24]. Patent protection and infringement is important enough for you to receive appropriate legal advice while developing and marketing a product—this advice applies to engineers, designers, marketing, and management.

Contract law is another area with which each of us should also have some acquaintance, especially business alliances and memos of understanding (MOUs). Again, you (engineers, designers, marketing, and management) should receive appropriate legal advice while doing business [24].

Our goal is to reduce liabilities to an acceptable level for our business or market. We also need to understand whom it affects and your degree of responsibility. See References 23 and 24.

1.16.3 Economics

Ultimately, when something does not work or fails, who is liable? Maybe another way to ask is, "Who pays? Does it affect you?"

One thing for sure, no one wants to undergo a product recall. Recalls are expensive; they cost your company money, and they damage the company's reputation for years. Success is short-lived; failure is long-term.

1.17 Priorities

Ok, some of you are probably saying, "Well, this all looks good, but I can't give the same attention to all of it." That is one of the big problems with design. The rest of this book is devoted to case studies to help you see where you might set your priorities and what remaining activities might not be as important.

1.18 Summary

This chapter sets a template for each of the case studies that occupy the vast majority of this book. The template should aid you to compare and contrast development issues in different products. Here is an outline of that template:

- Concept and market
- People and disciplines
- Architecting and architecture
- Phases
- Scheduling
- Documentation
- Requirements and standards

- Analyses
- Design trade-offs
- Tests
- Integration
- Manufacturing
- Support (training, logistics, maintenance, and repair)
- Disposal
- Liability

References

1. Ganssle, J. and Barr, M., *Embedded Systems Dictionary*, CMP Books, San Francisco, CA, 2003.
2. Fowler, K.R., *Electronic Product Development, Architecting for the Life Cycle*, 2nd ed., Oxford University Press, Oxford, UK, Chapter 2, 2008 (to be published).
3. Laplante, P., *Real-Time Systems Design and Analysis*, 3rd ed., IEEE Press and Wiley-Interscience, A John Wiley & Sons, Inc. Publication, Piscataway, NJ 2004.
4. Oshana, R., *DSP Software Development Techniques for Embedded and Real-Time Systems*, Newnes, an imprint of Elsevier, Boston, MA, 2006.
5. Williams, R., *Real-Time Systems Development*, Butterworth-Heinemann, an imprint of Elsevier, Oxford, UK, 2006.
6. Christensen, C.M., *The Innovator's Dilemma, When New Technologies Cause Great Firms to Fail*, Harvard Business School Press, Boston, MA, 1997, p. xv.
7. Rechtin, E. and Maier, M., *The Art of Systems Architecting*, CRC Press, Boca Raton, FL, 1997, p. 254.
8. Moody, J.A., Chapman, W.L., Van Voorhees, F.D., and Bahill, A.T., *Metrics and Case Studies for Evaluating Engineering Designs*, Prentice Hall PTR, Upper Saddle River, NJ, 1997, pp. 2–7.
9. Colleague, Personal communication, August 2, 2006.
10. Ganssle, J., Managing embedded projects, *Embedded Systems Conference*, November 1998.
11. Schoff, G.H. and Robinson, P.A., *Writing and Designing Manuals: Operator Manuals, Service Manuals, Manuals for International Markets*, 2nd ed., Chelsea, MI, 1991.
12. EMC Standards, *Compliance Engineering*, Vol. 21, No. 1, 2004 Annual Reference Guide, pp. 75–82.
13. ESD Standards, *Compliance Engineering*, Vol. 21, No. 1, 2004 Annual Reference Guide, pp. 103–105.
14. Telecom Standards and Regulations, *Compliance Engineering*, Vol. 21, No. 1, 2004 Annual Reference Guide, pp. 139–142.
15. Product Safety Standards, *Compliance Engineering*, Vol. 21, No. 1, 2004 Annual Reference Guide, pp. 165–172.

16. Wiegers, K., 10 Requirements Traps to Avoid, *Software Testing & Quality Engineering*, January/February 2000.
17. Reinertsen, D., In Search of the Perfect Product Specification, *IEEE Instrumentation & Measurement Magazine*, Vol. 3, No. 2, June 2000, pp. 28–31.
18. Allen, F., The Letter that Changed the Way We Fly, in *American Inventions, A Chronicle of Achievements that Changed the World*, Barnes & Noble Books by arrangement with American Heritage, a Division of Forbes, Inc., New York, NY, 1995, pp. 102–109.
19. Pazemenas, V., Rapid Development for Medical Products, *IEEE Instrumentation & Measurement Magazine*, Vol. 3, No. 2, June 2000, pp. 32–37.
20. Bogatin, E., *Signal Integrity: Simplified*, PTR Prentice Hall, Upper Saddle River, NJ, 2004.
21. Fowler, K., *Electronic Product Development, Architecting for the Life Cycle*, Oxford University Press, New York, NY, 1996, p. 373.
22. Freeman, H., Software Testing, *IEEE Instrumentation & Measurement Magazine*, Vol. 5, No. 3, September 2002, pp. 48–50.
23. Konold, W., Tittel, B., and Frei, D., *What Every Engineer Should Know about Patents*, CRC Press, Boca Raton, FL, 1988.
24. Silver, C., *The Pocket Lawyer For Engineers*, Elsevier Science-Newves, Boston, MA, 2008 (to be published).

2

Variations on the Theme—Considerations for Mission-Critical Equipment and Medical Devices

---

2.1 Development Processes

There are markets and products that require rigorous and well-defined development processes. Examples include industrial equipment, cars, trucks, tractors, military equipment, avionics, spacecraft, and medical devices. These types of products are called mission-critical equipment or, sometimes, safety-critical equipment.

Mission-critical means that the system or equipment must operate in a satisfactory fashion that is necessary for a larger overall operation. Should the equipment either fail or operate in an inappropriate manner, it compromises the overall operation or mission. An instrument on a spacecraft is an example of mission-critical equipment; its failure would seriously impair the space mission.

Safety-critical is very similar to mission-critical. The one distinction is that it affects human life. Should the equipment either fail or operate in an inappropriate manner, it puts human life and limb at risk.

This chapter will suggest some general processes used in design and development for mission-critical and safety-critical systems. It will highlight variations in the processes outlined in Chapter 1; these processes include standards, regulations, certification, documentation, development phases, plans, procedures, reviews, configuration control, archiving, and traceability.

2.1.1 What, Why, Who, When, Where, and How

You must develop mission-critical and safety-critical systems thoughtfully and carefully, with traceable evidence for every detail. Like a good journalist, you and your team must establish the "what, why, who, when, where, and how" in everything that you do.

What: You are establishing or operating a quality system that supports development of mission-critical and safety-critical equipment. It encourages rigorous and correct operation of procedures to help you build an appropriate and correctly operating product. Part of the quality system is a project plan, which describes the particular activities that must occur during development.

Why: A quality system should provide a complete audit trail throughout development. Anytime during development, the quality system should be able to demonstrate the entire "what, why, who, when, where, and how" of your project.

Who: Relevant personnel include anyone involved at any point in the development of mission-critical and safety-critical equipment. Most of the focus in this book will be on designers and engineers, but the "who" includes folks in management, administration, purchasing, manufacturing, and plant maintenance as well.

When: The project plan describes the phases of development when particular activities occur.

Where: The project plan describes where particular activities occur.

How: The essence of any quality system is to iterate through a simple five-step process in each stage of design and development. The steps are plan, execute, review, report and update (PERRU). This five-step process repeats at every level of effort, from preparing the simplest component to the highest-level architectural specification.

- Planning includes sketches, doodles, and notes in engineering notebooks, discussions with colleagues, meetings, and documentation of projected effort and responsibility.
- Execution means that you carry out the plans.
- Review includes code inspections, design reviews, tests, analyses, and simulations.
- Report means that you document the results of the three previous steps for immediate use (i.e. update) in the next iteration and for traceability later.

2.1.2 Economics

Mission-critical, safety-critical, and medical equipment all tend to be specialized, lower-volume products that generally have higher profit margins (ratio or difference between the price of the final product and the cost to develop, manufacture, and distribute it). One of the main reasons is the expense of non-recurring engineering (NRE) for certification and government approvals.

2.2 People and Disciplines

Quality systems include everyone on the team and in the company: engineers, quality assurance (QA), marketing, management, procurement, manufacturing, and administration.

2.3 Architecting and Architecture

2.3.1 Process

Most mission-critical, safety-critical, and medical projects operate under a form of process control. Many companies use the International Organization for Standardization, or ISO, standards and certification. More recently, some favor the Capability Maturity Model Integration, or CMMI, form of process control. More on ISO and CMMI appears below.

The design and development for these projects often use the modified Waterfall process, called the V-model, as shown in Figure 1.4. This is because any type of approval or certification does not allow further modification or upgrade to the design of the product. What you certify is what goes on the market. You cannot change it later without another round of approvals. Furthermore, I would venture to guess that the V-model process might be more traditional and its traceability better understood (even though traceability can be used with any process).

Two different sets of standards and regulations from the United States can guide the architecting and development of mission-critical, safety-critical systems, and medical devices. For medical devices, I will quote from the Food and Drug Administration's (FDA's) *Design Control Guidance for Medical Device Manufacturers*, March 11, 1997. You can find it at Reference 1. For mission-critical systems, I will quote from the U.S. DO-178B for avionics [2].

These standards give frameworks for development. They do not specify how each step is to occur. They only provide guidance with phrases like "represents current thinking on this topic" and "establishes a framework."

2.3.2 FDA Design Control Guidance

Intent, Purpose and Scope: " The regulation does not prescribe the practices that must be used. Instead, it establishes a framework that . . . provides manufacturers with the flexibility needed to develop design controls that both comply with the regulation and are most appropriate for their own design and development processes. . . . This guidance is intended to assist manufacturers in understanding the intent of the regulation" [1]. The purpose of the FDA Design Control Guidance is, " to assist manufacturers in understanding quality system requirements concerning design controls. . . . Design controls are an interrelated set of practices and procedures that are incorporated into the design and development process, that is, a system of checks and balances. Design controls make systematic assessment of the design [1]." The scope is, "Design controls are a component of a comprehensive quality system that covers the life of a device" [1].

Application of design controls: Figure 2.1 outlines the general application of the design controls. "Each design input is converted into a new design output; each output is verified as conforming to its input; . . . [1]." Basically, you can use the five-step process (plan, execute, review, report and update) and cover this intent. Figure 2.1 is similar to the V-model of development, but it can be applied to other forms of development too. "Manufacturers should use processes best suited to their needs" [1].

Design input: "*Design input* means the physical and performance requirements of a device that are used as a basis for device design" (21 CFR 820.3(f)) [1]. Design inputs are a complete, consistent, and unambiguous set of requirements. They specify intent, the "who, what, when, where, why," and indicate function, performance, and the interfaces. They should also give quantitative limits with measurement tolerances, characterization of the operational environment, and complete, relevant citations. Design inputs are *not* the following:

- Marketing memoranda or concept documents
- Prototypes
- "How" of design and development

Marketing memoranda or concept documents can provide the starting point for design inputs, but they do not provide complete coverage. Prototypes lack safety and ancillary features.

Design output: "*Design output* means the results of a design effort at each design phase and at the end of the total design effort. . . . The total

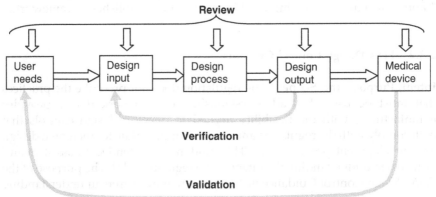

FIGURE 2.1

Outline of FDA design controls. (© 2002 by Kim Fowler, used with permission. All rights reserved.)

finished design output consists of the device, its packaging and labeling, and the device master record" (21 CFR 820.3(g)) [1]. The device master record contains the production specifications (e.g. assembly drawings, bill of materials, installation procedures, packaging, and labeling) and descriptive materials (e.g. risk analysis results, software source code, and the results of verification). You must document, review, and approve design outputs before release.

Design review: *"Design review* means a documented, comprehensive, systematic examination of a design . . ."* (21 CFR 820.3(h)) [1]. The design review describes a variety of reviews, procedures, and reviewers. The results of all reviews are contained in the design history file (DHF); it is " a compilation of records which describes the design history of a finished device" [1].

Design verification: Design verification has two components: verification and validation. Verification means that the device meets the quantitative requirements. Verification is an objective comparison of test results to metrics in the specifications. Validation means that the device satisfies the intent of user desires. Validation is a more subjective measure than verification, but validation addresses the system's overall function.

Design Transfer: Design transfer is a set of "procedures to ensure that the device design is correctly translated into production specifications" (21 CFR 820.3(h)) [1]. A design transfer contains items like the device master record, training materials, and manufacturing aids.

2.3.3   DO-178B

Purpose and scope: DO-178B is "to provide guidelines for the production of software for airborne systems and equipment that performs its intended function with a level of confidence in safety that complies with airworthiness requirements" [2]. The scope is, "those aspects of airworthiness certification that pertain to the production of software for airborne systems and equipment used on aircraft or engines." The document, "does not provide guidelines concerning the structure of the applicant's organization, the relationships between the applicant and its suppliers, or how the responsibilities are divided" [2].

Like the FDA *Design Control Guidance*, DO-178B recognizes that you can develop software in many different ways. It states, "This document recognizes that the guidelines herein are not mandated by law, but represent a consensus of the aviation community. It also recognizes that alternative methods to the methods described herein may be available to the applicant" [2].

Definitions: DO-178B categorizes failure under five headings:

- Catastrophic
- Hazardous/severe-major
- Major

- Minor
- No effect

The level of the software is then defined according to its potential failure conditions [2]:

- Level A for potential catastrophic failures
- Level B for potential hazardous/severe-major failures
- Level C for potential major failures
- Level D for potential minor failures
- Level E for potential no effect failures

Defining your software accordingly will then affect what level processes you use under DO-178B.

Processes: "DO-178B is primarily a process-oriented document. For each process, objectives are defined and a means of satisfying these objectives are described" [2]. The processes include [2]

- Software planning
- Software development
- Verification of outputs of software requirements
- Verification of outputs of software design
- Verification of outputs of software coding and integration
- Testing of outputs of integration
- Verification of verification process results
- Software configuration management
- Software quality assurance
- Certification liaison

2.3.4   Process Control

The development of mission-critical and safety-critical systems relies on some form of process control. The standards and regulations leave the choice to you with the intent that you cover important issues and concerns thoroughly. Process control includes configuration management and quality assurance. Components within configuration management include version control, archiving capabilities, and audit trails.

Commercial software tools are available to help you with process control and configuration management. These tools can provide various modules that provide object modeling, requirements management, version control, archiving facilities, and capture of test results for comparison to the requirements.

What follows are selected issues within process control. These are some of the things that you must do to develop mission-critical and safety-critical systems.

Performance verification: Another critical element of systems engineering is ensuring that every requirement is objectively measured and met. Therefore, the systems engineering team must develop a thorough compliance matrix that identifies the verification method for each requirement. The verification may be performed through one or more of the following four methods:

- *Test*: Most requirements and specifications will be verified by test or supported by quantitative test data. Testing will take place at various levels that may include component, board, module, assembly, or system.

- *Inspection*: This approach applies to requirements and specifications that describe a design characteristic or method. An example of a requirement that would be verified by inspection is "the application software shall be coded in C."

- *Demonstration*: Verification by demonstration is applied to some qualitative requirements and specifications that cannot be tested. For example, a requirement such as "a failure of one instrument shall not prevent successful operation of all other instruments" would be verified by demonstration.

- *Analysis*: Some requirements cannot be verified by test alone because of cost or physical limitations. In this case, analysis and simulation may help ensure that the system will meet its requirements. This is the effort of last resort.

System Validation: A critical element of systems engineering is ensuring that every requirement matches use or customer intent. Therefore, the systems engineering team must confirm that each requirement matches the expressed intent and has traceable specifications.

2.3.5 Project Risk Management

One of the most important things that a good process must do is accommodate risk management. Risk is the concern over the potential inability to stay within defined cost, schedule, performance, or safety constraints.

The FDA Design Control Guidance states, "Risk management is the systematic application of management policies, procedures, and practices to the tasks of identifying, analyzing, controlling, and monitoring risk" ([1], p. 5). Risk management continues throughout the development life cycle; it identifies risk, develops risk mitigation plans, then tracks and closes out those identified risks. Risk management includes planning for risk, assessing risk areas, developing risk-handling options, monitoring risks to determine how risks have changed, and documenting the overall risk management program.

Formal development activities to identify risks include Event Tree Analysis (ETA), Fault Tree Analysis (FTA), and Failure Modes and Effects Analysis (FMEA). Another less obvious activity is margin management, which helps to constrain and reduce risk as the project proceeds.

ETA helps you understand how your product responds to each potential and possible circumstance. It is a procedure to establish the relationship between an event and the causes of that event.

FTA is a top-down analysis. It is the process of establishing the relationship between a fault—an undesirable behavior or system state—and the causes of that event. FTA examines the causes of a bad result.

FMEA is a bottom-up analysis; it might be viewed as the reverse process of FTA. FMEA determines what can go wrong from a given cause. The FMEA picks up where the FTA leaves off at the component and unit level. In general, the FMEA considers all credible failure cases (e.g. high, low, or intermittent) for each component. The FMEA will help ensure that no unexpected single-point failures exist in the system.

Risk analysis establishes the likelihood of problems and the attendant severity. Margin management uses these metrics to then manage the system margins. It can help you avoid unexpected problems encountered during the system development.

2.3.6 Architecture

Mission-critical, safety-critical, and medical products require careful design. Dependability is a major topic in that design. Careful design to achieve the various stages of dependability can use a number of techniques, including (but not limited to) stress margins, redundancy and error checking, interlocks, fail-safe, trapdoors, limp-home, and fault tolerance. These can all help make devices and equipment that survive faults, failures, and problems, or at least degrade gracefully. Unfortunately, they are accompanied by a disproportionate increase in cost and development time.

2.3.7 Interfaces

Interfaces are important for partitioning a design into appropriate units and modules. Partitioning the development along physical, software, and functional lines should be a considerable part of the architecture and early design effort.

Among the interfaces and design decisions, the human interface is critical and very difficult to design. The interface design is the system to the user. It should be appropriate and as intuitive as possible. Both ETA and FTA should consider the many ways humans can misuse an instrument.

2.4 Phases

Mission-critical and safety-critical systems follow five development phases (Table 2.1). Each has many recorded activities that are traceable and may be audited by government agencies. The case studies will each give detailed lists of activities within each phase. Case Study 7 in Chapter 10, in particular, has nearly identical outlines for the first three phases that follow. Please note that the first four phases described here are in terms of the design review that marks the end of each phase.

2.4.1 Concept

Conceptual Design Review

The Conceptual Design Review concludes Phase 1: Concept; it should present the mission goals, objectives, and constraints. It should demonstrate that the requirements of the project are understood and that the proposed approach will meet these requirements. Example items, from a satellite subsystem, to be addressed in the CoDR are

- Program organizational structure, organizational interfaces, schedule, cost, policy
- Review mission objectives
- Requirements
  - Mission: environment, host resources, experiment requirements
  - Performance: technical characteristics
  - Major system function and interfaces
- Research—literature, patent searches
- Design constraints and major trade studies performed
- Requirements process and management

TABLE 2.1

Examples of the Five Phases of Development within Mission-Critical and Safety-Critical Systems

Phase	Medical Device	Satellite Subsystem
1	Concept	Concept
2	Planning and scheduling	Preliminary design
3	Design and development	Critical design
4	Controlled release	Production (manufacturing)
5	Commercial release	Operation, maintenance, disposal

- System architecture
  - Concept
  - Hardware components
  - Software components
  - Operations concept
  - Support systems and logistics
- Planned integration and test program
- Development drivers
- Risk assessment

The output of the CoDR will constrain the baseline design following the closure of any action items resulting from the review. Long lead items, development support equipment, breadboard parts, and materials can be purchased following the successful completion of the CoDR.

2.4.2   Preliminary Design

Preliminary Design Review

The Preliminary Design Review concludes Phase 2: Planning and Scheduling or Preliminary Design. It is the first major review of the detailed design and will be held prior to the preparation of most of the formal design drawings and software code development. The PDR is held when the design advances sufficiently to begin either some breadboard testing or fabrication of engineering models.

A PDR presents the design and interfaces through block diagrams, signal flow diagrams, schematics showing logic diagrams, first interface circuits, packaging plans, configuration and layout sketches, preliminary analyses, modeling, and any early test results. The PDR should present the estimates of weight, power, volume, and the basis for the estimates, as well as the mechanical, power, thermal, and electronic designs with load, stress, margins, and reliability assessments. A PDR should specify the software requirements, design, structure, logic flow diagrams, computational loading, design language, and development systems.

An example PDR, for a satellite subsystem, may cover the following items:

- Technical objectives, requirements, general specification
- Closure of actions from CoDR, completion of research, trade-offs, and feasibility
- Requirements—function, performance, interface
- Analyses
  - Mechanical/structural design, analyses, and life tests

       – Weight, power, data rate, commands, electromagnetic interference (EMI)/electromagnetic compatibility (EMC)

       – Electrical, thermal, mechanical, and radiation design and analyses

- Software requirements and design
- Support equipment design
- System performance budgets
- Design verification, test flow and test plans
- Host interfaces and drivers
- Parts selection, qualification
- Risk analysis—ETA, FMEA, and FTA
- Risk margin and management
- Contamination requirements and control plan
- Quality control, reliability
- Materials and processes

The completion of the PDR and the closure of any actions generated by the review become the basis for the start of the detailed design effort and the purchase of parts, materials, and equipment.

2.4.3   Critical Design

Critical Design Review

The Critical Design Review concludes Phase 3: Design and Development or Critical Design. It will be held near the completion of engineering evaluation using the breadboard model of the project. It will be held prior to any design freeze and before any significant fabrication activity begins.

The design at CDR should be complete and comprehensive. The CDR should present all the same basic subjects as the PDR, but in final form. An example CDR, for a satellite subsystem, should include all of the items specified for a PDR, but updated to the final present stage of development process, plus the following additional items:

- Closure of actions from the PDR review
- Changes from the PDR review
- Final parts list
- Final implementation plans including engineering models, prototypes, flight units, and spares
- Final software design and process implementation
- Engineering model and breadboard test results
- Design margins

- Completed design analyses
- Qualification and environmental test plans and test flow
- Safety requirements
- Operations plan
- Updated risk analysis—ETA, FMEA, and FTA
- Updated risk margin and management
- Test
  - Plans
  - Status of procedures and verification plans
  - Test flow
- Schedule
- Documentation status
- Test history of the hardware
- Product assurance
- Previous anomalies, deviations, waivers, and their resolution
- Identification of residual risk items
- Plans for shipping containers, environmental control, and transportation

Results from the CDR and resolution of all the action items generated by it constitute the baseline design.

2.4.4　Production or Manufacturing

Production Handoff

The Production Handoff concludes Phase 4: Controlled Release or Production; it occurs prior to manufacturing. The purpose of the PHO is to ensure that the design of the item has been validated through the environmental qualification and the acceptance test program, that all deviations, waivers, and open items have been satisfactorily closed, and that the project, along with all the required support equipment, documentation, and operating procedures, is ready for production. Here are some example items from a satellite subsystem:

- Rework/replacement of hardware, regression testing, or test plan changes
- Compliance with the test verification matrix
- Measured test margins versus design estimates
- Demonstrate qualification/acceptance temperature margins
- Trend data

- Total failure-free operating time of the item
- Could-not-duplicate failures should be presented along with assessment of the problem and the residual risk that may be inherent in the item
- Project assessment of any residual risk
- Review of shipping containers, monitoring/transportation/control plans
- Ground-support equipment status
- Post-shipment plans
- System-integration support plans

2.4.5   Logistics, Maintenance, and Disposal

This is the market or service phase of the product. It usually is the longest phase. Its length depends on both the product's longevity in service and the development effort. Medical devices might last 10 or 15 years, while automobiles might go twice as long. Buses and long-haul transport may run for 50 years or more. Certification and government approval usually are long, arduous, and expensive processes that significantly increase the development effort and prevent rapid market cycles.

This phase is the most varied in its requirements depending on the product, its uses, and its customers. This phase does not have a general outline of main considerations that were found in the previous phases; it simply varies too much from product to product.

2.5   Scheduling

Mission-critical, safety-critical systems, and medical devices have activities that need scheduling in addition to the ones already listed in Chapter 1. Here are some activities that you should include in your planning:

- Meetings with potential customers, focus groups
- Research of regulations and standards
- Meetings with the appropriate regulatory bodies
- Integration and test (can be greater than 50% of the development effort for larger systems with many subsystems)
- Certification—tests at specialized facilities
- Technical and operational evaluation (for military projects)
- Clinical studies (for medical devices)
- Documentation

I repeated documentation because it takes on new dimensions in volume and complexity for medical devices and for mission-critical, safety-critical systems.

2.6 Documentation

2.6.1 Purposes

Besides the stated purposes in Chapter 1 (to record the specifics of development, to account for progress, and to instruct the extent of functionality), documentation serves several other purposes in mission-critical and safety-critical systems. They provide the only real form of traceability and the basis to address audits. You must have a documentation system to be accountable to certifying bodies and to government organizations. Documentation and archives of documents are a major effort in mission-critical and safety-critical projects.

2.6.2 Types

Documentation must cover the entire "who, what, when, where, why, and how" of a project. Table 2.2 lists examples of documents and when and why they are prepared during mission-critical and safety-critical development.

Documents generally should not contain redundant information. Each should fill a specific purpose. If they overlap, one document should control the wording and the other should point to the first document without repeating the information. This practice significantly improves configuration control by reducing ambiguity; you do not have to update information simultaneously in separate documents.

2.7 Requirements and Standards

2.7.1 Markets

Each application area for mission-critical systems—automotive, military, aerospace, and medical—has its own set of standards. I will only point you in the direction of some of these standards here. The case studies in later chapters list examples of compliance standards that you will encounter in specific areas [3–6].

Many products, if not most, require certification through Underwriter Laboratories (UL) or have the Conformite Europeene (CE) mark for the European Union. Certification comes from evaluation testing in approved test laboratories.

TABLE 2.2

Example Listing of Some Documents for Development of Mission-Critical and Safety-Critical Systems

Who, What, When	Comments	Plan (Who, What, When, Where)	Design (How, Why)	Execute	Review
1. Project-Plan	Scheduling and specifying all other plans	2. Development plan System plan Software plan Hardware plan Parts control plan Test/support equipment plan		Parts inventory list	
		3. Configuration management plan		Traceability Design history file Version description Document Document control forms	
		4. Requirements and specification plan	Requirements and specifications		
		5. Test plan Laboratory tests Prototype tests Field tests V&V	Test procedures	Test results	Code and module inspections Corrective action reports
	Code standards, style guide	6. Documentation plan	Product description User manual Product design	Software source listings	Technical reviews CoDR, PDR, CDR

(Continued)

TABLE 2.2

Continued

Who, What, When	Comments	Plan (Who, What, When, Where)	Design (How, Why)	Execute	Review
			Documents System design Software design Hardware design Support equip. design doc. Software design Hardware design System design Technical communications— letters, memos, notes		Functional configuration review (Validation)
	Early phases	7. Risk analysis plan— FMEA, HA			Risk analysis report
	Later phases	8. Training plan	Training procedures	Training materials— manuals, brochures, video, interactive S/W	
	Later phases	9. Design transfer plan		Design transfer document	

2.7.2 Government and Market Standards

Automotive: Many standards exist for the automotive market. There are Corporate Average Fuel Economy (CAFE) regulations for fuel economy in the United States. The California Air Research Board (CARB) has standards for emissions reductions, while the European Association for Emissions Control has standards for emissions for vehicles sold in Europe. The National Highway Transportation Safety Administration (NHTSA) has standards for vehicle safety in the United States. You can find more information in References 7–10 and in the case study of Chapter 7.

In Europe, the OSEK standard defines characteristics for operating systems and in-vehicle communications protocols. OSEK stands for "Offene Systeme und deren Schnittstellen fur die Elektronik im Kraftfahrzeug," which means "open systems and corresponding interfaces for automotive electronics" [11].

Military: The U.S. government has been moving away from military standards over the last decade. Now they tend to put exactly what they want in a statement of work (SOW) or a procurement specification (PS) instead of calling out specific standards. You can find more information in the case study of Chapter 9.

Aerospace: General processes in the aerospace business can follow AS9100. Avionics follow DO-178B/C for software development and documentation. The Society of Aerospace Engineers (SAE) provides a CD costing US $2915 that has 2400 documents that can be applicable to aerospace [12].

Medical: For medical devices in the United States, the FDA classifies devices into three categories: Class I, II, or III. Class I devices must supply premarket notification, registration, prohibitions against adulteration and misbranding, and rules for good manufacturing practices. Class II devices have the same requirements as Class I plus they must meet performance standards. Class III devices have the same requirements as Class II plus they must gain premarket approval from the FDA. Some of these devices may be eligible for 510(k) status, which speeds their approval. A 510(k) status means that the device is substantially equivalent to a device in commercial distribution before May 1976. You can find more information in the Reference 13 and in the case studies in Chapters 12, 13, and 14.

UL: Underwriters Laboratory or UL is a certifying body for safety of products in the United States. UL has more than 800 Standards for Safety [14].

CE: The CE mark, now known as CE marking, is a compliance mark for products sold in Europe, in particular, and around the world in general. The certifying body documents compliance with specific standards. The CE marking on products serves as a "passport" into the European marketplace of 18 countries (also known as the European Economic Area, or EEA). The letters "CE" are the abbreviation for "Conformite Europeene" [15,16].

Most companies need help getting the CE marking. Third-party certification to certify products for the European Union are called "notified bodies"; they are authorized by European countries to serve as independent test labs and perform the steps called out by product directives. Notified bodies may be either a private company or a government agency.

ISO: ISO is the acronym for International Organization for Standardization. The goal of ISO is for companies to build and maintain processes that ensure quality products. Compliance with ISO standards is a significant undertaking for any company. A good book for getting a quick understanding of ISO implementation is Reference 17.

CMMI: CMMI is the acronym for Capability Maturity Model Integration. Its goal is for groups to build and maintain processes that ensure quality products, particularly software. I recommend Reference 18 for better understanding CMMI. Reference 19 is a good book for quickly understanding CMMI and comparing it to ISO.

2.8 Analysis

2.8.1 Event Tree Analysis

Event Tree Analysis (ETA) helps you understand how your product responds to each potential and possible circumstance. It is a procedure to establish the relationships between an event and the causes of that event through the use of an event tree. An event tree gives the sequence of hardware, software, and operator functions that make up a scenario. ETA gives both failed and successful paths through scenarios; it can also handle multiple failures simultaneously.

ETA is a combination of graphical and tabular techniques. ETA can be a more powerful tool for analyzing a system than FMEA. Figure 2.2 illustrates just a small example of one event tree that might be used in an ETA. (Reference 20, [pp. 175–176] is a good book for more detail on ETA and the following analyses of FTA and FMEA.)

2.8.2 Fault Tree Analysis

Fault Tree Analysis (FTA) is a top-down analysis. It is the process of establishing the relationship between a fault—an undesirable behavior or system state—and the causes of that event. FTA examines the causes of a bad result ([20], pp. 166–173).

FTA is a graphical technique. It begins with a fault or failure. It then traces backwards through successive layers of subsystems and components to identify the potential cause. It can reveal multiple causes for a particular fault or failure. Figure 2.3 illustrates just a small example of one fault tree that might be used in an FTA.

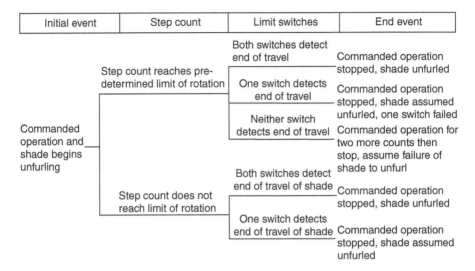

Initial event	Step count	Limit switches	End event

FIGURE 2.2
Example of an ETA diagram for part of the operation of unfurling a solar shade on a satellite. It uses a stepper motor to unfurl the shade and limit switches and counter circuitry to detect the end of travel for the shade. (© 2007 by Kim Fowler, used with permission. All rights reserved.)

2.8.3 Failure Modes and Effects Analysis

Failure Modes and Effects Analysis (FMEA) is a bottom-up analysis; it might be viewed as the reverse process of FTA. FMEA determines what can go wrong from a given cause. The FMEA picks up where the FTA leaves off at the component and unit level. In general, the FMEA considers all credible failure cases (e.g. stuck high, stuck low, or intermittent) for each component. The FMEA will help ensure that no unexpected single-point failures exist in the system.

FMEA is a "what-if" analysis. It attempts to list all the possible failure types or modes of a component and then list the outcome—the effect—of each. A FMEA worksheet should be rigorously filled in. It should identify the system and subsystem being analyzed, who is doing the analysis, and when they did it. Figure 2.4 illustrates just a small example of a chart that might be used in an FMEA.

Dunn cautions that FMEA should be conducted early in design. Merely tacking on a FMEA does not meet the spirit of its utility. Problems pointed out by FMEA are much cheaper to correct early in design, rather than late in design ([20], pp. 159–166). The same thing holds true for ETA and FTA; they should be done early in design.

2.8.4 Risk Analysis and Margin Management

Margin analysis establishes and manages system margins; it can help you avoid unexpected problems encountered during the system development.

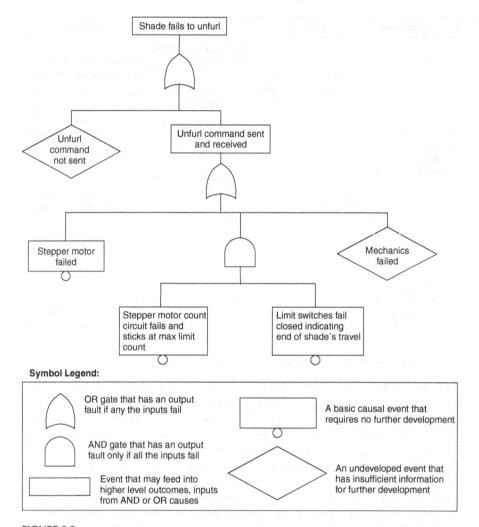

FIGURE 2.3

Example of an FTA diagram for part of the operation of unfurling a solar shade on a satellite. It uses a stepper motor to unfurl the shade and limit switches and counter circuitry to detect the end of travel for the shade. (© 2007 by Kim Fowler, used with permission. All rights reserved.)

The following parameters from a satellite subsystem are examples of items with margins, which must be managed:

- Mass
- Power consumption
- Data rate
- Data storage
- Critical CPU computational throughput

System: Solar Shade **Subsystem:** boom drive	**Operating** **Mode:** Unfurling	**Page:** 1 of x **Date:** 3 February 2007
Component	**Failure Mode**	**Failure Effect**
Stepper Motor	OFF	Solar shade does not unfurl. Stepper motor count circuit detects no rotation. Limit switches detect no closure.
Stepper Motor	ON	Solar shade unfurls and possibly over-extends to mechanical fracture of boom. Stepper motor count circuit detects continuous rotation. Limit switches detect closure at end of travel.
Stepper Motor count circuit	Stuck at count	Solar shade unfurls. Limit switches detect closure at end of travel. Computer commands motor off.
Limit switch 1	Stuck at OFF	Solar shade unfurls. Limit switch 2 detects closure at end of travel. Stepper motor count circuit detects rotation and maintains step count. Computer commands motor off.
Limit switch 1	Stuck at ON	Computer detects switch set, assumes its failure and continues operation by relying on redundant operations in switches and motor count. Solar shade unfurls. Limit switch 2 detects closure at end of travel. Stepper motor count circuit detects rotation and maintains step count. Computer commands motor off.
Limit switch 2	Stuck at OFF	Solar shade unfurls. Limit switch 1 detects closure at end of travel. Stepper motor count circuit detects rotation and maintains step count. Computer commands motor off.
Limit switch 2	Stuck at ON	Computer detects switch set, assumes its failure and continues operation by relying on redundant operations in switches and motor count. Solar shade unfurls. Limit switch 1 detects closure at end of travel. Stepper motor count circuit detects rotation and maintains step count. Computer commands motor off.
Boom breaks	. . .	. . .
Boom sticks	. . .	. . .

FIGURE 2.4

Example of an FMEA diagram for part of the operation of unfurling a solar shade on a satellite. It uses a stepper motor to unfurl the shade and limit switches and counter circuitry to detect the end of travel for the shade. (© 2007 by Kim Fowler, used with permission. All rights reserved.)

TABLE 2.3

Example of Risk Attribute Categories

Attribute	Value	Description
Likelihood	Very likely (v)	>50% chance of occurring
	High (hi)	10–50% chance of occurring
	Moderate (mod)	1–10% chance of occurring
	Low (lo)	0.01–1% chance of occurring
	Remote (r)	<0.01% chance of occurring
Consequence	Catastrophic (C)	Cost impact exceeds project reserves
		Schedule slip that affects launch date
	Severe (S)	Loss of mission
		Cost impact exceeds planned reserves
	Important (I)	Schedule slip affecting critical path but not delivery
		Major loss of capability in instruments or experiments
	Non-critical (N)	Cost impact smaller than element cost reserves
		Slip reduces slack to 1 month or <50% of remaining schedule
		Minor loss of capability or design/implementation work-around
		No impact to cost reserves
		Slip but slack greater than 1 month or >50% of remaining schedule
		Loss of capability/margin but all mission requirements met

- Critical processor memory (code and operational memory) loading
- Critical nonvolatile memory loading
- Data bus throughput

On balance, excessive margins can make a system more complex and expensive than necessary.

Margin analysis has two parts: the *likelihood* of failing to achieve a desired result and the *consequences* of failing to achieve that result. The likelihood of occurrence ranges from remote to very likely. The consequence of occurrence ranges from noncritical to catastrophic.

Table 2.3 is an example from the satellite subsystem of the criteria for specific levels that define risk consequence and likelihood. The risk severity is established by combining the consequence and likelihood as shown in Table 2.4.

Risks that fall in the light gray boxes (diagonal) and dark gray boxes (upper right corner) of Table 2.4 are tracked in the risk management database. Risks in the clear boxes (lower left corner) are tracked at the discretion of the program manager.

TABLE 2.4

Example of Risk Severity–the darker gray indicates greater severity

Consequence		r	lo	mod	hi	v
	C					
	S					
	I					
	N					
			Likelihood			

2.8.5 Numerical Simulations

Numerical simulations play a large role in both space and military operations. Often these missions or systems cannot be fully tested directly; so development teams will devote large efforts to considering many different scenarios. Some of these simulations can be updated to accommodate new principles, information, and understanding. For instance, changing diplomatic relationships may require new arenas of military operations. Or magnetic or gravitational anomalies may affect the navigation of spacecraft in different orbit orientations.

Unfortunately, entire simulations almost never can be tested against actual scenarios. Therefore, development teams must simulate modules and subsystems and then estimate the potential interactions within the complete system.

2.8.6 Testing

Prototype field testing is an important part of development for military equipment and medical devices. Advanced prototypes often undergo technical evaluations in military environments. Prototypes of medical devices usually undergo extensive laboratory testing. Medical devices in final commercial form also undergo clinical investigations to demonstrate efficacy and safety.

2.9 Design Trade-Offs

2.9.1 Architecture

Every mission-critical or safety-critical system usually has a primary objective. This means that these systems have a small set of primary requirements that address the objective. The requirements often focus on one of the following objectives: efficacy, performance, safety, or dependability. Selecting one of these objectives will drive the design of your device.

2.9.2 Dependability

Dependability refers to the quality of service provided; it has various components ([21], p. 104):

- Reliability
- Availability
- Fault tolerance or performability
- Testability
- Maintainability
- Safety

Dependability has analytical definitions and a theoretical basis. I recommend the books cited in References 20 (pp. 175–176) and 21 for further study of dependability.

Reliability: Reliability "is the probability that the system operates correctly throughout a complete interval of time. . . . Reliability is most often used to characterize systems in which even momentary periods of incorrect performance are unacceptable, or it is impossible to repair the system" ([21], pp. 4–5). You specify reliability in hours of continuous operation (i.e. without failure). Reliability might be in tens of thousands of hours for military equipment, or in years of continuous operation for satellites or implanted medical devices.

Complexity, vibration, shock, wide temperature variations, corrosion, and material aging all reduce reliability. Calculations for reliability do not accurately predict time intervals of correct functioning; they are better at indicating where problems might reside, such as in weak components or circuits or operations. Calculations for reliability have merit when comparing different design approaches.

Availability: Availability is the probability that a system is operating correctly at a specific instant in time. This is different from reliability. A system may be considered highly available even though it experiences frequent periods of nonoperation, which are extremely short in duration. "In other words, the availability of a system depends not only on how frequently it becomes inoperable but also, how quickly it can be repaired. Examples of high-availability applications include time-shared computing systems and certain transactions processing applications, such as airline reservation systems" ([21], pp. 4–5).

Fault tolerance or performability: Fault tolerance or performability means that a system continues operating in the face of a component failure or operational fault, but the performance may decrease. The difference between reliability and fault tolerance (performability) is that reliability defines the likelihood that *all* of the functions perform correctly, while fault tolerance (performability) estimates the likelihood that a *subset* of the functions performs correctly. Graceful degradation can be an aspect of fault tolerance (performability) whereby the system automatically reduces its performance during a fault ([21], pp. 5–6).

Fault tolerance or performability is an operational concern. It does not provide higher reliability; it provides a more robust approach to surviving faults and failure. Typically, fault tolerance will even lower reliability because it uses more components, subsystems, and operations to monitor, diagnose, and survive faults.

You would tend to use fault tolerance in situations where immediate repair or maintenance are not available. Examples of fault-tolerant systems would be a banking system or a server farm, where continued operation is critical to business. The "limp-home mode" already mentioned in automobile engine control is another example.

There are three primary ways to achieve fault-tolerant design: careful design, testable architecture, and redundant architecture. Each of these will be described in more detail a little later. Design to the application. Do not be overly conservative, which can drive costs up.

Testability: Testability defines the ease of test for certain attributes within a system ([21], pp. 5–6). Testable architectures generally do not provide continuous monitoring and they usually are not automatic either—either they must periodically trigger testing or an operator must initiate the testing. Finally, testing also requires that the calibration standard be known and understood.

Maintainability: Maintainability defines the ease of system maintenance, as well as the repair for a failed system. Quantitatively, it is the probability that a failed system or one down for maintenance will resume operation within a set period of time ([21], pp. 5–6).

Testability, built-in-test (BIT), diagnostics, and repairability all are components of maintainability. Furthermore, completing maintenance and repairs more quickly means that the system is more available than a similar system that is not maintainable. Ultimately this means that the system is more dependable.

Safety: "safety is the probability that a system will either perform its functions correctly or will discontinue the functions in a manner that causes no harm" ([21], pp. 4–5). Safety measures often use dissimilar operations or subsystems to check operations. Examples include external safety devices, interlocks, hardware to check software, and software to check hardware.

2.9.3   Trade-Offs for Dependability

These definitions help show that you must make trade-offs when designing dependable systems. Fault tolerance (or performability), for instance, does not improve reliability. Fault tolerance, such as redundant design, generally increases complexity, which lowers reliability. A side-effect of both redundancy and fault tolerance is that they make fault diagnosis generally more difficult. Hence, testability becomes more difficult.

Mission-critical, safety-critical, and medical products require careful design. Design trade-offs for dependability can use a number of techniques, including (but not limited to) stress margins, redundancy and error checking, interlocks, fail-safe, and fault tolerance. These can all help make for devices and equipment that survive faults, failures, and problems, or at least degrade gracefully. Unfortunately, they are accompanied by a disproportionate increase in cost and development time.

Careful design: Careful design uses conservative design practices, provides margin for overstress and abuse, and avoids faults and problematic situations. Here are examples of careful design:

- Reduce large thermal gradients
- Prevent or reduce overvoltage
- Eliminate leakage currents
- Prevent or reduce stray charge and protect against electrostatic discharge (ESD)
- Prevent or reduce shock and vibration
- Prevent incorrect hookups (e.g., keyed connectors)

The very best fault tolerance is fault avoidance. It is always better to avoid a problem than to suffer the fault and then try to recover from it. Examples include uniquely keyed connectors that prevent inadvertent plugging of cables into the wrong places. When you cannot avoid the fault, then you might use some of the following techniques.

Stress margins: Design should allow margin in the face of situations that stress the circuit or system. What you are trying to do is not operate components at the limit of their capabilities. An example of providing margin would be to avoid running a transistor near its rated power dissipation. Other examples of providing margin include dampening vibration in a system, draining charge in a controlled fashion, filtering out overvoltage transients, and dissipating heat to avoid large thermal gradients that can eventually pull physical connections apart through expansion and contraction cycles.

Avoiding mechanical and thermal stress is a good form of fault avoidance. Turning a component on and off frequently or cycling it through extreme environments is often stressful. Flicking a light bulb on and then off so that it cools between each lighting will drastically shorten its life because thermomechanical expansion and contraction accelerates fatigue fractures in the filament. Steady-state operation is often much more benign than cycled operations.

Error checking and data redundancy: Error checking can detect memory and communication failures. Error checking can use parity checks, checksums, and cyclic redundancy checks to find data upsets. The trade-offs in choosing a particular form of error checking revolve around simplicity, computational power required, and the depth of error coverage.

Data redundancy reduces certain types of failure through parallel transfers of data, preferably through different types of channels. Comparing datastreams and finding any discrepancies can indicate a problem or failure, speeding diagnostics and improving testability and maintainability.

Interlocks: "A safety interlock is a hardwired device employed in a system to inhibit effector motion when external conditions make effector action unsafe. Interlocks are commonly used in home appliances: the door switch on the microwave oven and washing machine that prevent operation when the door is open . . . safety interlocks should always be able to perform their safety function regardless of any failure that might occur in the system" ([20], p. 143).

One example is a thermostatic switch that opens should temperature exceed a maximum limit. If this switch connects to power and if the temperature exceeds the limit, the switch will interrupt power and shut down the device to prevent a failure due to overheating.

An interlock can use a hardware or mechanical function as a safety check or a limit for the software operation. One example is a watchdog timer, which is a simple hardware check of software operation. The timer continually counts up (or down); it requires periodic clearing (or resetting) from a software-induced signal; otherwise, if it is not cleared, it times out and generates an interrupt or a reset. Hopefully, the reset corrects the problem in the software's operation or at least signals a problem.

Fail-safe, trapdoor, and limp-home: Fail-safe operation, trapdoor, and limp-home mode are all responses to detecting a fault. The interlocks mentioned in the previous section are a quite rudimentary form of fail-safe. Generally, fail-safe, trapdoor, and limp-home responses are more complex operations than an interlock; they provide some sort of minimal functionality after a failure.

An example of fail-safe operation might be in a drug pump that can only deliver a single squirt of a drug on a timed basis. Should a fault cause a command for continuous high dosage, the pump is limited to cycling only a squirt of drug during each fixed interval. It cannot empty itself quickly; it is restricted to a maximum dose that is not harmful to the patient.

An example of trapdoor operation might be a survival mode for a spacecraft when a high-gain antenna fails. The trapdoor operation assumes minimal communications from low-bandwidth recovery communications and goes into an semi-autonomous mode that seeks basic commands and only runs the most important functions in the spacecraft: orientation toward the earth or the sun, maintaining attitude to ensure the solar panels keep receiving light and generating power, maintaining heater controls to keep the electronics above the minimum temperature required for operation, and powering the low-bandwidth communications with the low-gain antenna.

An example of limp-home mode might be the operation of an engine control module (ECM) for automobile engine when a crucial sensor fails, such as an oxygen sensor. The ECM can no longer provide fully optimized control for the engine, but it can estimate conditions and run in a reduced capability.

Testable architectures: Testable architectures provide means to test components, circuits, and subsystems. The purpose is to reduce repair time by locating problems quickly. Two simple examples are a collection of test points for technicians or automatic test equipment. Another example is built-in-test (BIT), which reduces or eliminates disassembly of equipment during diagnosis. BIT adds additional circuitry to the circuit or subsystem to test that subsystem. Testable architectures allow stimulation of circuits and recording of responses; then the circuit or an operator compares the results to the expected (or normal) response. Testable architectures add complexity, which reduces reliability; this trade-off allows quicker diagnosis of faults. Testable architectures do not monitor functions continuously—that is the purpose of redundant architectures. The following example discusses one such system.

Redundant architectures: Redundant architectures have multiple copies of circuitry and software that continuously self-check between functions. They detect faults and failures immediately. These are complex systems that are justified only when downtime for repair and maintenance cannot be tolerated.

The most common architectures are dual-redundant designs that can detect problems but not correct them. They are good for alerting operators to the problem. The following example discusses one such system.

Triply-redundant designs use three identical systems that have voting modules on the output to detect and override single-point failures. They are often discussed but rarely implemented. They are generally far too complex; the voting modules are a source of single-point failures that often put the entire system back in the same place of a unitary system without redundancy. These redundant designs often underscore the notion that it is better to avoid faults than be tolerant of them.

Example: The Arc Fault Detector (AFD)

Arcing faults might be called high-impedance short circuits between power bus bars. They generally occur in power supply systems;

they conduct sufficient current to sustain an arc but remain below the trip threshold of the circuit breakers that are supposed to protect the system. Arcing faults generate white-hot heat that consumes the metal in power switchgear in a few seconds [22].

Many years ago, I worked for the Johns Hopkins University Applied Physics Laboratory and with the Navy to design a system, called the Arc Fault Detector (AFD), for detecting and extinguishing arcing faults in switchboard cabinets in ships. The AFD would trip the appropriate circuit breakers to extinguish an arc upon detection. An important consideration was that the Navy required that the AFD generate no false alarms—even at the cost of potentially missing some actual events [23].

We designed the AFD with redundancy and testability to improve both the availability of the system and the chances of correctly detecting arcing faults. The architecture uses functional redundancy in the form of two different types of sensors, a photodiode detector and a fast-acting pressure sensor, to detect arcing faults and avoid false alarms. The AFD also has BIT, which is regularly exercised by Navy enlisted personnel to check functionality and locate failures if they occur. BIT improves availability by reducing the downtime of the system during repair.

AFD systems have operated for over 1000 ship-years. They have detected and extinguished at least nine arcing faults and have not generated any false alarms. Only through functional redundancy and testability has the AFD achieved this level of dependability.

2.9.4 Performance Margins

Performance is a very different concept from performability. Performance speaks of throughput, computational power, and capacity while performability addresses fault tolerant operation in the face of failure.

Both processing and memory constrain performance. Design for performance needs to provide margin for throughput data and memory capacity.

I suggest that you plan a project to use only about 25% of the memory capacity. The remainder should be available for upgrades or fixes that inevitably show up later in a project. The goal is to finish the project and release it in final form with less than 50% of the memory actually used. Most of you will need portions of that memory for changes after the final release.

I also suggest that you plan a project to use only about 50% of the rated throughput. Rate monotonic analysis has shown a theoretical limit at about 70% of a processor's rated computational throughput for a variety of independent tasks [24]. I try to steer clear of that 70% limit. If your software and data flow design exceeds 50% of the rated throughput, then you should seriously consider one of several options:

- Get a bigger processor
- Add more processing power (additional processors or components such as field programmable gate arrays [FPGAs])
- Revise the requirements to drop excess demands

Adding more processors is fine until the communications and control between processors become a bottleneck. Another problem with adding more processing resources is cost—more components not only increase the bottom line (i.e., cost more), they also require additional design time and effort. Conversely, tweaking software to run on a processor that is 90% loaded for throughput is terribly time consuming and less than satisfying in the end.

Sometimes a good trade-off is to incorporate a discrete analog circuit rather than a digital solution and software. Occasionally, an analog filter is more effective for either cost or performance than burdening the processor with additional signal processing. The down side is that analog circuitry does not change easily, can have noise sensitivities, and can consume too much power.

2.9.5   Cooling

Cooling, while sometimes necessary, contributes complexity (possibly lowering reliability) and increases cost. It becomes a part of the analysis, such as FTA, FMEA, and ETA, before detailed design. Realize that, for long-term installations, active cooling involving fans, forced liquid cooling, or refrigeration becomes a maintenance concern. Filters over fans need cleaning or replacement. Bearings on motors eventually wear out. Refrigeration fluids become a disposal hazard.

Many mission-critical systems do not allow an exchange of gases between the system enclosure and the ambient environment. This type of situation makes cooling more difficult and rules out fans to pull air through the enclosure. More esoteric solutions, such as heat pipes, thermoelectric coolers, and refrigeration, must then be used.

2.9.6   Power

Power conversion and distribution is a chronic source of problems. In the 1970s and 1980s, 20% of all military electronic failures traced back to the power converters [25]. My own experience in spacecraft systems is that many problems, if not the majority, fall into the realm of the power conversion and distribution subsystems. Some problems are as simple as incorrect connectors or wiring. Other problems arise from load transients when a subsystem turns on. Understanding the source and the loads in a system is critical to successful design of a power subsystem.

Many mission-critical and safety-critical devices use batteries as either the primary or the backup source of power. Voltage is not constant. Batteries have discharge curves that depend on their chemistry and on the current demand. Battery-powered systems require power conversion modules to provide appropriate and constant levels of voltage to circuits. Furthermore, the energy capacity of batteries is quite limited; for implanted medical devices, the battery life often determines the operational lifetime.

Spacecraft typically use solar cells to recharge batteries. As a satellite goes in and out of the earth's shadow, power from the solar panels surges and shuts off. Solar cells also degrade over time. Considering these issues, the recharging and power regulation becomes an important design effort in itself.

Even the more plebeian applications that use line or mains power still must consider the power design. Raw line power is not constant; it is noisy, spiky, and has periods of over- or undervoltage. All these things can adversely affect your equipment and must be accommodated.

2.9.7   Software

The development of mission-critical and safety-critical devices absolutely requires good design processes for the software! Good processes include code inspections, design reviews, unit tests, and integration tests. Processes for software really cannot be separated from the hardware and certainly not from the system. Adhering to a standard, such as CMMI, will go great distances to helping you and your development team achieve quality, functioning software.

2.9.8   Hardware vs. Software

Safety interlocks for software are an example of how hardware may complement software. Conversely, software can check and verify hardware function. One example of this is BIT, where the software exercises the hardware by stimulating circuits and monitoring the responses. Correct responses generate a nominal indication; incorrect responses generate an indication of fault.

2.9.9   Buy vs. Build

Many mission-critical and safety-critical devices are custom-designed for a specific application. Companies building larger pieces of equipment, however, might buy circuit boards, chassis, and subsystems and incorporate them into the final product. Often these, commercial off-the-shelf (COTS) components are specialized for a particular market—possibly ruggedized or certified for an extreme environment or qualified as appropriate tools.

Usually, the main concern is whether the COTS component fits the application or if modification must be made. This concern is compounded by a long learning curve for complex components. At some point, it becomes easier to build the component yourself than to purchase it. (See Chapter 1 and Section 1.10.8 on buy vs. build.)

2.10   Tests

2.10.1   Formal Processes

Mission-critical and safety-critical devices require formal processes for every aspect of development, including testing. All tests, whether laboratory tests, inspections, hardware subsystem tests, or software unit tests, must be recorded and maintained in a device history file. These files are subject to audits from certifying bodies or from customers.

Everyone involved in any type of testing, anywhere along the development timeline, must understand the processes and adhere to them. These people include the design engineers, test engineers, technicians, and personnel at test facilities.

Chapter 1 covered all of these tests. Here I will focus in more detail on two types of tests and review—the design review and test simulators.

2.10.2   Design and Peer Review

Effective review of the project should have a stereotyped format including agenda, checklists, and minutes. Complete minutes of a review include

- The date of the review
- The review agenda
- Who attended
- Who presented the design
- The lead and independent reviewers
- What major decisions were made
- What action items were generated, their due dates, and who is responsible for each

Reviews should generate action items to ensure that identified issues are addressed. All action items should be tracked in a database. Each action item should have the following fields:

- Unique identifier number
- Status (open, closed, in work, and in sign-off)

- Date opened
- Brief summary
- Response summary
- Requestor
- Assignee
- Due date

Design reviews: For larger programs and projects, the design review should have independent reviewers who are not directly associated with the project. The review committee for each formal review should consist of at least four members plus a designated chairman; none of whom should be members of the project team. I have to say this rarely happens—it is the ideal, but most companies do not have the people, time, or resources to devote to independent review.

Most companies put together design reviews with members of the project team presenting and reviewing. Sometimes they will ask a customer or client to attend and critique. This form of review is still effective.

Regardless of the format, you should send a review package to each reviewer about 2 weeks ahead of the review. The review package should contain a copy of all the slides to be presented at the review along with appropriate background material.

Code inspections and system reviews: Software code walk-throughs are a legitimate form of review, as well. They are a form of static testing—but highly effective. These forms of peer review are an excellent way to encourage proper designs and good development processes. For some types of products, such as medical devices, they are an important part of the formal development. You should still have procedures (yes, they can be simple and straightforward) for recording notes or minutes and then maintain a database of action items.

I would suggest that this same sort of review would be good for early system integration. If done regularly by team members, these reviews would reveal more problems earlier and give you a better chance at ironing them out sooner.

2.10.3   BIT, BITE, and Simulators

Often we need automated help to provide routine test coverage of subsystems under development. This test coverage can go on to become an integral part of the system and product. Built-in-test (BIT), built-in-test equipment (BITE), and simulators provide portions of that automated help during test.

For some projects, such as avionics and spacecraft, test support equipment is crucial to their development. It bridges most gaps between test and integration for larger systems; I will address this in the next section.

BIT and BITE: "BIT can generally be described as a set of evaluation and diagnostic tests that uses resources that are an integral part of the system under test" [26]. BIT measures, diagnoses, and estimates the state of health of the system. The subtle difference between BIT and BITE is that BIT is entirely self-contained within the product that it tests while BITE includes any set of external components needed to provide full coverage, such as wrap-around cables that connect outputs to inputs to allow the BIT to test those functions.

You need to specify BIT in terms of the amount of coverage to detect faults. Here is an example from a military avionics program: the BIT had to "detect 95% of all faults, isolate 95% of any fault to one of over 50 replaceable assemblies, and allow false alarms for less than 2% of all detected faults" [27]. For wide coverage, you will expend significant engineering time and resources to design the BIT and incorporate it in the system. A benefit of a comprehensive BIT is stated by Steinmetz, "This information is as useful in the factory as it is in the field. A system with so much internal measurement data is often referred to as self-instrumented" [27].

Simulators: Simulators can be an excellent form of dynamic testing for software. They represent other system components with which the primary instrument interfaces. The closer the simulator comes to replicating the interface or coupled subsystem, the better the test coverage and the higher the final confidence in the integrated system. Simulators can fill the void while the various subsystems are being developed. The downside is that the closer a simulator gets to replicating the actual subsystem, the more expensive and the longer development it becomes.

I have worked on spacecraft instruments where the ground support equipment (GSE) served dual purposes as both test support during design and later ground support during space flight. This makes a lot of sense; you get parallel development of necessary subsystems and you get to exercise both early and often.

2.11   Integration

2.11.1   System Behavior

System integration is an important phase of most mission-critical and safety-critical devices. A clear understanding of the operational interfaces is necessary between the various subsystems and operational groups— including the hardware, software, human, and environmental components. Integration is the planned combination of these operational groups in measured steps, with the goal to reveal all interactions, understand consequences and potential consequences, and limit undesirable operations. Good systems integration provides both verification of requirements and validation of design intent.

Too often people do the "big bang" form of integration, where they connect all the components and then hope for the best. It never works. Even if it appears to work, something may be amiss but obscured.

The best form of integration is to bring together a minimal number of subsystems and limit the number of variables. This technique closely mirrors the spiral form of development (Figure 1.3b). For more complex systems, simulators that imitate various subsystems are mandatory. As a subsystem matures and migrates to integration, it replaces its simulator.

A good example of an integration test bed was Boeing's 777 Systems Integration Laboratory (SIL). "The 777 SIL included all the electrical power systems, electromechanical systems, avionics, environment control systems, propulsion systems, and a portion of the payload electronics. The integration testing included realistic simulations of flight modes to support verification and validation of production equipment *before* the first flight of the aircraft. It also provided support for certification and validated the correct performance of both the physical and functional interfaces in the electrical and electronic systems during concurrent operation of multiple subsystems and failure" ([28], p. 14). The Boeing 777 SIL was incredibly complex. Some of its particulars were as follows ([28], pp. 14–15):

- 40,000 airplane wires in 1,000 bundles
- 28,000 lab-unique wires
- simulate 4000 signals and data buses
- record 1000 signals and data buses
- Support for all the LRUs (line-replaceable units) on the actual airplane
- Cooling for the LRUs
- Antennas for radios, global positioning system (GPS), and navigation systems
- 800 kW of power

Integration requires the close cooperation of the design team, test team, manufacturing team, and integration team. Depending on the complexity of the system, integration may require its own separate facility, such as Boeing's 777 SIL.

2.11.2 Environmental

Mission-critical and safety-critical devices often encounter some sort of extreme environments: spacecraft encounter vacuum and radiation in space, aircraft encounter temperature and pressure swings during flight, and medical devices encounter the hostile milieu of the human body. For these extremes major subsystems, if not the entire device, must undergo environmental testing. These tests have two purposes: to shake out problems and to certify a device as capable of surviving the expected environments.

Environmental tests are primarily mechanical in nature: thermal, shock, vibration, pressure, and vacuum. Spacecraft instruments and subsystems undergo thermal–vacuum tests and shock and vibration tests. Sometimes, the entire spacecraft is tested in large thermal–vacuum chambers or on shake tables. Automobile manufacturers will subject new model cars to extreme weather, particularly cold and hot, for extended periods of time.

Some environmental tests are more unusual, such as condensation or EMC tests. Some military systems must endure salt-spray and condensation testing. Most products do have to pass standard EMC tests.

So far I have described tests that aim at certifying a product. Environmental tests that stress test prototypes have a different purpose; they try to uncover incipient faults. Stress in the form of thermal or power cycling, vibration, shock, or condensation applied in extreme forms can reveal weaknesses. High temperature can precipitate or accelerate diffusion processes on silicon die, the oxidation of fractures, and reduce timing margins. Temperature cycling will expand and contract interconnections, such as solder joints and ball bonds. Elevated humidity that causes condensation can promote corrosion and the breakdown of electrical isolation [29].

These environmental stress tests go by a variety of names: Accelerated Stress Test (AST), Accelerated Environmental Stress Screen (AESS), and Accelerated Life Test (ALT). They can also have the prefix of H to indicated "highly" accelerated, as in Highly Accelerated Stress Test (HAST), Highly Accelerated Stress Screen (HASS), and Highly Accelerated Life Test (HALT).

2.11.3 Field Tests

Field tests are particularly important for complex systems. You really need to see the product in use to understand its operation and utility. Some examples are clinical tests for medical devices (which are very formal field tests that support certification), test track trials for cars, and military technical evaluations followed by operational evaluations of equipment on the battlefield.

2.11.4 Certification

Many mission-critical and safety-critical devices must be certified by a regulatory agency or designated service. Certification, while a separate activity, might be considered either a form of integration or a form of acceptance test. Not understanding the specifics of certification and regulatory requirements will waste your time and your company's money.

Standards Organizations: Various organizations can provide different types of certification. A notified body, a third party designated by authorities, can be either a commercial firm or a government organization that assesses the compliance for safety, performance, intended use, and risk analysis. Underwriters Laboratory (UL) in North America, for example, provides much product-safety certification. Manufacturers must submit product samples and information, as well as meet the applicable safety standards for UL certification. Within the European Union, the certification is through CE marking, which assesses compliance with appropriate directives [13].

Standards: These notified bodies test to the appropriate standards for the selected market. Section 2.7.2 begins to list some of these possible sets of standards. The case studies, found later in the book, give more standards for specific markets.

Safety Evaluation: Standards or portions of standards often focus on safety. Equipment must be safe in two different ways:

- Normal conditions, which are situations *likely* to occur
- Single fault conditions, which are situations that *could* occur, for example, failure of a component

Certification for safety includes the actual equipment, its markings and labels, software, biocompatibility, and EMC. The equipment in the evaluated system must meet the following conditions to receive certification:

- Fit the scope of the standards
- Include connected equipment
- Identify potential hazards in normal use and abuse
- Verify power requirements and fusing

For certification, you must provide the following to the notified body:

- Insulation diagram
- Documentation for components with UL or American National Standards Institute (ANSI) standards
- Illustrations of components
- One or more samples of the equipment

Documentation from Safety Evaluation: The safety evaluation generates three primary documents: the UL report, an informative test report, and a certified body (CB) report. The UL report is a product description and a test report; it authorizes you to apply the UL mark to the product. The informative test report is a complete record of meeting all the requirements of the applicable standard and contains the insulation diagram, illustrations, and markings. The CB report contains both an informative test report and a certificate from the issuer. The CB report, which is recognized internationally, helps you obtain third-party certification marks [13].

Common Mistakes: People repeat several types of mistakes in failing certification. The most common mistake or noncompliance is in the documentation, in particular, leaving out a required inclusion [13].

The next most common mistake or noncompliance is selecting a power supply that is not UL certified. Some of the problems with these units are mechanical spacing, leakage current, and the wrong mains component. Use a UL-certified (2601-1) unit to avoid some of these types of problems [13].

The third most common mistake or noncompliance is using the incorrect colors for indicator lights. Use red lights or light emitting diodes (LEDs) only for warning and nothing else; do not use red LEDs to indicate power-on. Use yellow LEDs only for caution [13].

2.12   Manufacturing

Manufacturing of mission-critical and safety-critical devices often has its own set of special requirements and quality standards. For small quantities, a company might try to outsource to a special contract manufacturer where the work is often hand-assembled. For large quantities of a product, a company might develop the necessary resources and capability to manufacture in-house.

A company building small quantities of a special mission-critical or safety-critical device might have to develop expertise in-house to assemble the product. Many companies will outsource components, modules, and subsystems but do the final assembly and test in-house. Often they will outsource the circuit board fabrication and assembly. They will also outsource the production of specialty mechanics and enclosures.

Typically, trained personnel or technicians will perform most tests for manufacturing quality. The cost of using or programming automatic test equipment, or ATE, is prohibitive in these small quantities and for these demanding special applications.

2.13   Support

2.13.1   Fielding

Fielding of mission-critical and safety-critical devices can take many different forms. Some devices, such as implanted medical devices, are replaced and the failed unit returned to the manufacturer for diagnosis and disposal. Other products, such as military equipment, will have modules called line-replaceable units (LRUs) that allow swapping-out in the field by trained personnel or operators; they then send the failed LRU back to a depot for repair.

Sometimes a company finds a systematic fault or failure pattern in a particular product line. When that happens, they must implement a recall program. A recall exercises the entire network of logistics, distribution, service, inventory, and disposal, expending great effort by many people and costing lots of money. Other times, a company may institute a low-level program to fix a problem with low public visibility. The automotive industry has the somewhat euphemistically-named "service programs" to update and modify subsystems whose behavior has nearly risen to the level of a recall.

Repair often requires skilled field personnel to diagnose and fix problems. Companies will train either technicians or representatives to handle problems and failures. They almost never use general repair shops for repairing or maintaining mission-critical and safety-critical devices. Some companies providing specialty equipment employ field engineers to do the detective work.

2.13.2   Logistics and Maintenance

Logistics include concerns such as regular replenishment of consumables, maintenance, repair, and disposal. Do not confuse the terms maintenance and repair or use them synonymously. Maintenance is the regular checking of components and replenishing of consumables, such as lubrication. Repair is the replacement of broken or failed components. Diagnostics can be common to both maintenance and repair. Maintenance might lead to repair if a component is detected as marginal in operation.

Maintenance: Most of us are familiar with regular maintenance like changing the oil in our automobile engines or rotating the tires. Some equipment or devices may need periodic checks to confirm calibration or alignment. Some devices might need filters cleaned or replaced; some industrial equipment operating in a dusty environment can quickly clog the filters over their cooling fans.

Replenishment: Replenishment of consumables can take on many forms. Ink-jet cartridges in printers are consumables that need replenishing. Recharging or replacing batteries in portable devices is another form of replenishment. Regular lubrication of moving parts, shafts, and bearings is another form of replenishment.

Inventory: If the product has replaceable components or subsystems, then some number of spares must be stored and available. This requires warehouse space and a distribution system, which all costs money and takes manual labor. Reducing the requirement for inventory can decrease the life-cycle costs of most systems and products.

2.13.3 Repair

Assuming the product is not a "throw-away" appliance, repair is needed when it breaks or fails. First, someone must diagnose the problem and determine its cause. Then a technician must repair the problem, usually by replacing a failed component or by adjusting or tuning the operation.

New forms of diagnostics are transforming certain sectors of industry. Condition-based maintenance, for instance, automatically monitors the operation of a piece of equipment and raises an alarm if operating signatures exceed established bounds. Diagnostics are beginning to use artificial intelligence and may operate from one or more different approaches ([30], pp. 16–20):

- Rule-based
- Model-based
- Learning
- Fuzzy reasoning
- Neural networks
- Hybrid

Rule-based approaches incorporate the experience of designers and skilled maintenance personnel in the form of rules. These rules often are conditional statements, such as, "IF signature (or condition or symptom) THEN consider particular fault (or take course of action)." A rule-based approach might have hundreds or thousands of rules. A rule-based approach can be simple, but it suffers from difficulty in acquiring sufficient knowledge, the inability to deal with novel situations, and system dependence ([30], p. 17).

Model-based approaches overcome some of the deficiencies of rule-based approaches through approximate representations of the actual system. A model-based approach predicts faults using observations of the actual system and then generating and testing and discriminating hypotheses. Behavioral modeling of circuits is an example of a model-based approach. The disadvantages of model-based approaches are that computational effort climbs with system complexity, good models are difficult to develop, and information on failure mechanisms may not be available ([30], p. 17).

Learning approaches go beyond the previous two approaches, which are restricted to fixed levels of performance. Learning can use case-based reasoning to store experiences of past solutions, use an analogous case for a new situation, and use past successes and failures in diagnosis to improve performance. Such case-based reasoning, plus the methods of fuzzy logic and neural networks, are at the forefront of research into more powerful diagnostics ([30], p. 18).

2.13.4 Technical Support

Technical support takes on new meaning for specialized mission-critical and safety-critical devices. It can range from training and explaining basic operation and maintenance to supporting customers implementing a function. In the latter situation, companies employ field representatives. For implanted medical devices, the field reps can be highly educated and trained to advise in certain medical procedures; some are even doctors.

Technical support can handle a number of unusual situations. Component obsolescence becomes important for products with a long service life, for example, motors, industrial process control, military equipment, and medical devices. Technical support can help find replacements for obsolete components. Technical support can also perform field studies to understand really "sticky" problems and then prescribe an appropriate corrective action. These are the folks that probably initiate a recall or "service program" in appliances or automotive systems.

2.14 Disposal

Few companies escape concerns with Restriction of use of certain Hazardous Substances (RoHS), waste from electrical and electronic equipment (WEEE), and recycling. If anyone does escape these concern, it is the small quantity, specialty mission-critical and safety-critical devices, such as military equipment or spacecraft. Medical device companies, on the other hand, not only must insure proper disposal, but also must record the serial number and keep a database and log of every device throughout its life cycle.

2.15 Liability

Liability is important for mission-critical and safety-critical devices. Everything mentioned in Chapter 1 on liability applies to mission-critical and safety-critical devices, just in greater magnitude. The attendant legalities also increase. The economics and safety concerns arising from failure can cripple a small company producing a mission-critical or safety-critical device.

2.16 Priorities

Everything is important with mission-critical and medical devices; that is why they are mission-critical or safety-critical. At the top of these concerns are good processes—from methods to implementation; everything must be carefully considered and recorded. Documentation plays a big role in good process. Most important are the people involved; they need to be self-motivated to do a good job.

2.17   Summary

This chapter focuses on mission-critical and safety-critical devices. The development of mission-critical and safety-critical devices requires an understanding of the appropriate standards for each market and specific trade-offs that might be made in developing them. Process for mission-critical and safety-critical devices is rigorous and must be carefully planned.

References

1. U.S. FDA, *Design Control Guidance for Medical Device Manufacturers*, March 11, 1997, relates to FDA 21 CFR 820.30 and sub-clause 4.4 of ISO9001. http://www.fda.gov/cdrh/comp/designgd.pdf. pp. i, 1, 2, 4, 5, 8, 13, 19, 23, 37, 43.
2. *Software Considerations in Airborne Systems and Equipment Certification*, RTCA/ DO-178B, December 1, 1992. RTCA, Inc., 1828 L Street, NW, Suite 805, Washington, D.C. 20036. pp. A-2, A-3, 1, 7, 68–77. You can purchase DO-178B at http://www.rtca.org
3. EMC Standards, *Compliance Engineering*, Vol. 21, No. 1, 2004 Annual Reference Guide, pp. 75–82.
4. ESD Standards, *Compliance Engineering*, Vol. 21, No. 1, 2004 Annual Reference Guide, pp. 103–105.
5. Telecom Standards and Regulations, *Compliance Engineering*, Vol. 21, No. 1, 2004 Annual Reference Guide, pp. 139–142.
6. Product Safety Standards, *Compliance Engineering*, Vol. 21, No. 1, 2004 Annual Reference Guide, pp. 165–172.
7. http://www.nhtsa.dot.gov/
8. www.arb.ca.gov (California Air Resources Board).
9. US Environmental Protection Agency, Office of Transportation and Air Quality, publication EAP420-B-00-001, February 2000.
10. www.aecc.be/en/european_legislation.htm (European Association for Emissions Control).
11. Absmeier, J., Dundar, B., and Wilcutts, M., *Embedded Controller Hardware and Operating System Selection for MoBIES Powertrain Testbed*, Vehicle Dynamics Laboratory, University of California, Berkeley, October 31, 2003. http://vehicle.me.berkeley.edu/mobies/powertrain/reports/Controller_Hardware_and_Operating_System.doc
12. http://store.sae.org/webcd.htm
13. Marcus, M. and Biersach, B., Regulatory Requirements for Medical Equipment, *IEEE Instrumentation & Measurement Magazine*, Vol. 6, No. 4, December 2003, pp. 23–29.
14. http://www.ul.com/info/standard.htm
15. http://www.ce-mark.com/cedoc.html
16. http://ts.nist.gov/Standards/Global/pg17.cfm
17. Monnich, Jr., H., ISO 9001:2000 for Small and Medium Sized Businesses, *ASQ Quality Press*, Milwaukee, WI, 2001.
18. Kasse, T., *Practical Insight Into CMMI*, Artech House, Boston, MA, 2004.

19. Persse, J.R., *Process Improvement Essentials*, O'Reilly Media, Inc., Sebastopol, CA, 2006.
20. Dunn, W.R., *Practical Design of Safety-Critical Computer Systems*, Reliability Press, 2002, Solvang, CA, pp. 143–176.
21. Pradhan, D.K., *Fault-Tolerant Computer System Design*, Prentice Hall PTR, 1996, pp. 4–6, 104.
22. Land, H.B., Eddins, C.L., Gauthier, L.R., and Klimek, J.M., Design of a Sensor to Predict Arcing Faults in Nuclear Switchgear, *IEEE Transactions on Nuclear Science*, Vol. 50, No. 4, August 2003, pp. 1161–1165.
23. Land III, H.B., Sensing Switchboard Arc Faults, *IEEE Power Engineering Review*, April 2002, pp. 18–20, 27.
24. Laplante, P., *Real-Time Systems Design and Analysis*, 3rd ed., IEEE Press and Wiley-Interscience, A John Wiley & Sons, Inc., Publication, Piscataway, NJ 2004, pp. 94–96.
25. Military-power-supply failures give rise to unofficial MIL standard, *EDN*, July 6, 1984, pp. 49–57.
26. Drees, R. and Young, N., Built-In-Test in Support System Maintenance, *IEEE Instrumentation & Measurement Magazine*, Vol. 5, No. 3, September 2002, p. 25.
27. Steinmetz, M., Built-In-Test Instrumentation and 21 Rules of Thumb, *IEEE Instrumentation & Measurement Magazine*, Vol. 5, No. 3, September 2002, p. 31.
28. Lansdaal, M. and Lewis, L., Boeing's 777 Systems Integration Lab, *IEEE Instrumentation & Measurement Magazine*, Vol. 3, No. 3, September 2000, p. 14–15.
29. Chan, H.A. and Englert, P.J., (eds.) *Accelerated Stress Testing Handbook, Guide for Achieving Quality Products*, IEEE Press, Piscataway, NJ 2001, p. 71.
30. Fenton, B., McGinnity, M., and Maguire, L., Fault Diagnosis of Electronic Systems, *IEEE Instrumentation & Measurement Magazine*, Vol. 5, No. 3, September 2002, pp. 16–20.

3

Tools of the Trade

3.1 Introduction

You will need a basic set of tools to develop a real-time embedded system. This chapter is not an exhaustive survey of tools; it may not cover all the necessary ones. The tools that are covered in this chapter are for illustration only; they indicate what you might use to implement good processes.

3.2 Tools for Estimation and Feasibility

3.2.1 Spreadsheet

A simple spreadsheet can provide a reasonable "bottom-up" estimate of time, effort, and cost for your projects. The reference Web site has a template spreadsheet that you may download and modify to suit your projects [1].

Many different activities make up a project. An accurate estimate requires accounting for all of them. Here are some activities that you should include in your planning:

- Planning and meetings
- Travel
- Analysis and simulation
- Hardware design
- Software design and coding
- Prototyping and field tests
- Test and review
- Fabrication, manufacturing, and assembly
- Installation and technical support
- Documentation

Table 3.1 is an example of some simple spreadsheet calculations for activities within a project.

TABLE 3.1

Example of a Spreadsheet Checklist with Some of the Activities to Schedule (© Kim Fowler, 2006,)

Concept Phase (Engineering Models Prepared)

Activity	Number of Meetings	Number of People	Average Time Spent (h)	Total Effort (h)	Procurement and Travel Cost ($)	Engineer and Program Management Cost ($)	Designer Cost ($)	Total Cost ($)
Planning	3	4	1	12		1,560		
Review and status	2	4	1	8		1,040		
Formal presentations	1	4	4	16		2,080		
Presentation preparation	1	3	4	12		1,560		
Travel	1	3	24	72	3,000	9,360		
Simulation		1	16	16		2,080		
Human interface design		2	40	80		10,400		
Environmental standards		1	16	16		2,080		
Circuit design		1	24	24		3,120		
PCB design		1	80	80			8,000	
Assembly		1	40	40	6,600		4,000	
Software design		2	80	160				
Coding		1	40	40				
Packaging design		1	80	80	1,000	10,400		
Prototyping		1	80	80			8,000	
Memos		3	10	30		3,900		
Requirements and specifications		2	40	80		10,400		
Design documents		2	20	40		5,200	4,000	
Test plans		1	8	8		1,040		
Manufacturing transfer plan				0				

Support and training documents	1	8	8				
Inspection	1	1	1	100			
Test		1	0				
Approvals	3	1	8	390			
Procurement	1	8	107	800	1,620		
Program management	1	107	107	13,943			
Technical support			0				
Contingency			40	4,000	5,200		
Minimum calendar time (h)			619				
Minimum effort (man-h)			1061	$12,220	$83,753	$28,900	$124,873 (Minimum)
Maximum calendar time (h)			929				
Maximum adjusted effort (man-h)			1592	$18,330	$125,629	$43,350	$187,309 (Maximum)
Minimum estimated effort (man-month)			6.3	Calendar (months)	3.7		
Maximum estimated effort (man-month)			9.5	Calendar (months)	5.5		
Engineer overhead ($/h)	130						
Designer overhead ($/h)	100						
Risk factor (%)	50						

3.2.2   Gantt Charts

Scheduling software can also help you develop reasonable estimates of time, effort, and cost for your projects and then format it into a Gantt chart. Figure 3.1 shows an example Gantt chart. As in the previous section with a spreadsheet checklist, you must be honest and diligent to fill in all activities needed by the software—it cannot do that for you. The reference Web site has a template Microsoft Project® file that you may download and modify to suit your projects [1].

You can find more information about Microsoft Project® in Reference 2. Please remember that this discussion only introduces the concept of the Gantt chart; other software packages mentioned later also have the capability to produce these kinds of charts.

3.2.3   Estimating Feasibility

There are some software packages available on the market to help you estimate feasibility of a proposed project. One such package from Galorath Associates is the SEER® software. QSM provides another set of tools. Both companies have databases with many complete software projects to allow metric comparisons and assessments to make better estimates. These are two examples of sets of tools on the market; other packages also exist.

The software packages from Galorath Associates can help you plan and control projects with these modeling tools [3]:

- SEER-SEM is for systems engineering
- SEER-H is for hardware development
- SEER-DFM is for manufacturing

QSM has a set of tools called Software LIfecycle Management (SLIM) that support decision making during the software life cycle. The tools can estimate, track, benchmark, and analyze metrics. Each tool can stand alone or operate as part of a suite of software tools. QSM's tools are as follows [4]:

- SLIM-Estimate helps estimate the time, effort, and cost to satisfy a set of software requirements.
- EstimateExpress is a software project estimating tool for projects with smaller requirements.
- SLIM-Control has statistical process control techniques to provide status on a project. It can compare the project plan against project actuals and forecast completion.
- SLIM-Metrics and SLIM-DataManager can preserve project history, assess competitive position, identify bottlenecks, quantify the benefits of process improvement, and defend future project estimates.

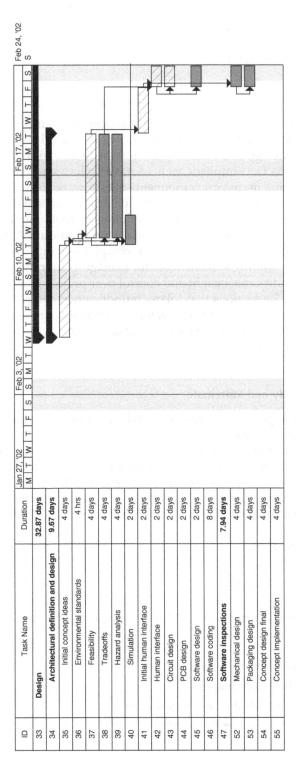

FIGURE 3.1
Example of a Gantt chart with some scheduled activities. (© Kim Fowler, 2006, used with permission. All rights reserved.)

3.3   Tools for Project Control

3.3.1   Overview of Various Software Tools

Managing projects and having adequate control over them is very important to developing mission-critical and safety-critical devices. Management and control starts with generating specifications and high-level requirements, moves through design and test, and ends with review and documentation. Obviously, this is a simplistic view of project phases and cycles but serves to indicate some necessary components for software tools.

References 5 and 6 have nice outlines of tools for project management. Within Reference 5 is a short outline with a sampling of tools found in Reference 7; Table 3.2 lists some of these tools and their example attributes. Please note that this listing is neither inclusive nor complete—it only serves to whet your appetite for finding suitable tools for managing your projects.

While these tools provide some insight and control over projects, they are not complete packages. The sections that follow introduce two commercial packages that attempt to cover the entire development process. These tend to be very large, expensive, complex packages that take significant corporate commitment to buy and use.

3.3.2   Telelogic Rhapsody and Statemate

Telelogic has two software tools, Rhapsody and Statemate, to manage design and development processes [8].

Rhapsody is an environment for systems, software, and testing that Telelogic calls Model-Driven Development. It provides a graphical means to specify the design of the system and its software and to model the embedded system. Statemate supports modeling, analysis, performance, and code generation. The goal of Rhapsody and Statemate is threefold:

- To automate simulation
- To automate validation
- To output production code

Rhapsody and Statemate support many different aspects of embedded system design:

- Requirements modeling
- Lifecycle traceability and analysis
- Team collaboration with document generation
- Design for testability and test generation

TABLE 3.2

A Small Sampling of Tools for Project Management—These Only Serve as Examples, This is Neither a Definitive Nor Exhaustive Listing

Software Tool, URL	Attributes
Adaptable Process Model (APM), www.rspa.com/index.html	APM provides a process flow in a hypertext document, descriptions of many software engineering tasks, document templates, and checklists
Artemis Views, www.artemispm.com	Artemis Views Project Management provides a collaborative environment for project and resource management that has role based access to project data through a web browser. It provides both structured data, such as dates, resources, and costs, and unstructured data, such as documents, risk, e-mails, and reminders
Microsoft Project®, www.microsoft.com	The standard against which most commercial packages measure themselves. It has various views including Gantt, resources, loading, and summary
Milestones, www.kidasa.com	Project management program for projects of all sizes. It can create presentation reports in a variety of formats including Earned Value, Resource, Summary, Milestone, and Gantt. It has many features and is Microsoft Project® compatible
Open Plan, www.deltek.com	An enterprise project management system that provides multiproject analysis, critical path planning, and resource management
Project KickStart, www.projectkickstart.com	Project management program for small to medium-sized projects; it both plans a project and creates a schedule. It guides users to identify goals, obstacles, and resources and suggests a strategic plan
TurboProject, www.turboproject.com	Project management program for small to medium-sized projects; it can create calendars, timelines, and Gantt charts, assign resources and tasks, and check them off upon completion. It has tools for resource and activity views, automatic over allocation warnings, split and recurring activity scheduling, and presentation templates. Microsoft Project® compatible

- Rules-based code generation for C, C++, Java, Ada for 8-, 16-, 32-, and 64-bit targets
- Adapts to a commercial RTOS

3.3.3 Rational Unified Process

The Rational Unified Process®, or RUP, is a set of tools to manage design and development processes [9]. RUP comprises RequisitePro® (also known as ReqPro), ClearQuest®, ClearCase®, TestManager®, ProjectConsole®, and SoDA®.

RequisitePro® helps create and manage requirements throughout the life cycle of a project. It not only captures requirements but also manages the changes in requirements. It can help manage the scope of a project through clarifying requirement attributes and traceability.

ClearQuest® tracks change requests during development. Its goal is to unify defect and change tracking so that team members can manage changes in system development.

ClearCase® is a version control tool. It helps the development team to ensure the accuracy of software releases, to build and patch previously shipped products, and to organize an automated development process.

TestManager® organizes the testing process by connecting and communicating between the test plan, test cases, requirements, and test records. It links test cases to requirements. It uses Crystal Reports to provide graphical and text reports on test metrics, test results, pass–fail status, and load performance.

ProjectConsole® helps developers monitor the status of their projects by creating a project metrics Web site based on data collected from RUP or third-party products and presenting the results graphically. It can update on demand or on schedule to provide an up-to-date view of the project.

SoDA® creates and manages documentation from the RUP suite of tools. Completeness and consistency are its goals, as well as easy update. SoDA® has over 70 templates for documents and supports both the Mil-Std 498 and DoD 2167A formats. It also allows you to format your own proprietary documents.

3.3.4   Version Control

Though it is simple, one of the best things that you can do to improve your processes and ultimately the quality of your products is to exercise strict version control. Knowing the latest version of a document or software file quickly becomes confusing if two or more people have any involvement in its preparation. Occasionally, we want to go back and view the evolution of a particular file, as well.

Many software products are available to archive and control your documentation. These systems allow only one team member at a time check out a particular file. They also keep a clear record of the evolution of each file, maintain all updates to the file, and store its latest version.

Rational's ClearCase® is one example of a version control tool, while Microsoft's SourceSafe® is another example. Please note these are only two examples, many other competent products are available. The most important thing to do is to get a version control tool and use it religiously.

3.4   Tools for Design

3.4.1   Simulators

Simulation can be a good start in designing a product. Please recognize that simulation is just that—a simulation; it is not real life. Simulations are based on simplifying assumptions and can never possibly account for all situations. Ultimately you will have to test the physical design.

With that said, many different software simulators can help with initial design. These simulator tools include system models and software generators (already mentioned earlier), electronic circuit—both digital and analog—thermal environments, and mechanical structures.

Circuit simulators take several different forms. Digital circuits can use behavioral or register transfer level (RTL) models. SPICE modeling can be used to analyze analog circuits and even digital circuits. It is possible to use electromagnetic and transmission-line models to approach the analog signal environment. The more sophisticated the tool, the higher its cost, the more computational power needed, and the slower the simulation.

Simulators for thermal environments can help locate "hot spots" and map forced convection flows. Simulators for mechanical packaging and mechanisms can help identify stress and load points. These types of simulators tend to find a home in environments where either many different products are developed or where testing is very expensive or impossible (some spacecraft instruments can only be simulated–they are not tested until space flight).

3.4.2   Computer Aided Design

Computer aided design (CAD) tools help out with documenting specific aspects of a product, particularly the mechanical package, the circuit design, and the circuit board layout. Many competing products are available on the market and are adequate for the task.

Most circuit design packages can also provide a layout tool for circuit boards. While most advertise automatic routing features for the circuit boards, you will still have tune the design by moving traces manually. This is especially true to drive the layout toward self-shielding and for high-speed design.

3.4.3   Software Design Tools

Software design tools include at least some of the following:

- Text editor
- Compiler
- Assembler
- Linker

The text editor is a simple word processing program that can produce a simple hex file. The compiler can be from one of several different sources: commercial, open source, or certified. Debate rages as to the most effective form of compiler. A certified compiler eases the approval a bit for mission-critical or safety-critical devices. Open source is free, but you must provide the technical support. A commercial compiler often has some sort of technical support. Most commercial compilers automatically assemble and link the compiled code; you probably would not have to worry about doing it yourself. Typically, you will be concerned about assembling and linking modules only when you write subroutines in assembly language to speed up a section of code.

Another tool is a static code checker that reviews the program structure and provides comments. These can be effective in catching some problems.

Another tool type includes modeling packages, such as MatLab®, which not only simulates a system but can also generate code [10]. One concern for automatic code generation is that you need to understand the assumptions that underlie the code structure; sometimes, confusion or ignorance about these assumptions can lead to unforeseen consequences.

3.5   Laboratory Equipment

3.5.1   Instruments and Tools

Electronic development, particularly for embedded systems, needs a basic set of equipment in the lab. These basic tools include the following (Figure 3.2):

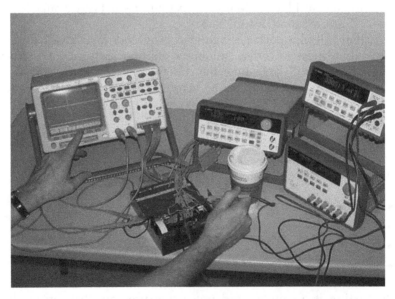

FIGURE 3.2
Examples of lab instruments used in developing and diagnosing hardware design: power supply, waveform generator, DMM, and a combination oscilloscope and logic analyzer. (© 2004 by Kim Fowler, used with permission. All rights reserved.)

- Several good-quality power supplies or a power supply with multiple outputs
- Multimeter that has V, I, and Ω settings
- Signal generator—preferably an arbitrary waveform generator
- Oscilloscope
- Logic analyzer

The multimeter should measure down to microvolts and microamperes. The signal generator should have at least 2–5 times the bandwidth of the expected signals that your design might encounter. The oscilloscope should have a bandwidth greater than 20 times the clock frequency of your design. The logic analyzer needs to acquire signals and glitches that are a fraction of the clock period—again 20 times the frequency is a lower bound. For any kind of wireless design, a spectrum analyzer will also be needed.

Cutting and connecting wires and traces will always be a part of circuit design and debug. You will need diagonal cutters, needle nose pliers, wire strippers, a good soldering iron, and various gauges of wire (Figure 3.3). Even with a fairly simple set of tools, you can do some fairly sophisticated developments.

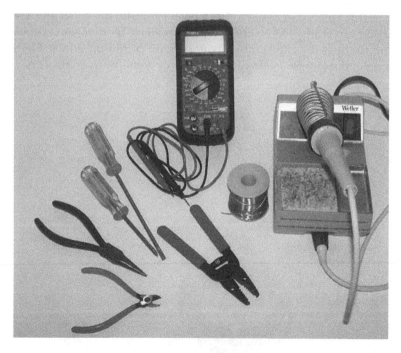

FIGURE 3.3

Examples of basic tools for an electronics lab—nothing fancy, just necessary. (© 2007 by Kim Fowler, used with permission. All rights reserved.)

3.5.2 Development Systems

Today, you can find a number of development systems to help with embedded system development. These are usually simple "pods" that connect to your desktop computer and then plug into your system.

Traditionally, microprocessor emulators would be plugged in place of the microprocessor and allow a complete view of all the internal registers in real-time or in a single-step mode. These were always expensive and depended on "bond-out" chips from the manufacturer; these are not available as they once were. Now emulators tend to be EEPROM emulators to allow you to download and monitor program flow without using an expensive or exclusive microprocessor emulator.

Another form of development system is the Joint Test Association Group (JTAG) analyzer. These plug into the JTAG ports on a circuit board and provide a measure of visibility into the component operation. They also allow program uploads if the circuit board is designed to do so.

3.5.3 Evaluation Boards

Another very effective form of development is the evaluation board. This is a relative cheap board from a manufacturer of a selected component, such as a processor, microcontroller, or an analog-to-digital converter (ADC), on it. The evaluation board tends to have been carefully designed and laid out so that you can study and evaluate the component. It often has rudimentary software routines to control the chip along with communications. Figure 3.4 shows several example evaluation boards for some microcontrollers.

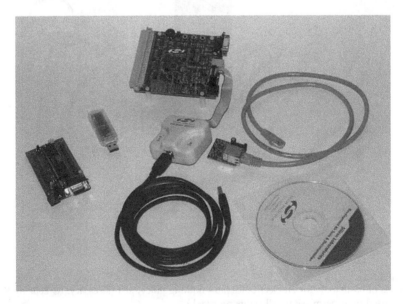

FIGURE 3.4
Examples of a few different evaluation boards. (© 2007 by Kim Fowler, used with permission. All rights reserved.)

With larger systems where you might be using a single-board computer (SBC) or multiple SBCs, you should buy an SBC and use it as a development board. If you are evaluating which SBC to buy, I would suggest the following steps:

1. Review the competing products; ask for a users' manual for the product—pay for it if you must.

2. Call technical support at each company and ask a slate of prepared questions (e.g., power consumption—maximum and typical values and under what conditions, the recommended software environment, throughput—under what conditions, etc.)

3. From Steps 1 and 2, select two or three competing products from those companies who competently responded to you.

4. Buy and evaluate the SBC from each of the competing vendors. Do not try to get a loaner system; these usually loan on a very short-term basis and may only be prototypes anyway.

References

1. Website with some basic templates: www.cool-stream.com
2. Gantt charts and scheduling: http://office.microsoft.com/en-us/project/default. aspx
3. Estimation and project control: http://www.galorath.com/
4. Estimation and project control: http://www.qsm.com/products.html
5. Listing of project management tools: http://www.projectreference.com/#PM Software
6. A directory of program management tools: http://home.houston.rr.com/interplan/
7. Short outline of tools: http://commercial-solutions.com/pages/pmsoftware.html
8. http://www.telelogic.com/Products/focalpoint/index.cfm
9. http://www-306.ibm.com/software/rational/offerings/scm.html
10. MatLab: http://www.mathworks.com

4

Case Study 1—Major Appliances

4.1 Concept and Market

4.1.1 Who, What, Why, How, Where, and When

This chapter focuses on major consumer appliances such as kitchen ovens or washing machines. I will use the development of a kitchen oven as the focus of this case study. Figure 4.1 shows some examples of these products. Costumers are primarily distributors who then resell to consumers.

Interestingly, at least one company and most likely more follow the practices of mission-critical development for these appliances. They need to get it right the first time because once a product goes out the door, it will not be coming back for repair or upgrade. Furthermore, these appliances are expected to last somewhere between 20 and 40 years!

4.1.2 Economics

Between 3 and 5 million appliances are sold every year. These appliances are cost-sensitive and yet are expected to last for a long time. Clearly, cost and reliability are prime concerns for the company. For most of us, walking the line between cost and reliability is a mutually exclusive analysis; as reliability increases so does cost, or as cost is driven down, reliability can suffer.

Profit margins are slim in major appliances and consumer white goods. Anything that can lower cost and still maintain reliability and functionality is worthy of a significant amount of research and effort.

4.1.3 Incremental Evolution

Changes in design and features of major appliances and consumer "white goods" are incremental or evolutionary, at best. The use of microprocessor controls is only now becoming prevalent. The battle between cost, functionality, and reliability has delayed the entry of this technology into major consumer appliances. This delay for consumer appliances has lagged behind other markets and applications.

FIGURE 4.1
(a–e) Examples of white goods consumer products. (© 2006, photographs by Kim Fowler, used with permission. All rights reserved.)

4.2   People and Disciplines

A number of people are involved in developing a major consumer appliance. Marketing defines the implementation for a particular appliance. The product-development team includes a chassis-performance team with mechanical, electrical, software, chemical, sensor, and food engineers.

For a high-end kitchen oven, the engineering team typically might include four software engineers, three hardware engineers, one computer-aided design (CAD) designer, and one technician. A group of controls engineers provides the appropriate control algorithms, such as proportional integral differential (PID) control of the oven temperature. A team that handles Underwriter Laboratories (UL) certification also provides part-time support to the development.

4.3   Architecting and Architecture

4.3.1   Process

As mentioned, major appliances and consumer white goods appliances must be exceedingly reliable. The engineers and designers use a combination of V-model and spiral development in a mission-critical fashion.

The combination of waterfall and spiral development for appliance design goes through five phases: concept, prototype design, laboratory design, manufacturing, and launch (Figure 4.2). Neither the hardware nor the software changes once finalized; therefore, spiral development of new features finishes in the laboratory design.

4.3.2   Analysis

Marketing performs a feasibility analysis for each product idea. They send the idea to a "toll gate" where the decision is made to go forward or to stop the development. After an affirmative decision, the product development team kicks off the product development.

The product development team uses Fault Tree Analysis (FTA) to analyze functionality and fault tolerance on all possible single-point failures and Failure Modes and Effects Analysis (FMEA). The engineers check for the effects of flipped bits in memory and in the software.

4.3.3   Architecture

The product manager and the product development team determine the features desired and prepare a specification for a new kitchen oven.

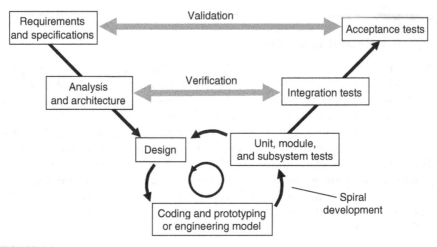

FIGURE 4.2
A modified combination of V-model and spiral development processes. (© 2007 by Kim Fowler, used with permission. All rights reserved.)

The user or human interface is the most critical item. The primary constraints on design are cost, time (to develop), and features (see Figure 4.3).

The team first establishes the necessary parameters for the new oven. Some of the parameters include air temperature, airflow volume and rate, evenness of browning, and maximum current draw (which must be under 50 A for the United States).

Next, the team considers the new features; displays, for example, are an important concern. The team evaluates the usability of the display to customers. For instance, how would customers turn on the oven? What makes sense?

Beyond "touch and feel" and intuitive operation, the team considers how these features affect software and hardware. For touch displays, they evaluate materials for front panels, such as glass vs. Mylar. Furthermore, they attempt to reuse previously developed modules in both software and hardware. These modules have already been proven through testing and long scrutiny; they represent significant savings in development time and money when incorporated in new products.

4.3.4   Interfaces

The user or human interface is the most complex arena of design. It is difficult to test all possible combinations of events—even now as manufacturers move to using liquid crystal displays (LCDs). A curious factor in user interface design is that consumers want each feature to be one button; consumers do not want menus with nested levels of selections.

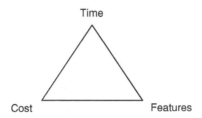

FIGURE 4.3
Three primary constraints for product development in major appliances.

4.4   Phases

Most companies have a proprietary development cycle. Most use the same types of phases as I have described in earlier chapters: Concept, Prototype Design, Laboratory Design, Manufacturing, Product Launch, and Logistics–maintenance–disposal. During each phase they update the schedule, add or revise features, and study the impact of those changes. They also review the software, its quality, correctness, and progress, at least once per phase.

4.5   Scheduling

Marketing uses top-down planning to set a target date for product launch. The product development team then uses bottom-up planning and sets the resource allocation in both dollars and staffing effort. Finally, both teams, marketing and product development, negotiate features and development so that the separately derived target dates move into alignment with each other.

4.6   Documentation

Most companies will use all the standard forms of communication and documentation, both formal and informal. These include notebooks, e-mail messages, letters, memos, project documents, manuals, brochures, and presentations.

The documentation important to the product development team consists of specifications, design documents, analyses, and safety concerns. These all help a team survive a quality or safety audit or should a product recall occur.

4.7   Requirements and Standards

4.7.1   Standards and Regulations

Most companies certify their appliances to UL and CSA regulations. Most also have proprietary standards for designing appliances. In one instance, the CAD team has rules and guidelines for designing circuit boards and modules, which include items such as trace spacing, location of test points, and layout for electromagnetic capability (EMC).

4.7.2   Preparing Requirements

Both marketing and product development teams help prepare the requirements. Marketing maintains contact with customers, consumers, and distributors to follow current trends. Once marketing develops a concept for an appliance, called an "innovation product", it goes to the "toll gate" for a GO/NO GO decision toward development. A GO decision leads to a product kickoff, where the product development team prepares the requirements and specifications.

4.8   Analysis

A number of different analyses can be performed: innovation product feasibility, heuristics, calculations, and prototype testing. These are primarily performed by the product-development team.

Teams study feasibility of an innovation product by studying, comparing, and contrasting new feature sets. The staff develop heuristics for reusing hardware and software modules. They use FTA and FMEA calculations to define fault tolerance and robustness. Finally, they use prototype field testing with customers to verify and validate concepts.

4.9   Design Trade-Offs

4.9.1   Hardware

The market for consumer white goods and major kitchen appliances such as ovens places special demands on electronic hardware. Here are some examples of those demands:

- Integration of components must be high to reduce assembly costs and still achieve the desired functionality. The designers use the 8-, 10-, or 12-bit analog-to-digital converter (ADC) inside the micro-controller for reading analog signals from sensors; they do not use a separate, stand-alone component. The analog subsystems and ADCs, in turn, drive the preference toward 5 VDC powered microcontrollers, which have adequate voltage swings to provide noise margin. They also do not want to change between models because of development legacy. Microcontroller vendors, however, are forcing them to move to 3.3 VDC, which reduces the analog noise margins.

- They need a freeze on integrated circuit (IC) fabrication to avoid EMC problems. When a vendor shrinks the design rules for line widths and transistor size to improve their fabrication yields, it causes faster edges on the digital pulses over the original IC design. This, in turn, increases the amplitude and number of harmonic components for the same clock frequency, which then leads to more EMC radiation and interference.

- Flash memory retention needs to be much greater than 20 years (remember, many of these appliances are expected to last as long as 40 years). Right now, no IC vendor is willing to specify retention beyond 20 years.

- Engineers prefer single-layer circuit boards for the lowest cost. They begrudgingly go to two-layer circuit boards if EMC becomes a concern.

- The CAD designers that lay out the circuit boards have developed their own set of rules to control EMC. This decoupling of knowledge from the design engineers can cause problems and force iteration in design to achieve all the objectives.

4.9.2 Power

For kitchen stovetops and ovens, the input power is 220 VAC, 60 Hz in the United States. For many international markets the input power is 240 VAC, 50 Hz.

Companies must ensure operation and tolerance in the face of changes by vendors in their manufacturing. One way that this is done is to use two different terminal fuses of the same specification from different vendors in a redundant circuit to reduce the effects of unexpected changes.

Another concern is power quality. One company tests their appliances for 52 different AC line conditions.

4.9.3 Software

For the kitchen oven in this case study, software engineers write software in C for appliances. They also design their own custom real-time operating system (RTOS) for hosting the software on 8-bit microcontroller systems. They use a custom time-slice RTOS that guarantees timeliness and is amenable to analysis. They use a custom pre-emptive priority, interrupt-based RTOS for 32-bit systems. In both types of systems, engineers reuse software modules that have been proven with time and testing. This reduces cost and effort.

Software has to be correct when the appliance comes off the production line; it cannot be upgraded or revised once shipped. Software engineers use careful development processes. They conduct code reviews, do extensive, carefully-planned tests, and perform field tests. The code reviews are held at least four times before release; this reduces to one or more reviews per phase.

4.9.4 Hardware vs. Software

The constant battle in designing appliances is desired features vs. either reliability or utility. Just because a feature is possible, does not mean that it satisfies the incremental gain in utility or the potential decrease in reliability.

One interesting trade-off is in the control panels. Most consumers want each feature to be governed by a single hardware button; they generally do not want menus with nested levels of selections, which implements easily in software. As fashions and features change in devices used in all parts of life, people adjust to new ways of operation. The invasion of menu-driven devices and appliances, such as cell phones, personal digital assistants (PDAs), and computers, is pushing the acceptance of menus in appliances.

4.9.5 Buy vs. Build

Because major appliances sell in high-volume, low-margin markets, nearly every subsystem that one company uses is custom-designed. These custom designs are highly optimized and tightly integrated for cost, functionality, and reliability.

4.9.6 Manufacturing

Cost is of paramount importance. Companies use both design-for-manufacture (DFM) and design-for-assembly (DFA) to reduce manu-facturing costs. Shipping is another consideration; whatever reduces the cost of packaging the appliance and distributing it goes straight to the bottom line. If the cardboard carton used for shipping can be revised in such a way to cut pennies of cost while maintaining an acceptable level of protection, it will be done.

4.9.7 Test and Maintenance

Appliance companies test for manufacturing quality; those tests are described in Section 4.12. Otherwise, test points on circuit boards and hardware modules aid diagnostics in the laboratory to debug problems before manufacturing.

Maintenance is not planned for appliances. Consumers are not good about maintaining equipment; therefore, engineers design for reliability and avoid maintenance.

Repair is typically by replacement of the failed module. A good technician can run the appropriate diagnostics to isolate faults and failures, but good technicians are not common. Still, companies maintain ongoing research into diagnostics tools and techniques.

4.10 Tests

Some companies use a large number of different tests to develop a product. The product development team uses field tests and laboratory tests to refine features and implementation. They use subsystem tests for both hardware and software to verify function.

4.11 Integration

It is rather arbitrary to separate integration from the tests described in Section 4.10; I am only doing it to provide a consistent format between case studies. Hardware and software subsystems are integrated and tested for functionality. System tests are prescribed to verify specifications, after which the product is sent to certification testing for UL approval.

4.12 Manufacturing

4.12.1 Electrical, Electronic, and Mechanical

Some companies do all the manufacturing in-house; this includes all the printed circuit boards used in appliances. These are usually one- or two-sided cards to maintain cost. One appliance company also manufactures much of the chassis, doors, hinges, and panels for its appliances.

Most companies only purchase the smallest or most basic components from vendors: integrated circuits, electronic displays, wires, electrical cords and plugs, heating elements, and fans. One company also purchases the materials but builds the glass windows for the oven compartment and the nonstick glass cooktops.

4.12.2 Assembly

Electrical assembly uses programmable machines to form the cable harnesses that connect the display panel, buttons, and processor module.

Electronic assembly uses automatic pick and place machines and soldering reflow ovens to build the circuit boards. Mechanical assembly uses programmable machines to attach the metal panels to the chassis.

Most assemblies are automatic, with programmable machines to reduce assembly costs. Manual assembly is unavoidable in areas where operations are too complex for assembly-line equipment and robots.

4.12.3 Tests

Assembly-line testing also strives to reduce the need for different equipment or operations. A case in point is that one company uses one tester interface for all products within a specific market, for example, for this case study, all cooker products.

Manufacturing personnel run proprietary protocols on all units that come off the assembly line. They test

- All loads (heating elements)
- All connections
- All front-panel buttons

They also perform a visual inspection of every display.

Automatic test cross-checks for the version of the software load. Reuse of hardware means that the same circuit board may end up in 15 or 20 different products. The only distinguishing difference is the version of software load; that version must be confirmed.

4.13   Support

4.13.1   Maintenance and Repair

There is no planned maintenance, other than occasional cleaning by users. Most companies generally do not plan for software upgrades; it would be too costly. Third-party companies and local shops repair these appliances by replacement of parts or modules only.

4.13.2   Technical Support

Technical support covers basic operation, repair concerns, and customers attempting to try out a function new to them. Although failures are reported back from the field, it is still difficult to understand the exact problems without engineers being there to question users. This is a basic weakness with all consumer appliances.

4.14 Disposal

All manufacturers are concerned with Restriction of use of certain Hazardous Substances (RoHS). On the horizon from Europe is the specter of recycling, which will place further burdens on manufacturers.

4.15 Liability

The biggest safeguard against liability is careful design and documentation. Recalls for faulty products are costly in terms of time, money, and company reputation; building it right the first time is the only way to avoid recalls. Careful design is the only way to build it right the first time.

4.16 Summary

4.16.1 Emphases

- Developing major appliances, such as a kitchen oven, requires mission-critical design processes.
- Designing consumer appliances requires a constant struggle to balance time, cost, and features.
- The user interface, or the graphical user interface (GUI), is a major focus of any appliance. Much effort is expended studying and researching the most effective and desired operations.
- The design of major appliances must account for longevity of operation, potentially up to 40 or 50 years.

4.16.2 Gotcha's

- Current microcontroller chips have unknown or limited retention for flash memory.
- IC vendors can and do change fabrication processes, which usually means smaller features sizes that lead to sharper digital pulse edges. This increases the frequency of the signal harmonics and affects EMC.

Acknowledgment

The material for this chapter comes from interviews with a colleague in the major appliance industry. We could not get approval to use any particular company's name.

5

Case Study 2—Telecom Products

5.1 Concept and Market

5.1.1 Who, What, Why, How, Where, and When

This chapter focuses on products in the telecommunications market that move and transfer data. These products include DSL (digital subscriber line) systems, remote access servers (colloquially termed "ISP in a box"), and VOIP (voice-over-internet-protocol) boxes. Figure 5.1 shows some of these products.

The company is Patton Electronics, located in Gaithersburg, MD, USA; it sells products internationally. Costumers include end users, distributors, integrators, and value-added resellers (VARs). Patton will "rebadge" products for overseas carrier deployment; this means that the company will put a customer's logo on each unit to make the unit unique to the customer.

5.1.2 Economics

The folks at Patton Electronics focus on "high-mix, low-volume" products. This means that they design, develop, manufacture, and sell many different products but in fairly small volumes. Patton's total average output for all products is about 15,000–20,000 units per month.

Patton picks markets within its expertise and develops advanced products. They tend to be early adopters of technology, but they avoid the "bleeding edge" in the state-of-the-art. An example of their mode of operation is their remote-access servers. They entered the market in 1997, fairly late in the game, then competed with and outlasted some other companies. A big competitor left the business in 2001 and turned over their product line to Patton. Patton's goal is to persist in a market.

5.1.3 Market Definition

The people at Patton Electronics struggle continually to define their markets. They discuss focus and priorities, channels of products vs. sales regions vs. applications, and whether to segment product lines or not. Market definition is rather more amorphous than crisply defined goals.

(a)

(b)

(c)

(d)

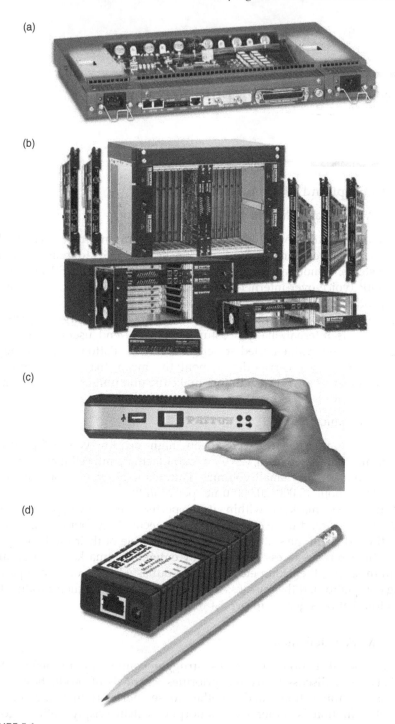

FIGURE 5.1
(a–d) Examples of telecom products. (© Patton Electronics 2006, used with permission.)

There tends to be a lot of angst and hunger for better definition while remaining "customer-driven."

5.2 People and Disciplines

Patton Electronics uses several different types of teams to develop a product. One team, called the product review board (PRB), helps define products and markets. The PRB comprises the president, executive vice president (VP), VP of product management, VP of manufacturing, VP of product development, and the director of technical support. The director of technical support, who may seem to be an incongruous member of the PRB, provides the insight as to whether the company will be able to support and service the new product or not. Another team does the engineering work; these teams are usually very small, typically one or two people. Manufacturing is done in cells or groups by a variety of skilled technicians.

5.3 Architecting and Architecture

5.3.1 Product Definition

Anyone within the company may generate a product idea. Once a month the PRB collects product ideas and discusses them. If a particular idea merits further consideration, someone designated by the team then writes a marketing requirements document (MRD), which establishes the business case. The MRD estimates potential revenues, costs, and margins. Once the MRD is written and signed off, the idea becomes a product nomination. All nominations go to a subcommittee for further selection; the executive VP, VP of product management, product managers, head software engineer, and head hardware engineer comprise that subcommittee.

5.3.2 Product Development

Engineers within Patton Electronics use both waterfall and spiral development processes (Figure 5.2). Hardware designers use the waterfall process to develop a platform for a new product. Software developers use spiral development that iterates many times far beyond the initial sale and deployment.

The waterfall development of hardware goes through three or four phases: specification, preliminary design (which sometimes combines with critical design), critical design, and design verification. Hardware platforms seldom change once designed and produced, and so spiral development does not apply to designing the hardware. Patton Electronics does some

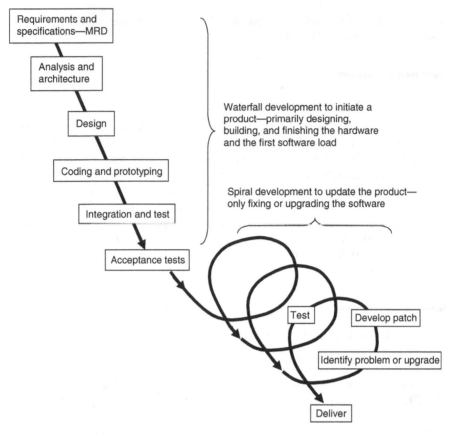

FIGURE 5.2

Patton Electronics uses both a waterfall process and a spiral development for products. The waterfall process applies to the hardware platform and the first software load. Spiral development of the software then continues throughout the life of the product. (© 2007 by Kim Fowler, used with permission. All rights reserved.)

field testing, and when changes do occur they are generally implemented in software.

Software development is different from hardware development. In the beginning of a project the software group meets with the product manager to determine the course of software development. Most products receive a basic software load upon initial deployment. Then Patton generates frequent updates, about once per month, to increase the functionality and to patch bugs. This business model of frequent updates and releases works well for a number of telecom products.

5.3.4 Architecture

The chief engineers for software and hardware define the data flow, the data buses, and the memory size. The software engineer defines the software core and how it will grow from the alpha version to the beta version to the follow-on updates after product release.

5.3.4 Interfaces

Most of these telecom products are embedded within other equipment and are seldom stand-alone for direct operation by users. So, other than the data buses and ports, the only other interface for human interaction is the Web page that a network manager uses to configure or change the operation of the unit. A manual is not necessarily needed.

5.4 Phases

5.4.1 Specification

Once the MRD is completed and the product selected for development, it enters the specification phase. It usually spends about 1–2 months in this phase. The specification review defines its exit from this phase.

A product manager and a software engineer typically write the specifications. The executive VP, VP of product management, VP of manufacturing, VP of product development, and the director of technical support often comprise the review team for the specifications.

5.4.2 Preliminary/Critical Design

Next is the preliminary design phase. Sometimes it combines with the critical design phase for simpler products. This phase spans anywhere from 3 months to one year. Sometimes, the team develops a prototype and submits it to field testing to find the necessary constraints and concerns for final development.

The team usually has one or two engineers, most often software engineers. If the project is big enough, the team may have a hardware engineer full time, otherwise they receive some part-time help from a hardware engineer to develop the hardware platform. The product manager, VP of product management, VP of manufacturing, VP of product development, and the director of technical support typically make up the review team to complete the phases.

5.4.3 Manufacturing

This phase is primarily for the hardware platform. The software engineer or team continues work on the first software load. This is where pilot versions of the product are built and tested.

Patton has manufacturing facilities and staff. They can assemble circuit boards, install power supplies, build wire harnesses, stuff enclosures, and test, all in-house. Only fabrication of the circuit boards and certification or compliance testing is performed outside of Patton.

5.4.4   Launch

A successful review determines product launch. The folks involved in that decision include the product manager, VP of product management, VP of manufacturing, VP of product development, and the director of technical support.

5.4.5   Logistics, Maintenance, and Disposal

The software engineer or team enters a spiral development process for revising and updating the software. New releases are frequent and can occur monthly.

RoHS compliance and regulations require that Patton collect and dispose of product and manufacturing equipment at the end of its life.

5.5   Scheduling

Scheduling and product management follow a visual work flow. Each product nomination receives a "ticket" that contains a project name, the name of the assigned product manager, and any notes particular to its development. These enter a product funnel for selection; the selected products then flow out of the funnel into a timeline with the other products in development. At each stage along the way, notes may be added to the ticket or checked off. This allows team members to quickly assess the progress of the project (Figure 5.3).

Scheduling is bottom-up, with deadline dates set along the timeline as a product percolates through its phases.

5.6   Documentation

5.6.1   Hardware

The documentation for the hardware platforms includes the following:

- Hardware specification
- Design verification report

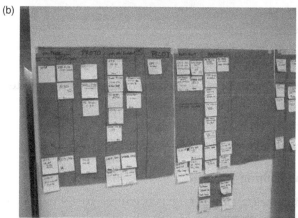

FIGURE 5.3

(a–c) Schematic diagram of project flow and progress visualization at Patton Electronics. This deceptively simple scheme allows engineers to gather around and at a glance assess the state of all projects. (a) Nominations and product funnel. (b) Prototypes and pilot runs. (c) Compliance and certification. (Photographs © 2006 by Kim Fowler, used with permission. All rights reserved.)

- Parts list
- Assembly instructions
- Qualified manufacturers lists for parts
- Production test procedure
- Compliance test reports
- Regulatory certificates

The hardware engineer generates all of these documents except the last two. Outside test firms generate the test reports and the regulatory certificates. A component engineer maintains the database of manufacturers' declarations pertaining to compliance and regulation.

5.6.2   Software

The software team generates the following documents:

- Software specification, which becomes the release notes
- Design verification report
- Version control
- Bug management database

5.6.3   Manufacturing

Manufacturing has a number of procedures to follow, all controlled by various software programs. Manufacturing also keeps a database that contains customer identification, key generator, software load for the unit, and the "pack out" bar codes, which confirm the Internet address of the unit. The challenge in all this is to maintain quality in isolated applications and try to integrate the manufacturing environment.

5.6.4   User Manuals

Patton Electronics maintains Web pages for products that serve as user manuals. These Web pages are written by the software engineer who developed the unit; the Web pages serve a very small number of users—system operators that manage the equipment.

5.7   Requirements and Standards

The primary regulation that this telecom equipment must meet is the FCC (Federal Communications Commission in the United States) part 68.

The products often must meet CE (Conformite Europeene) compliance for European and international markets.

5.8 Analysis

5.8.1 Feasibility—The Business Case

The primary effort to determine the feasibility of a new product is in the MRD. The MRD establishes the business case by estimating potential revenues, costs, and margins (the differential between costs and sale price). Implicit in these estimates is the technical feasibility.

An engineer or product manager assigned by the PRB writes the MRD. This happens early on in the process and becomes part of the product nomination.

5.8.2 Heuristics, Calculations, Definitions, Approximations, and Simulations

The rules of thumb and estimates usually pertain to the size of the software loads, needed memory, and rates of data throughput. Estimates for power consumption and cooling are also made.

An engineer or engineering team makes these estimates. These go into the MRD and help shape the design specifications that follow after selection for further development.

5.8.3 Field Tests

Occasionally, a product needs further definition through prototype and field trials. An example of this is any VOIP product; beta test units often spend extended time in the field. When this happens the engineering team develops the prototype with oversight from the product manager. The prototype is fielded and monitored by the engineering team. Field tests typically take between 2 and 10 months. These occur at a prospective customer's site or in the laboratory.

5.9 Design Trade-Offs

5.9.1 Hardware

Most of the hardware focuses on data flow; therefore, field-programmable gate arrays (FPGAs) are used extensively. Patton uses a variety of microcontrollers and microprocessors to control the telecom equipment; these include 8051 series microcontrollers running between 4 and 20 MHz, MIPS processors running at 400 MHz, and i960s. The software loads generally range between 1 and 5 MB of memory.

Typically, a hardware engineer works closely with the lead software engineer to determine the size of the memory and data through requirements.

5.9.2   Power

Input power is nearly always AC line, or mains, power. Patton uses power supplies that have universal input (100–240 VAC and 50–60 Hz) to allow for international applications.

The size, shape, and power consumption all factor into selecting the power supply. For a telecom product that uses a cPCI chassis, they will select a commercial off-the-shelf (COTS) cPCI power supply that slides into the chassis. For self-contained enclosures they will select an open-frame power supply. For lower-power applications in a small box they select a "wall wart" type of power supply.

5.9.3   Cooling

Patton does not use fans to cool the circuitry. They strive for low-power designs and avoid the maintenance concerns of fans and filters. They depend on forced air cooling provided by equipment chassis.

5.9.4   Software

The software engineers write most of their software in C. They do not use code reviews but rely on field tests and monthly releases to control the quality of the software. They write their own custom real-time operating system (RTOS) for embedding in their products.

I tend to think that the lack of peer review is a weakness in the software development process. Again a custom RTOS might not be the best way to go, unless Patton has developed a template for their custom RTOSes. The caveat in my comments is that I do not know the full extent of the software in Patton's telecom products.

5.9.5   Hardware vs. Software

The biggest trade-off is that the hardware platform must have sufficient memory and computational power to support the frequent software updates. Once designed and released, the hardware is not changed. Only software updates are used to enhance or modify any particular product. The lifetime of a typical product is between 3 and 5 years.

5.9.6 Buy vs. Build

Nearly all of Patton Electronics' telecom products are custom-built and manufactured for specific applications. Patton has in-house manufacturing facilities for assembling, testing, and shipping products. Consequently, most subsystems are custom-designed and custom-built.

The one exception is the power supply. Typically, Patton buys COTS power supplies for their products. The one place where they find it difficult to buy a COTS supply is for 1 U rack mount enclosures, where they need long skinny form factors for the power supply. A product in a 1 U enclosure will often dictate a custom design of the power supply.

Even when Patton does do custom design of a power supply, they still use standard types of pinouts for the DC-DC convertors. This allows them to swap out DC-DC modules to fit different applications.

5.9.7 Manufacturing

Patton Electronics designs and develops products with design for manufacturing (DFM) and design for test (DFT) in mind. This means that layout of components and subsystems aid the smooth and efficient assembly of product. Patton also uses DFT to do production testing before a product ships. Chassis usually have Joint Test Association Group (JTAG) ports for diagnostics.

5.9.8 Test and Maintenance

Patton Electronics does not maintain a test-engineering staff. The hardware and software design engineers develop the appropriate tests along with the production design. Maintenance is not a consideration because, once a product ships, it does not return for repair. If a unit fails, it is simply replaced.

5.10 Tests

5.10.1 Formal and Informal

Most testing at Patton is laboratory bench diagnostics or manufacturing qualification tests. Patton relies on component vendors to test and qualify components and subsystems; a components engineer maintains a database of compliance declarations by the vendors.

Patton does not peer-review software. Software engineers rely on frequent updates of the software in the field.

Field testing often is viewed as an integration activity. Patton uses prototypes and field-tests them to complete the specifications for a final product.

5.10.2   Manufacturing

Products are usually tested as final assemblies in manufacturing. Examples of tests carried out on the assembly line include power-on-self test, software build-in-test (BIT), and link tests. They also test the interfaces for meeting communications protocols such as T1, DSL, and ISDN. They test for bit error correction by inserting bit errors with a test article and then check for correction by the production unit.

The hardware and software engineers on the design team design the manufacturing tests. Technicians then carry out the tests on the production units at the end of the manufacturing line.

5.11   Integration

5.11.1   System

Integration brings the hardware and software together; the initial software programs are loaded on the hardware platform. Integration occurs during the two different phases: pilot and manufacturing release.

During integration, the system is validated. Most of the validation is devoted to software functionality.

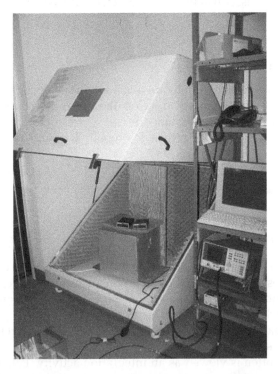

FIGURE 5.4
Anechoic chamber for precompliance testing. (Photograph © 2007 by Kim Fowler, used with permission. All rights reserved.)

The engineering team, both hardware and software engineers, performs the integration in both pilot and manufacturing release phases.

5.11.2 Environmental

Patton does do some precompliance testing for electromagnetic capability (EMC) to reduce the amount of time and effort spent at official test and certification sites (Figure 5.4). These tests check for radiated emissions and conducted susceptibility. Precompliance testing gives a measure of assurance before Patton goes to an outside firm to test and certify the design.

Patton tests and certifies to both the CE mark and FCC part 68 regulations.

5.12 Manufacturing

5.12.1 Electrical and Electronic

Patton Electronics has a fairly complete and flexible capability to manufacture telecom products. They have pick-and-place equipment to "stuff" circuit boards (Figure 5.5) and then solder reflow ovens to bond the

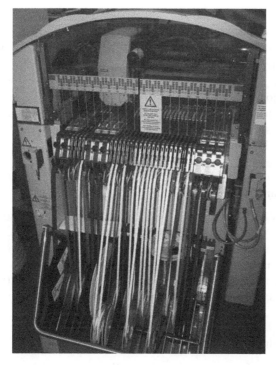

FIGURE 5.5
Pick-and-place equipment for "stuffing" circuit boards. (Photograph © 2006 by Kim Fowler, used with permission. All rights reserved.)

FIGURE 5.6
(a and b) Solder reflow oven for attaching components to circuit boards. (Photographs © 2006 by Kim Fowler, used with permission. All rights reserved.)

components to the circuit boards (Figure 5.6). Afterwards, a technician inspects the circuit boards for quality and then sends them onto assembly.

Any wire harnesses are prepared by hand assembly. In general, Patton avoids discrete wires and harnesses; they try to make connections directly to the circuit boards.

Patton does not fabricate its own circuit boards; it sends out board layout files to an outside vendor, who then fabricates and delivers the circuit boards back to Patton.

5.12.2 Mechanical

Most products use a standard form factor for their enclosures, including 1 U boxes and PCI chassis. Even stand-alone products use COTS enclosures. If the product is not a PCI chassis, then the circuit boards mount in the enclosures with a motherboard and daughter cards.

These enclosures are hand-assembled by technicians organized into groups, or cells. Patton also hand-assembles some components (Figure 5.7).

5.12.3 Assembly Control

Many different software programs run the manufacturing lines; it is not an integrated process because of the complexity of the processes in a

FIGURE 5.7
Assembly of components before inserting into a product. (Photograph © 2006 by Kim Fowler, used with permission. All rights reserved.)

"high-mix, low-volume" environment. Patton Electronics relies on skilled technicians to run the appropriate software to control manufacturing.

5.12.4 Testing

Patton Electronics does not maintain a test-engineering staff. The hardware and software design engineers develop the appropriate tests along with the production design. A skilled technician then tests the products at the end of the manufacturing line (Figure 5.8). These tests are for functionality and quality.

5.13 Support

5.13.1 Logistics and Maintenance

Products from Patton Electronics are generally not field repairable. Repair is by replacement of the unit if hardware fails. Otherwise, fixes and upgrades occur through frequent software releases.

The technical support team fields the initial inquiry. If technical support cannot immediately fix the problem, then the product manager or designated engineer provides the solution.

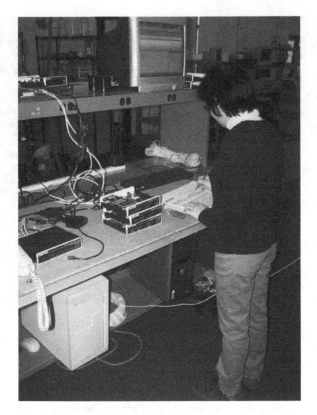

FIGURE 5.8
Testing products at the end of the manufacturing line. (Photograph © 2007 by Kim Fowler, used with permission. All rights reserved.)

5.13.2 Technical Support

Technical support is primarily instruction into using the Web page of the individual product. The Web page instructs the user how to configure the product.

5.14 Disposal

Patton Electronics, like other manufacturers, must deal with Restriction of use of certain Hazardous Substances (RoHS) and waste from electrical and electronic equipment (WEEE). They are contracting with third-party companies to recycle old equipment. They also have the option to pick up and recycle the equipment themselves, but have not done that yet. Implementation of WEEE recycling is not clear, and each country independently determines how it wants the recycling to be performed. There is some talk of a fee-based recycling based on product weight, but that is not decided as of early 2007 in any country.

5.15 Liability

These telecom units and market have fairly low liability; they are not anything like a medical device or automotive device or even a kitchen appliance. They do not really have major safety issues other than the AC power input, which many electronic products have.

The biggest concerns are the legalities surrounding intellectual property. Here Patton primarily deals with products that follow government and commercial standards. Intellectual property disputes, as with most products in any market, are the primary concern but very infrequent for Patton.

5.16 Summary

5.16.1 Emphases

Patton Electronics' forte is persistence in the telecom market. Their products generally remain behind the "bleeding edge" of technology but serve a niche market in data-transfer products. The company specializes in "high-mix, low-volume" products and have the design and manufacturing resources to support that type of operation.

5.16.2 Gotcha's

Patton Electronics does not perform peer review of its software. They also build their own custom RTOSes. Smaller size and extent of the software and reuse between products can mitigate these concerns.

Acknowledgments

My thanks to Craig Silver and Bryan Dubois at Patton Electronics for providing the information for this chapter and for reviewing it.

6

Case Study 3—Commercial Laboratory Equipment

6.1 Concept and Market

6.1.1 Who, What, Why, How, Where, and When

This chapter focuses on scientific instruments for both laboratories and high-tech manufacturing. These instruments and systems address manufacturing tests of electronics, process monitors, and scientific and engineering research. Figure 6.1 shows some of these products.

Keithley Instruments, Inc., of Cleveland, OH, offers about 500 products in dozens of markets. Keithley's primary markets for test equipment [1]:

- Manufacturing production test for electronic components
- Testing flat panel displays and optoelectronics
- Accurate, very sensitive, low-level measurements for research into superconductivity, semiconductors, metals, polymers, and insulators
- Device characterization for semiconductor wafer processing
- Testing telecommunications devices such as cell phones and digital switch systems

Joseph Keithley founded the company in 1946 in a small workshop in Cleveland, OH. Its products have evolved and multiplied in the arena of electronic test, measurement, and data acquisition for engineers and scientists. The company has grown into an international concern with sales worldwide [1].

6.1.2 Economics

Keithley Instruments, Inc., has more than 650 employees and yearly sales of nearly US$150 MM [1]. Keithley's annual report for 2005 gives the following

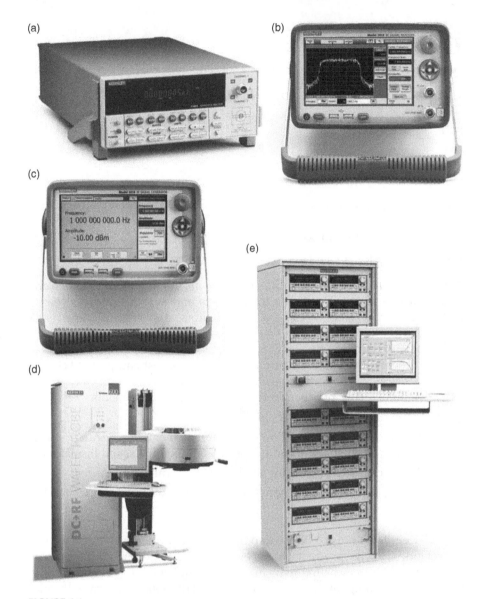

FIGURE 6.1
Examples of scientific instrumentation for the laboratory and production line. (© Keithley Instruments, Inc. 2006, used with permission).

breakdown of types and price ranges of their products [2]:

- Parametric test systems for semiconductor wafer manufacturing range from US$150K to $500K
- Semiconductor characterization system ranges from US$30K to $75K
- System instruments range from US$15K to $25K

- Bench-top, stand-alone instruments range from US$1K to $10K
- Switch systems range from US$2K to $50K
- Personal computer (PC) plug-in boards range from US$200 to $4000

Like the company in the last chapter, Keithley focuses on "high-mix, low-volume" products. This means that they design, develop, manufacture, and sell many different products but in fairly small volumes. Keithley's average output per month for all products is between 3,000 and 5,000 units with a total yearly output of between 45,000 and 60,000 units.

6.2 People and Disciplines

Keithley Instruments has five distinct groups of people working on every product [3]:

- Strategic marketing
- Research and development (R&D)
- Manufacturing
- Customer service and applications support
- Commercial marketing

Each group has specific responsibilities and tasks depending on the phase of the project.

Strategic marketing is responsible for the business case, financials, and planning for a new product. The folks in strategic marketing must understand customers' applications and requirements and the competition to gauge both market position and potential. They determine the product's fit with the goals and capabilities within the company for other products. They also develop and maintain key customers and partners who will participate in the development of the product [3].

R&D is responsible for generating technical innovation, assessing the technology risks, establishing major milestones during development, the product design, negotiating product requirements, and quality and reliability. The people in R&D establish external development partnerships, the schedule, and the resource plans. They also contribute to the *DFx* goals; DFx stands for Design-For-x, where x might be assembly, test, manufacturing, and so forth [3].

Manufacturing eventually builds the product, which means that manufacturing is responsible for the product assembly and test. They also finalize the DFx goals, conduct the test run pilot, and build demo units. Manufacturing also must make sure that material sourcing and material entry into the purchasing system all happen [3].

Customer service and applications support is responsible for planning and implementing the product service and support and for the training in the use of the new product. Commercial marketing is responsible for the launch plan, the sources of leads, and sales by geography [3].

6.3   Architecting and Architecture

6.3.1   Process

Keithley Instruments uses the stage-gate process to select and develop product ideas. Keithley's stage-gate has four sections or phases. In the first two phases, called the *fuzzy front end*, the product is defined and invented. These phases produce the concept and assess its feasibility. The final two phases are called the *concrete back end*; the product is implemented, designed, built and demonstrated in these two phases. Somewhere between Phase 2 (the fuzzy front end) and Phase 3 (the concrete back end) is a line in the sand. It represents the switchover from invention to implementation.

Part of the process and parallel to the fuzzy front end and the concrete back end is the product funnel (Figure 6.2). The people at Keithley Instruments, as in the company described in the last chapter, have more ideas than can be executed at any one time. They must set priorities and then narrow the number of candidates for further development. Keithley defines its product funnel through criteria, such as financial payback, risk, corporate strategy, and core competencies. Most of the selection occurs in Phases 1 and 2 to weed out product ideas, which leaves full resources devoted to product implementation in Phases 3 and 4.

Keithley is using a spiral form of development through these four phases, particularly the last two. It consists of a full integration of the current build of modules, both hardware and software, to test and verify the

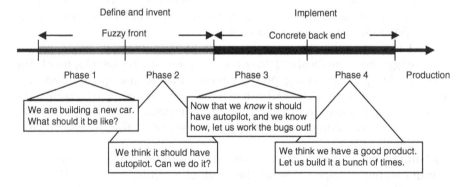

FIGURE 6.2
Typical resource balancing throughout the phase/gate process. (Used with permission, *IEEE Instrumentation & Measurement Magazine* [3] and Keithley Instruments, Inc.)

specifications and to validate the functionality of the instrument. The company is using spiral development to reduce risk in developing products. It is an ongoing effort that requires extra work and is still resisted by some engineers; the ultimate goal is to make it into a "best practice."

Part of Keithley's "best practices" is to localize risky concerns into modules and not to distribute them across multiple interfaces. Risky modules may be interchanged with more developed, less risky implementations until the technology has resolved the concerns.

Keithley has also established a unit called Keithley Labs to take high-technology risk items as much as possible off of the critical path of the projects. Their goal is to find high risk new technologies and intersect with projects that will begin to develop the new technology within 2–3 years followed by a new product in the marketplace in 4–5 years time.

6.3.2 Parameters and Analyses

This case study does not focus on a single product; it considers the general processes that Keithley Instruments uses to develop instruments and laboratory equipment. Consequently, these parameters and analyses generalize to all products.

The parameters and analyses tend to have significant business orientation and include

- Risk and financial payback
- Opportunity cost of resource usage
- Likelihood of emerging as a critical new industry measurement
- Alignment with corporate strategy and core competencies
- Importance of measurement for key customers
- Maintaining reputation and brand name (e.g. for high reliability)

In addition, considerations of technical issues may include

- Physical limits of measurement
- Stability of sources (e.g. voltage and current sources)
- Low-level noise (e.g. thermal noise overwhelming a low-level measurement or EMF of dissimilar metals at a junction)
- Power density and cooling
- Radio frequency (RF)—phase noise, distortion, and frequency ranges

Instruments destined for different industries or market segments have different emphases, hence they focus on different parameters. Instrumentation for RF testing focuses on phase noise, distortion, and frequency ranges, while instrumentation for direct current (DC) measurements focus on low-level signals and noise reduction. Equipment and instruments for measurements in semiconductor manufacturing operate round the clock, so they need reliability of the highest degree.

6.3.3   Architecture

Keithley Instruments uses "platform leverage" to help define the architecture of instruments. A platform is a collection of technology blocks or subsystems, such as power supply, digital signal processor (DSP) module, digital board, or display board. Platform leverage relies on reusing these technology blocks to develop new instruments quickly.

Until recently, Keithley has relied on *pull-forward leveraging*: Keithley defined a need for a new instrument, then the designers looked for appropriate technology blocks, already developed, to realize the final product. Pull-forward leveraging actually results in distinct instruments, each with distinct support and logistics needs. Because their technology blocks were pulled from various previous developments on different platforms, distinct instruments shared few common needs in support or logistics. Such product development create extra problems in the high-mix and low-volume world.

Keithley Instruments is moving toward *forward-looking platform management* and preparing roadmaps for future instruments, up to 5 years away, that share common technology blocks and development processes. The goal is to spin off new product variations and improvements from this common platform and still maintain a more efficient support infrastructure.

Manufacturing companies in the hi-tech arena are always looking for higher-density equipment racks that take less floor space. They also want increased throughput to reduce test time and cost. This means that parallel tests with multiple sets of source stimuli and receivers or probes are increasing in use. Furthermore, some manufacturers, such as in semiconductor fabs, operate continuously, i.e., 24/7, hence demand high reliability. This need affects the architecture of the test instruments on the assembly line.

Keithley has an internal module for many instruments called the Test Script Processor (TSP™). It is a combination of processing hardware and software that makes possible the fast execution of a sequence of commands. It has an easy-to-use programming language to generate the test scripts. The internal hardware accepts commands in batch fashion, executes the commands including decision blocks, calculates the result of a complex algorithm, and returns the answer. All of this can be done without continuous communication with the external computer for each command. The result is a much faster execution of a list of commands. Any instrument with TSP™ can act as either a master or a slave to expand functionality—only a simple local area network (LAN) cable connection is needed between units, and then all measurement channels look like they are in the same box. In this way, Keithley's TSP™ provides "seamless extensibility."

6.3.4 Interfaces

Instrumentation will always have several important interfaces: electrical, mechanical, and user. An instrument will have a main sensor or probe, which means sensing, conditioning, and conversion; all of which occur at an interface: transduction of phenomena to electrical signal, translation and isolation of the electrical signal, and translation from the electrical signal to the digital domain. An instrument will have a user interface—both display and manipulative input (e.g. buttons). Not all instruments have a display mounted directly on the enclosure, and some instruments use only a remote interface. Finally, if an instrument is a piece of manufacturing test equipment, it will have a control interface to maintain process control and data storage of manufacturing test results.

6.3.5 Branding

Keithley Instruments, Inc. has always focused on building a solid reputation with quality equipment. Careful control of the Keithley instrument appearance is one way to reinforce its brand and the perception of quality. They want every product to have a look and feel that immediately identifies the equipment as Keithley. Keithley pays special attention to the entire appearance and operation of the equipment, from the logo to the color scheme to the operation of the buttons and displays.

6.4 Phases

6.4.1 Overview

Figure 6.2 gives a good overview of the phases that Keithley Instruments goes through to develop a product. A product develops in the first four phases and then it enters the production and support phases. I will focus on product development in the first four phases.

6.4.2 Phase 1—Concept

Phase 1, or Concept, is part of the fuzzy front end when Keithley personnel prepare and refine a product concept. It is a very difficult stage to estimate and schedule. Product definition is highly iterative. Knowledge of both the market and the customer feeds into the current understanding of technology and then Keithley refines the concept.

This phase might be called the invention phase. The primary ownership is in the hands of the strategic marketing group, who receives technical and logistic input from the R&D group. Table 6.1 outlines some typical tasks in Phase 1 [3]. During this phase Keithley will gain greater understanding of

TABLE 6.1

Phase 1—Typical Activities (Used with Permission, *IEEE Instrumentation & Measurement Magazine* [3] and Keithley Instruments, Inc.)

Tasks to Complete	Typical Owner	Purpose	Risk with Deletion
Understand competitive offerings, customer applications, critical requirements, core contribution; positioning in market	Strategic marketing	Aid in positioning with respect to competition and defining product that will solve customer needs	Miss key definitional elements; competitive offering may overshadow new product; may not be prepared to aid customers pre- or postsales
Strategic fit with goals and capabilities; market potential; preliminary financials	Strategic marketing	Maintain synergy with other plans; assure this is financially the right investment	Disjoint product plans create inefficiencies; without solid financials, may over or underestimate payback
Preliminary key customer/partner list	Strategic marketing	Customer to aid in making the right product	May engage key customers too late in development to influence design
Technology assessment; list of high-risk areas, major decisions, and major milestones	R&D	Assess technical risk, prepare for major tasks to be completed in next phase	May enter into development with too much technology development on main path; schedule slippage
List of potential development partnerships	R&D	Early recognition of risk or cycle time benefit; make/buy decisions the need to be made	May pass up opportunity for significant development benefit
Preliminary schedule, resource plan, other goals (such as DFx)	R&D	Aid in financial analysis, organizational awareness, and commitment	Poor information input to financials; resources not available when needed; need goal/ direction for team
Key issues preliminary material sourcing plan	Manufacturing	Early recognition of major efforts	Schedule slippage due to surprises
List of key issues	Customer service/ applications support and commercial marketing	Early recognition of major efforts	Schedule slippage due to surprises

the customer applications sufficient for R&D to identify the key technical issues to be solved, and to begin their resolution. The key is to identify separable blocks in the architecture where the technical risk is within the block, and not within the interface. Breadboards of potential solutions for key technical issues will be produced and tested. At this stage, Keithley is taking a minimalist approach, building only those elements necessary to clearly identify the major issues and potential solutions.

An example of iteration through understanding the customer and the technology is Keithley's Model 2602 Source Measure Unit (Figure 6.3). It has features that addressed previously unidentified customer need. "Through careful understanding of the competitive offerings and customer applications, through complete technology assessment and significant breakthroughs, we produced a product that has exceeded expectations of our customers in rack density, extendibility to many channels, testing speed . . . and cost . . . " ([3], p. 16).

6.4.3 Phase 2—Investigation

Phase 2, or Investigation, is the other part of the fuzzy front end when Keithley eliminates major risks from the product concept: market, technology, and logistics. Phase 1 identifies the key risks and possible solutions, and Phase 2 beats these risks down. Engineers build breadboards and test solutions to all risky portions of the architecture or system. They only

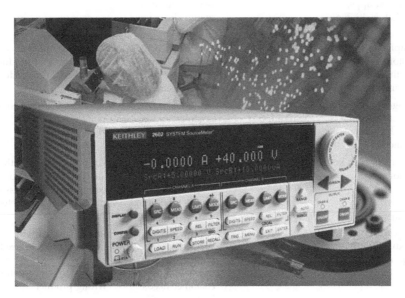

FIGURE 6.3
Keithley Model 2602 two-channel source measurement unit. (Used with permission, *IEEE Instrumentation & Measurement Magazine* [3] and Keithley Instruments, Inc.)

prepare a system breadboard if full integration is the only way to clarify a system concern, such as heat dissipation or electromagnetic capability (EMC). At the end of this phase, Keithley engineers have solutions to the most risky technical issues. Also, during this phase, the team at Keithley develops a plan for resources and project cost.

Ideally, Phases 1 and 2 should consume about 50% of the project's calendar time. If too little time is spent in the fuzzy front end, Keithley runs the risk of either not correctly defining the product or of taking too much technical risk into the next phases. If too much time is spent in these first two phases, time may be wasted and this could strongly affect revenue, especially in a rapidly moving market. Keithley strives to deliver the right solution to a customer's problem within an appropriate window of time.

Phase 2 is the responsibility of both the strategic marketing and R&D groups, but they receive significant help from manufacturing, especially if a significantly different manufacturing plan is needed for the project. Table 6.2 outlines some typical tasks in Phase 2 [3].

TABLE 6.2

Phase 2—Typical Activities (Used with Permission, *IEEE Instrumentation & Measurement Magazine* [3] and Keithley Instruments, Inc.)

Tasks to Complete	Typical Owner	Purpose	Risk with Deletion
Update on competition, customer apps, market positioning	Strategic marketing	Update definition of product and key reasons customer would chose your product	May have obsolete product before introduction
Completion of customer requirements documentation	Strategic marketing	Solidify agreement with R&D on deliverable as seen by customer	Without completion, stand good chance of product rework and project delays
Changes to other aspects of business plan including financials	Strategic marketing	Update organization on changes	May miss a key decision point if conditions changed
Agreement with key customers/ partners	Strategic marketing	Identify customers who will aid in key elements of product definition	Will not have in-depth analysis by one who uses equipment; may miss the mark in key parameters
Completion and agreement on product requirements	R&D	Turn customer requirements into product requirements— translation of customer needs into methods for achieving	May not have understanding with marketing on the need and method of taking the "need" to the "solution"; wrong product

(Continued)

TABLE 6.2

Continued

Tasks to Complete	Typical Owner	Purpose	Risk with Deletion
Decisions on architecture and leverage/reuse; complete invention—near elimination of high technology risks	R&D	Lock in on high level decisions that affect technology risk; get to point where you can plan for implementation	Uncertainty in technology employed and likely schedule slippage
Agreement with development partners	R&D	Understanding with all parties on who will do what development	Uncertainty in roles and responsibilities; schedule slippage
Completed resource plan, achievable schedule; ID of major milestones	R&D	Financial analysis, organizational awareness, and commitment	Poor information input to financials; resources not available when needed; intermediate goals for team not clear
Production process strategy Preliminary material sourcing plan Final DFx goals	Manufacturing	Define major elements of manufacturing process; influence design for manufacturing	Chance of major process change in direction late in project; miss opportunity to influence product; potential for schedule slippage
Resource plan	Manufacturing	Financial analysis, organizational awareness, and commitment	Poor information input to financials; resources not available when needed
Preliminary product service and support strategy	Customer service/ applications support	Estimate required budget, identify design-for-serviceability issues; identify customers' service requirements	Unexpected expenses; difficulties in servicing product; service strategy not accepted by customers
Preliminary launch plan and alignment with other plans; key assumptions such as lead sources, sales by geography, and so forth	Commercial marketing	Estimate scope and cost of promotion; set preliminary lead and opportunity objectives	Inadequate lead and opportunity generation; excessive promotional costs

At the end of this phase, they hope to have completed product definition. They should be finished with inventing and ready to implement the concept, but this is not always possible.

"This ideal of having a completed product definition at the end of this phase is not always possible. This is especially true in cases where one is working on cutting-edge technology, such as MEMS and nanotechnology, and the product needs are being defined interactively while a key customer is completing their technology investigation ... " (Figure 6.4). In new technology areas, Keithley works "tightly with a customer and alter[s] the product definition more deeply into the project cycle than the ideal" ([3], p. 19).

For the Model 6220 Current Source (Figure 6.5), an instrument used in the nanotechnology field, Keithley "learned progressively more about the customer needs, as their applications were refined. It was necessary in this case to live with a greater level of uncertainty in the product definition going into Phase 3. This uncertainty did, of course, produce a longer cycle time for the product owing to some level of rework on the product design" ([3], p. 19).

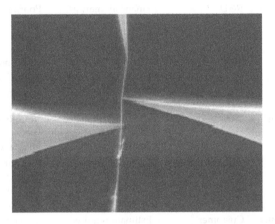

FIGURE 6.4
Measurement of carbon nanotube using Keithley Model 4200-SCS Semiconductor Characterization System and Zyvex S100 Nanomanipulator. (Used with permission, *IEEE Instrumentation & Measurement Magazine* [3] and Keithley Instruments, Inc.)

FIGURE 6.5
Keithley Model 6220 Current Source. (Used with permission, Keithley Instruments, Inc.)

At other times a new product arises while Keithley works with a customer and understands their application to such a level that they can give significant help in critical measurements. One such development was the Model 4200-SCS with Pulsed IV (Figure 6.6) which solved a specific semiconductor device problem; it integrated a pulse card, an oscilloscope, and applications software into the Model 4200-SCS Parameter Analyzer. "Although the solution had been proven with an external pulse instrument, one of the key risks in the project was to create and integrate a voltage pulse card into the Model 4200-SCS. Reduction of this key risk, clearly on the critical path of the schedule, could be accomplished through creation of a working circuit on the bench. This was necessary before we left the 'fuzzy front end'." ([3], pp. 19–20).

6.4.4 Phase 3—Development

Phase 3, or Development, is part of the concrete back end; it is where Keithley personnel implement and complete the design plan for the product and build a prototype to meet the critical specifications. R&D is most prominent in this phase. Table 6.3 outlines some typical tasks in Phase 3 [3]. In this phase, the team typically generates between one and three full system integrations to work out the issues and to guarantee a high-quality manu-facturable product. Upon exiting this phase Keithley expects no turns of updated board [meaning they do not plan to fabricate any revised printed circuit boards (PCBs)] before turning on full manufacturing; they also expect that full hardware quality audits have been performed and expect that all the critical firmware functionality has been incorporated into the instrument.

FIGURE 6.6
Keithley Model 4200-SCS with Pulsed IV. (Used with permission, *IEEE Instrumentation & Measurement Magazine* [3] and Keithley Instruments, Inc.)

TABLE 6.3

Phase 3—Typical Activities (Used with Permission, *IEEE Instrumentation & Measurement Magazine* [3] and Keithley Instruments, Inc.)

Tasks to Complete	Typical Owner	Purpose	Risk with Deletion
Changes to business case including financials	Strategic marketing	Update organization on changes	Miss a key decision point if conditions change
Create manufacturable product, completed features, no planned redesigns	R&D	Complete the design	Potential schedule slippage from reworking units in following stage; potential for not meeting quality, cost, yield goals
Verify hardware, firmware, and software quality and reliability	R&D	Tests to assure a quality product released to desired performance	Potential schedule slippage; potential for not meeting quality goals
Complete preparations for test run . . . training, materials list and mfg drawings, preliminary specs	R&D	Training of technicians, assemblers, testers, engineers; readiness for manufacturing processes	May have untrained people, material shortages for next phase; schedule slippage
Test run (pilot) build plan including demo units	Manufacturing	Number of units to demonstrate yield, and so forth during next phase	May not have good sampling for statistics
Preliminary production ramp plan	Manufacturing	Plan for first production run; stage material, and so forth	May not be ready with material for first production run
Complete product assembly and test processes, material entry into purchasing systems	Manufacturing	Readiness for processes to build samples during next phase	May not be representative test of production processes
Identify unique materials—set up special processes	Manufacturing	Special handling identified	Quality, cost, production cycle times or performance may suffer
Complete product service and support plan	Customer service/ applications support	Prepare plans to procure spares; identify required training	Inability to service product due to unavailability of spares or trained personnel
Verify adequacy of solution introduction plans	Commercial marketing	Verify optimal solution rollout into marketplace	Confused or inadequate market introduction

6.4.5 Phase 4—Pilot

Phase 4, or Pilot, is the other part of the concrete back end when Keithley personnel build a pilot run to test the product for cost and yield to performance specifications. They also test the manufacturing processes—material ordering, assembly, and test. Manufacturing owns this phase. Table 6.4 outlines some typical tasks in Phase 4 [3]. Quality audits will again be performed on the hardware, and the firmware is sent through a rigorous process of problem finding, reporting, and fixing. At the end of this phase, the manufacturing organization has confidence that they will be able to fulfill a customer order on time and with high quality.

TABLE 6.4

Phase 4—Typical Activities (Used with Permission, *IEEE Instrumentation & Measurement Magazine* [3] and Keithley Instruments, Inc.)

Tasks to Complete	Typical Owner	Purpose	Risk with Deletion
Changes to business case including financials	Strategic marketing	Update organization on changes	Miss a key decision point if conditions change
Agree on engineering transition plan for R&D engineers	R&D	How long or through what goals will R&D support manufacturing during first runs	No commitment from R&D to focus on release product until goals met
Aid in any issues that come up in mfg test runs	R&D	Assure that test runs go smoothly	Manufacturing engineering left to solve all problems, risk to schedule
Reverify hardware, firmware, and software quality and reliability	R&D	Tests on final product to assure a quality product released to desired performance	Possibility that final product with recent changes not represented by earlier verification
Build significant sample of product	Manufacturing	Check yields, process, material flow	May have line shutdown later due to inability to build to specs
Specification negotiation with marketing and R&D	Manufacturing	Use results of data taken in Phase 4 to set realistic specs balancing yield risk with market demand for specs	May not have product specs supported by manufacturing results
Updated production ramp plan	Manufacturing	Final check on material flow	May not ramp quickly enough to meet market need
Implement launch plan and sales plan	Commercial marketing	Ensure optimal solution rollout into marketplace	Suboptimal or inadequate market introduction

This phase is important for smooth transfer into manufacturing. Any changes during Phase 4 should be minor.

On a occasion Keithley has "tried to shortcut this phase. This is desirable when time to market is of the greatest concern. You may get a product out the door and into the market place sooner. But without a complete Phase 4, the struggles will [occur] in the first production runs, with orders on the books and customers waiting, and with possible exposure to the customer of quality and delivery problems"([3], p. 21).

6.4.6   Production and Support

This phase is where Keithley Instruments manufactures the product in bulk and ships it to customers as a regular catalog item. During the first months of production, the R&D team may still be somewhat engaged as volumes increase and issues are revealed. For this reason, manufacturing will typically enter production under a "transition agreement" with R&D engineers to secure time from the R&D engineers as needed. As volumes and experience with the product increase, the responsibility for continuous support transfers to the manufacturing engineers.

6.5   Scheduling

As can be seen from the previous section on phases of development, planning requires both top-down and bottom-up efforts. Obviously goals and schedule objectives are set in a top-down manner. The fuzzy front end in the first two phases can alter this; bottom-up planning then plays a role to readjust the schedule. This bottom-up planning produces a work breakdown structure and its schedule, which gives greater buy-in from the development teams. (The previous section gives examples of what can change the schedule during fuzzy front end.) And always, management will ask the team, "What can I do to help you be more successful? What can I do to help you improve the schedule, or improve the chances of hitting the needed performance? What do you need from me to either assure that you meet this agreement, or better yet, to improve on it?"

6.6   Documentation

6.6.1   Types

Keithley Instruments, like any company, has the standard types of documentation: notebooks, letters and e-mail messages, memos, project documents, manuals, brochures, and presentations. One major point about Keithley's documentation: they know that it is always written for somebody else, both external customers and internal owners of the product. Without this documentation, transfer of responsibility between internal owners would not be possible and interruption during the next development project would be likely.

Engineers who provide support for the product or do follow-on designs need its theory of operation in the documentation. This is a *gradient* type of documentation because it is a living document; it shows trends and gives the explanation so that upgrades, modifications, and support may continue. The documentation provides the basis for changes in applications or design evolution.

Manufacturing, on the other hand, needs a *snapshot* type of documentation—a picture of the product at a particular point in time. These documents include schematics and assembly instructions to build product.

One form of documentation that stands out at Keithley is their application notes and white papers, which help customers better understand the use of the instruments. The applications group, often called "apps" in the industry, works closely with customers to solve problems in test and measurement; they then publish application notes and white papers to help other customers through similar experiences.

Keithley Instruments used to give the customer everything—user manuals, explanations, schematics, and theory. Now they must reserve that kind of service for bigger customers who have their own calibration laboratories and repair shops. Protection concerns for intellectual property have made it more difficult for companies to give away so much information without losing legal protection for their designs and ideas. Keithley has to limit the documentation so as to "not give away the farm."

6.6.2 General Formats for User Manuals

A manual from Keithley Instruments is a carefully planned and executed document. Every manual has a similar format. The following outline combines outlines from the manuals for the model 6220 and model 2600S:

- Front material
  - Warranty information
  - Title
  - Manual print history
  - Safety precautions
  - Table of contents
  - List of illustrations
  - List of tables
- Getting started
- Introduction

- General information—warranty information, safety precautions, unpacking and handling, options, and accessories
- Front and rear panel familiarization—briefly describe buttons and displays and their operations
- Precautions—heat sinks and cooling vents
- Power-up
- Display menus
  - Editing controls
  - Password
  - Remote interface
  - Error and status messages
  - Default settings
- Output connections
  - Configurations
  - Guards
  - Connections to DUT (device under test)
  - Using a test fixture
- Operations and commands specific to each instrument
- Appendices
  - A: Specifications
  - B: Command scripts
  - C: Frequently asked questions
  - Index

Fonts, font sizes, subheadings, and indentation in every manual follow a company prescribed format and are simple and clear. The manuals use simple line drawings to explain the features—these are effective in communicating the information. Manuals provide screen shots for instruments that connect to a control computer where applicable.

6.7   Requirements and Standards

6.7.1   Markets

The markets for Keithley Instruments are scientific and engineering laboratories and high-tech manufacturing lines. Their products address accurate, very-sensitive, low-level measurements in semiconductors, metals, polymers, and insulators. Their products also go into manufacturing facilities to test electronic and optoelectronic components, flat panel displays, cell phones, and digital switching systems.

Keithley Instruments, Inc., designs, develops, manufactures, and markets complex electronic instruments and systems geared to the specialized needs of electronics manufacturers for high-performance production testing, process monitoring, product development, and research. Keithley has approximately 500 products used to source, measure, connect, control, or communicate DC, RF or optical signals. Product offerings include integrated systems solutions, along with instruments and PC plug-in boards that can be used as system components or stand-alone solutions. Keithley's customers are engineers, technicians, and scientists in manufacturing, product development, and research.

Keithley partners with customers to anticipate their current and future measurement needs. A thorough understanding of their applications coupled with Keithley's precision measurement technology enables Keithley to add value to customers' processes by improving quality, throughput, and yield of their products. These partnerships help Keithley determine which test applications they choose to serve. Keithley deploys its own sales and support employees throughout the Americas, Europe, and Asia, as opposed to relying on a contract sales force, because Keithley believes this serves customers much more effectively.

Keithley leverages its applications expertise and product platforms to other industries by concentrating on interrelated industries and product technologies. Keithley gains insight into measurement problems experienced by one set of customers and uses this insight to solve problems for other customers.

6.7.2   Standards

All of Keithley Instruments' products must meet Underwriter Laboratories (UL) standards for product safety. Beyond these standards, they must be CE compliant for European and international markets.

6.7.3   Preparing Requirements

Section 6.3 covers much of the preparation of requirements during phases of development. Strategic marketing works with customers to better understand the needs of the industry. R&D helps assess the technology to show feasibility. All this occurs in Phase 1.

6.8   Analysis

6.8.1   Feasibility

Section 6.3.2 introduced some parameters and analyses, including risk and financial payback, corporate strategy and core competencies, and the physical limits of measurement, that Keithley Instruments uses to determine feasibility. Also mentioned in Section 6.3.2 is that each industry has a different emphasis, and so Keithley focuses on different parameters and analyses.

The folks involved in determining feasibility are from the marketing group. The R&D group contributes insight and trade-offs to marketing's analyses.

6.8.2   Heuristics, Numerical Simulations, and Calculations

Keithley has proprietary heuristics and core competencies for which their numerical simulations and models are so good in some market segments, such as DC measurements, that they have (in the past) been able to predict product performance accurately before any pilot runs. These predictions from the theoretical calculations have been good enough for Keithley to fix specifications early. This has changed in the past several years as Keithley has standardized on a pilot phase for all projects, along with statistical setting of specifications. In other market segments, the physics are not as well understood, and the designers in R&D have to develop a pilot model. In emerging fields, such as nanotechnology and carbon nanotubules, Keithley R&D staff work alongside scientists and engineers at customer sites to develop theory that eventually will lead to usable models and calculations.

6.8.3   Testing

Breadboards and prototypes are important for developing cutting-edge scientific instruments. They are necessary to proving the concept and demonstrating feasibility, but Keithley Instruments recognizes that "typically only 10 to 50% of the work has been done by the time this first breadboard is done" ([3], p. 21).

A prototype only points the way. A final product must undergo design trade-offs, satisfy manufacturing and test concerns, and have ready support once in the field. Keithley personnel recognize that they are not just producing an invention; they are producing a supply chain. The development task is not complete until this supply chain is in place.

6.9   Design Trade-Offs

6.9.1   Architecture

Keithley Instruments begins any new product by examining architectural concerns. Their forward-looking platform management affects the form factor of all products; for instrumentation used in the manufacturing environment, Keithley has standard dimensions of full rack or half rack (for width) and 1U,

2U, 3U, or 4U (for height). Keithley strives for similar dimensions in their bench-top laboratory equipment. Their goal is encourage the smooth migration of thought and test from the R&D bench top to the factory floor.

All enclosures are custom and conform to internal company standards. Mechanical engineers design the enclosures to account for thermal dissipation, vibration, and acoustic noise. For many years, Keithley has incorporated a considerable amount of reuse in developing new instruments. As they move from pull-forward leveraging to using forward-looking platform management, they have some standard dimensions and interfaces for buttons and displays.

While Keithley's TSP™ can provide "seamless extensibility," not every customer application and not every instrument needs it. When it does not provide enough added-value, the TSP™ is left out of the instrument design to save component and manufacturing costs.

There is a balance to reuse and optimization; every module, whether hardware or software, cannot be designed and built with such flexibility that it can meet every possible future need or contingency. Even if it could meet all of the performance needs, the cost of this modularity and flexibility would probably violate other trade-offs in engineering. At some point, the instrument will need some unique capability to perform its ultimate task.

6.9.2 Hardware

Forward-looking platform management strives for standard sets of processors, memory, and processor boards. Common platforms save inventory costs by reducing the variety of different components. Common platforms also reduce manufacturing and assembly costs by having increased volumes. As stated in the previous section, a common platform may not suit a particular instrument; unique requirements may force custom designs to accomplish the task.

Circuit boards are another area for design trade-offs. Generally they are multilayer, usually between 4 and 16 layers and possibly more. Both EMC isolation and complex interconnects drive the design of the circuit boards, their complexity, and the number of layers.

Keithley engineers try to balance between theory and practice. The skill lies in understanding the point of "too much" versus "not enough" theoretical modeling and knowing when to do a quick turn PC board to test an idea. Keithley has a PCB milling machine that can turn out a double-sided PCB for prototyping circuits in less than an hour. This capability allows them to quickly test out ideas for circuits.

6.9.3 Power

Keithley products sell internationally, with about one-third of its sales in Asia, one-third in Europe, and one-third in America. Power supplies built into their equipment, therefore, have universal inputs; they take 120–240

VAC and 50–60 Hz line power. Nearly all power supplies within Keithley instruments require custom design, which is forced by sensitive, low-level measurements and electromagnetic interference (EMI) control within the instruments.

Most power supplies are switching supplies to gain efficiency and power density. Switchers require careful design of output filters and layout to control conducted and radiated internal noise EMI (a particular problem with low-level measurements) and to maintain EMC.

6.9.4 Cooling

Most instrument designs strive for passive convection cooling. If engineers have a concern about heat and dissipation, they do thermal modeling to analyze whether active cooling is needed or not. They will also use infrared (IR) cameras and thermocouples on prototypes to verify the thermal models and to modify designs as necessary. Keithley tries to avoid fans because of electrical noise problems that can disrupt measurements and the acoustic noise that can annoy an operator.

A fan immediately adds complexity and cost to an instrument. The power for the fan must have a separate circuit to isolate conducted EMI produced by the fan motor from the very "clean" power supply dedicated to sensitive, low-level analog circuits. Fans also have mechanical bearings that eventually wear out. Finally, dust can thermally insulate components and cause temperatures to rise above the expected—Keithley could install filters to trap dust but these need periodic cleaning. Fans are avoided, if possible.

6.9.5 Software

Much of the source software for operation of instruments is written in C. Keithley is moving to object-oriented programming and languages, such as C++, in new designs, particularly for the user interfaces, to help engineering personnel improve structure, documentation, and efficiency in developing software.

Older products used a "super loop" form of operating system and processing. Now Keithley uses real-time operating systems in virtually all new products. The choice of the real-time operating system (RTOS) depends on the trade-offs, such as speed and cost, but currently include OSE®, WinCE®, as well as others.

Keithley has efforts in several directions to improve the quality and access to improvements. First, Keithley engineers conduct organized technology forums—communities of similar disciplines such as software and firmware engineers—to share and develop best practices. On their own initiative, they have promoted processes such as code reviews and peer checking to improve quality of software production. Second, the company has set up a Web site to facilitate software upgrades once a product is in field. In addition, Keithley has created more rigorous software audits, better design for quality and reliability processes, and better checks and balances between R&D and manufacturing.

6.9.6 Hardware vs. Software

Keithley uses software to gain flexibility for upgrades and future expansion—assuming that performance is the same whether using a hardware implementation or a software module. DSP and field-programmable gate array (FPGA) engines in circuit boards help provide this kind of flexibility. A solution in hardware will usually result in a faster instrument; it is used when product speed performance is critical. Creating a solution in software (e.g. firmware, DSP, FPGA) will usually be faster to implement and reduces design cycle time while increasing flexibility. In addition, the user interface for test equipment is becoming more and more important for a company to distinguish itself. This interface is primarily, of course, controlled by instrument software.

The LXI consortium is a group of test and measurement companies putting together a common protocol for instruments over the Ethernet. They are defining user interfaces that avoid extensive hardware and put the functionality in the software with rich features through display Web sites. A page on a Web site can replace the two-line liquid crystal display (LCD) from previous generations of instrument front panels with many more varied features. The trade-off is speed—a Web site page is slower than an LCD (intelligent design of the web or instrument interface can minimize this difference in speed). For more information, see www.lxistandard.org.

6.9.7 Buy vs. Build

Keithley buys components and then custom-builds the vast majority of its modules and subsystems. The market for tightly specified and highly focused laboratory and test equipment requires custom design; high-margin products (60% gross margins—see Keithley Instruments' annual report [2]) support custom design.

Sometimes, they will buy a display subsystem for an instrument and then wrap a keypad around it. For large test systems built for manufacturing customers where Keithley might not have expertise in all the functions, they will buy a specific instrument to cover functions they lack. That instrument then acts as a subsystem in the larger test system.

One design trade-off is whether to buy standard integrated circuits or to build an application-specific integrated circuit (ASIC) to accomplish a particular function. For the high-mix, low-volume market, a major concern for Keithley is the initial investment cost. For an ASIC, the investment can be between US$500K and US$1M. If you assume a US$500K investment and an $80 savings per unit, as an example, it requires 6200 units to break even; this might be greater than the number of units sold for the entire product's market life. Justifying this level of integration based on volumes just does not make much sense in the high-mix, low-volume market; other factors, such as performance, determine the decision for Keithley.

Keithley custom-builds all its enclosures. This is a first line of effort in branding their products. Custom builds also allow them to achieve the desired mechanical strength and user interfaces.

6.9.8   Manufacturing

DFx is a general philosophy, where DF means "design for" and x stands for just about everything. Keithley strives for DFx in a number of ways. A subset of considerations is:

- Design-for-reliability (DFr), DFM (manufacture), DFA (assembly), and DFT (test) to assure quality in manufacturing
- Design-for-flexibility (DFf), this is part of their forward-looking platform management; hardware design may be updated every year and firmware may be updated every fiscal quarter
- Design-for-transfer (DFt), which allows them to move manufacturing to different sites or to contract manufacturing
- Design-for-improvements (DFi), which reduces cost and increases both yield and ease of assembly (in some ways it overlaps DFA)

Keithley is in the business of building laboratory instruments and specialized test equipment. Such a business is based on high-mix, low-volume, high-margin products. Consequently, designers look for ways to cut costs when the potential savings are hundreds of dollars per unit. Forward-looking platform management is a major part of this type of thinking—saving money by reusing a previously designed module or optimizing a function for a totally new instrument design. Unlike consumer appliances, where margins are low and production volumes are high, things like saving pennies on the design of a shipping container are not a concern for Keithley.

6.9.9   Test for Quality

There are various ways to insure quality in the circuit boards through testing of the assembly. Most commonly used methods are functional test, built-in-test (BIT), and in-circuit test (ICT). BIT and functional test are distinguished from ICT by the level of capital investment necessary for each circuit board. ICT can provide a higher level of test coverage along with more exact fault reporting. In all cases, specific design considerations must be taken into account during layout of the circuit board.

BIT requires little in the way of capital or equipment investment, and may be known by other names such as power-on-self-test (POST), diagnostics, or self-test. The test resides in the firmware internal to the operation of the instrument or in a software program. These methods make a great deal of sense when the volume of the product is low, since accomplishing BIT requires some software (SW) or firmware (FW) development and attention to circuit board layout. The layout requires test pads and the addition of components to the design to increase the test coverage.

There are several methods and types of ICT. The most common are boundary scan, flying probe, and full fixture, bed-of-nails ICT. Flying probe testers pin down on component leads, PC board traces, and test pads provided on the circuit board. This is the least expensive form of ICT in terms of fixture cost. Flying probe testing is much slower than a bed-of-nails test method and is usually used during prototype builds and for very-low-volume products. Boundary scan testing is used most effectively for digital designs. The components have to be selected for the capability to support boundary scan during design. A test connector or test pads on the circuit board provide test access. The test process can be run from a computer through a vendor's boundary scan module or accessed from the flying probe and bed-of-nails testers. ICT will usually require a "bed-of-nails" type fixture that may cost from US$20K to $30K. This fixture will need to be modified with each board modification, and it may need to be replaced if the board layout changes too much. For the high-mix, low-volume market, a major concern for Keithley is the initial investment cost. If you assume a US$50 savings per unit, as an example, it requires 600 units to break even; for products that might ship only a 100 per year, it does not make sense to incorporate ICT. BIT may be a better choice. As a result, Keithley carefully examines when to use BIT and when to use ICT; other factors, such as quality, help determine the decision for Keithley.

6.9.10 Maintenance and Repair

Keithley has centers around the world for repairing and maintaining their products. These centers are particularly important for feeding back information to Keithley's engineers on newly introduced products. Other, more-established products can have regional dependencies that determine how they are repaired and maintained. In some cases, calibration and repair are contracted to third-party firms. For large customers with their own facilities, Keithley provides the documentation and training so that the customer can do the work themselves.

6.10 Tests

Keithley Instruments has a full range of development tests—informal, formal, and laboratory. The more informal or lab bench tests include prototypes and breadboards, such as the two-sided PCBs milled out in short order to try out ideas. The formal tests include both inspection and peer review, promoted from within by the engineers. They also include subsystem tests for both hardware and software modules. The testing is typically done on nearly complete instruments, or on subelements of a system. Examples are tests of the preamps in a test head for automated parametric test (APT) equipment (for process control of semiconductor wafer fabrication). Keithley performs environmental tests on certain subsystems and on the final system. The tests use thermal cycling, shock, and vibration to exercise the product.

Engineers lead the development of test processes for quality during design development. Audit engineers contribute to assuring quality. Another group of test engineers are those who develop tests of the product itself. While engineers currently perform the testing, Keithley is moving toward using trained technicians.

6.11   Integration

As mentioned in Section 6.3.1, Keithley Instruments uses cycles of full integration to build up functionality—spiral development. In some instruments with much software, the design team focuses the integration builds on the software system. An important point: the designers and engineers try to localize risk into modules and not across interfaces. This practice eases test and integration by reducing and focusing the trouble-shooting of the problems. The integration team comprises all engineers on the project. Engineers responsible for a given module subgroup help insert that module into the larger system. The verification of the performance of the system also involves engineers from the manufacturing design teams.

Integration testing takes hardware and software modules and configures them into a system that resembles the final instrument. System tests exercise the interactions between modules; they also exercise the full system functionality. Eventually, Keithley runs environmental tests with thermal cycling, shock, and vibration to exercise the entire instrument.

Once an instrument completes full integration testing in the laboratory, Keithley will often subject units to field tests at customer sites. Field tests take place with greater frequency and intensity, working closely with the customer when Keithley tackles fundamentally new areas of measurement. The field tests are carried out with key customers that are identified early in the project. People involved in the tests are the customer, a few key project engineers and people from the strategic marketing organization.

6.12   Manufacturing

As outlined in Section 6.9.8, DFx is a general philosophy for quality manufacturing. Besides DFM, DFA, and DFT, Keithley does design-for-reliability, DFt, and DFi. DFt allows Keithley to move manufacturing to different sites, including to contract manufacturing to use manufacturing resources most efficiently. DFi provides the "hooks" in the design that allow future changes that reduce the cost of components, improve the yield, and ease the assembly effort.

Although Keithley buys components, it fabricates and manufactures most of the electrical and electronic subsystems that go into its products. These are custom designs because of the special nature of low-level measurements.

Keithley assembles but does not fabricate its own PCBs for its products. Control and cleanliness in the assembly of PCBs are very important to avoid interference in low-level signals and very sensitive physical measurements. For circuits that do not have quite the stringent requirements on control, Keithley will go to a contract PCB fabricator to build some boards if Keithley's facilities are busy.

Keithley works with vendors to fabricate and manufacture the mechanical elements of subsystems and enclosures, too. Keithley will then assemble most subsystems on site (Figure 6.7). Branding and the nature of custom design tends to keep much of the assembly in-house.

Keithley assembles its circuit boards and its products. The primary reason, again, is quality control to achieve instruments that consistently function correctly while measuring low-level signals. Parts of the assembly and manufacturing environment are inspection and test; Keithley personnel inspect subsystem and final assemblies and perform both manufacturing tests with dedicated test rigs and BIT, if BIT is incorporated in the product.

FIGURE 6.7
Keithley personnel assembling a product. (Used with permission, Keithley Instruments, Inc.)

6.13   Support

Section 6.9.10 mentions Keithley's repair and maintenance centers around the world for servicing their products. Repair can be any one of several modes: swapping out units, replacing modules, or even replacing specific components.

Support in the high-technology arena, however, means more than shipping product, supplying a user's manual, and maintaining service centers. Technical support in the form of personal service, application notes, and white papers is a critical part of selling, using, and operating sophisticated measurement instruments. The applications (apps) group works closely with customers to solve problems in test and measurement. They also help potential customers understand problems and concerns with low-level measurements and the significance of particular techniques. Keithley's Web site has a number of application notes and white papers immediately available for downloading to serve customers and potential customers [1].

6.14   Disposal

Keithley Instruments will meet both the RoHS (Restriction of the use of certain Hazardous Substances) and the WEEE (waste from electrical and electronic equipment) directives. Although it will take a great deal of work to comply with the regulations and reduce risk—some components currently used in measurement instruments just cannot be free of lead—discussions with peer companies in test and measurement have shown that Keithley is leading many others in preparing for the new regulations. The good news for all test and measurement companies is that they still have more time to meet these regulations. The test and measurement community has received relief from RoHS until 2008 or 2009.

6.15   Liability

While the company's product manuals have prominent warnings about safety, Keithley's level of concern is probably less than that of most appliance manufacturers. Most customers of their products have a reasonable understanding of the dangers of line power and of certain measurements. Keithley is more concerned over the protection of intellectual property or IP; it is the primary asset in Keithley's business. The economics of losing ground to a competitor who copies a Keithley design can be quite significant.

The possibility of lawsuit has, in the past, been of little concern to Keithley. They are, however, concerned with their obligation to their customers to maintain an excellent metrology lab and traceability to National Institute of Standards and Technology (NIST). Keithley remains very active in this effort.

6.16   Summary

6.16.1   Emphases

High-mix, high-margin, and low-volume products, such as those designed, manufactured, and sold by Keithley Instruments, Inc., are custom designs. Custom design can better achieve both quality (i.e., meeting all the specifications) and branding. Keithley uses the philosophy of *forward-looking platform management* to capitalize on both risk reduction and cost reduction through reuse. For further enhancement of quality and risk reduction, they do DFx, where DF is "Design For" and x represents a variety of different concerns: A for assembly, T for test, M for manufacture, R for reliability, i for improvement, and Tr for transfer. The company uses spiral development and goes through cycles of full integration and test to evolve functional and high-quality products.

6.16.2   Gotcha's

Reuse might use *pull-forward leveraging*, which is a backward-looking way of retrieving (or pulling in) previous subsystem designs. It does not incorporate a product roadmap to predict trends in commonality between platforms; this can result in disparate modules combined together into distinct instruments. While the modules have a common background, the architecture of the platform is unique, thereby increasing the burden on support and inventory.

Acknowledgment

My thanks to Larry Pendergrass, VP of New Product Development at Keithley Instruments, Inc. for providing the information for this chapter.

References

1. www.keithley.com/company
2. Keithley Instruments, Inc., 2005 Annual Report, p. 7.
3. Pendergrass, L., Climbing the Commercialization Hill, The Four Phases of Product Development, *IEEE Instrumentation & Measurement Magazine*, Vol. 9, No. 1, February 2006, pp. 12–21.

7

Case Study 4—Automobile Engine Controller

7.1 Concept and Market

7.1.1 Who, What, Why, How, Where, and When

Electronics invaded the engine compartment of automobiles in the early 1980s. The introduction of engine control modules (ECMs) improved the performance of automobile engines and decreased pollutant emissions. Engine tune-ups virtually disappeared. In the three decades since the first ECMs, development has continued and all aspects of engine operation and performance have improved. Figure 7.1 shows an example of one engine compartment and ECM on a recent model automobile.

Electrical, mechanical, and software engineers all participate in the development of ECMs. The design cycle for an ECM usually takes about 2 years; it continues for about 5 years after the first model introduction with various upgrades and improvements. Source code for the ECM is then archived and preserved for at least 10 years after the last unit is produced; this action is required by government regulations in the event of a recall that requires reprogramming the ECM.

With more performance being squeezed out of internal combustion engines and environmental concerns growing, the design of ECMs will increase in importance. The recent introduction of hybrid vehicles only further emphasizes the importance of the ECM—now it's not just the control of the internal combustion engine; it also involves the entire power train, inverters, power electronics, battery control modules, motor controllers, and energy/power management through the electric motor and batteries, as well (Figure 7.2).

7.1.2 Economics

Automobile manufacturers in the United States produce about 12 million cars, light trucks, and light commercial vehicles each year. The United States produced 4.3 million cars and 7.2 million light trucks and light commercial vehicles in 2005. NAFTA—the United States, Canada, and Mexico—produced 6.7 million cars and 9.1 million light trucks and light commercial vehicles

FIGURE 7.1
An automobile engine in a recent model car—the ECM is on the left side between the white plastic bottles. (© 2005 by Kim Fowler, used with permission. All rights reserved.)

FIGURE 7.2
The ECM for this hybrid car is under the large silver cover just right of center. (© 2004 by Kim Fowler, used with permission. All rights reserved.)

in 2005 [1,2]. Detroit and automotive plants around the world produce varying volumes of the different models of cars. Some assembly lines produce as few as 5,000 cars for special editions and hybrids, while production for a particular model of small car can be as high as 200,000 vehicles per year.

Small cars have very little margin for profit. Big, expensive cars and sports utility vehicles have good profit margins. Yet every car has an ECM controlling the engine. (Some cars even have a second electronic control unit [ECU] just for control of the throttle plate to precisely meter air to match the fuel provided to the injectors for stoichiometry and a target torque/power value and to control emissions.)

7.2 People and Disciplines

Teams that design ECMs include electrical engineers, software/firmware engineers, and mechanical engineers. Mechanical engineers comprise a significant portion of these teams because they understand the problem domain of the internal combustion engine and its control, although more computer engineers, software/firmware engineers, and electrical engineers are entering the field now. The size and composition of the design teams vary according to the company and supplier.

7.3 Architecting and Architecture

7.3.1 Process

Designing and manufacturing ECMs is a mission-critical process. While each company has its own process, most rely on a waterfall or V-model process. Most do not do spiral development because the ECM design generally must be correct when it goes to commercial production.

The automotive industry does have one major departure from pure process—it is called "calibrations." Many parameters are not understood completely until commercial production, and so designers build in hooks or calibration points (called configurations) that are set at production release.

Software or firmware development does continue beyond production release. When changes need to be incorporated into the ECM, they are for:

- Emergency defect mitigation
- Late requirements
- Future development of new ECM products
- Emission-related campaigns (sometimes this is a euphemistic phrase for a recall)

7.3.2   Analysis

All ECM design teams (either working for original equipment manufacturers—OEMs—first-tier suppliers, or working for second-tier suppliers) consistently do Fault Tree Analysis (FTA) and Failure Modes and Effects Analysis (FMEA). Some teams do these analyses better than others. The better teams use these analyses to drive design decisions; they do not just tack on the analyses to satisfy industry regulations.

7.3.3   Architecture

A typical instrumentation block diagram might represent and describe an ECM (Figure 7.3). It has input from sensors and the driver's foot pedals; the sensors monitor shaft position, temperature, air temperature, coolant temperature, air pressure, and oxygen in the exhaust. The ECM diagram has a central processing block, which is what most of us think of as the ECM box. Finally, the ECM has output actuation that controls the spark plugs, various valves, and even the electric fan on the radiator.

ECMs have circuits based on highly integrated microcontrollers that contain a number of peripherals, such as timing processors and analog converters. Manufacturers are moving toward common platforms for ECMs to reduce the cycle time of development through the functional reuse of hardware and software modules. Configurable hardware (e.g., FPGAs) is beginning to play a role in developing ECMs, while application-specific integrated-circuits (ASICs) have been used for nearly two decades.

Automobile engine control is hard real-time embedded control at its best. It cannot miss a processing deadline without important, if not severe, consequences.

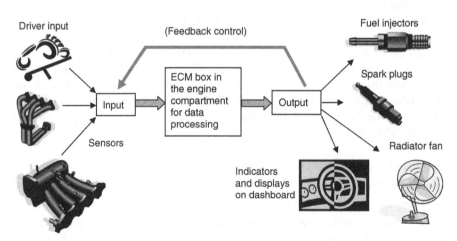

FIGURE 7.3.
An instrumentation block diagram for an ECM. (© 2006 by Kim Fowler, used with permission. All rights reserved.)

Example

An eight-cylinder automobile engine might allow a maximum of 8000 revolutions per minute (RPM). One spark plug out of four fires each revolution of the crankshaft. This means that the engine has two ignition sparks per revolution; at 8000 RPM this is 133 revolutions per second or 267 sparks per second. The maximum time the ECM has to process between sparks is the reciprocal of this or 3.75 msec.

The timing resolution on the sparks is even more stringent. The ECM must time and initiate the spark and control it to within 0.4° of the crankshaft angle, which resolves to 900 positions per revolution. At 133 revolutions per second, the maximum time is 7.50 msec. The time resolution = (minimum time to compute/revolution)/ (positions resolved/revolution) = is (7.50 msec)/(900) = 8.33 μsec/ position. This type of high resolution in timing needs a dedicated peripheral to do hardware timing on the microcontroller chip.

In a large, fast microcontroller, 3.75 msec can be a long time if spark timing were the only algorithm it ran, but that is not the case. The firmware must take input readings from both the sensors and the driver control. It must process the data to give the optimum balance between performance, fuel economy, and pollutant emissions. Finally, the firmware must actuate a number of outputs: firing the sparkplugs, pulsing the fuel injectors, setting the throttle opening, running the radiator fan, maintaining communications (e.g. CAN, Keyword/KWP, UDS, LIN), and preparing diagnostics, which can consume 30% or more of the processing capabilities and ROM storage.

Many ECMs use a 32-bit processor or microcontroller. A larger data path can provide for more sophisticated algorithms than smaller, simpler, and cheaper microcontrollers. As mentioned, the firmware in the ECM must perform many functions including high-speed calculations. One way to speed up operations and decrease the time of calculations is to use large look-up tables (LUTs); the trade-off is speed, gained by a large LUT for memory size, which increases cost and decreases reliability.

7.3.4 Interfaces

The interfaces are electrical, mechanical, and chemical at both the inputs and the outputs. (Figure 7.3 shows some of these inputs and outputs.) The sensors detect mechanical values, such as position or rotational velocity, physical quantities, such as temperature or air pressure, or chemical concentrations, such as oxygen. The sensors and transducers translate the physical phenomenon into electrical signals for transmission to the ECM "box."

Within the ECM box, the hardware converts the electrical analog signal to a digital data value. Most inputs use the internal analog-to-digital converters (ADCs) built into the microcontroller. These signals have resolutions of 10 or 12 bits. Some signals, such as pressure, require fairly high resolution, which means that separate 16-bit or 18-bit or even 20-bit converters are sometimes used. (There can also be external smart sensors, which communicate through serial buses such as SPI or I^2C.)

Software then uses these data values. The software also has calibration points that need setting during or before manufacture on the assembly line. The ECM must have suitable interfaces to memory and these calibration points.

7.4 Phases

ECMs go through four or five phases of development plus follow-on support:

1. Concept
2. Prototype design
3. Iterative testing/validations/development trips
4. Manufacturing development
5. Commercial production
6. Logistics, maintenance, disposal

A slight difference exits between automotive design for ECMs and the general phases laid out in Chapters 1 and 2. The names of the phases vary and indicate more of an environment conducive to prototype and field test in the automotive world. The prime differentiator is "calibrations," which are the parameter thresholds or base values that are set at the beginning of development, refined during development, and eventually specified just before—and sometimes after—the ECM goes to full-scale manufacturing.

7.5 Scheduling

Scheduling is top-down. Every activity flows down from the deadlines set for development. In the automotive world, meeting schedule is the primary thrust. Remember, "Everyone wants to do the right thing, but schedule always wins" (quote from Section 1.6.1).

The design team fixes the requirements once schedule is set. After that they must trade-off between cost and quality. Quality is that elusive goal that indicates how well the requirements are met (assuming that the requirements were set correctly and that they accurately represent intent).

The design cycle time for ECMs is 2 or 3 years, depending on the company, the team, and the particular ECM. After production release, only firmware changes to the ECM are allowed for the first 2 or 3 years of production. Three to six years following production, the design team might modify the hardware in the ECM, if needed.

7.6 Documentation

Like all mission-critical project developments, full documentation, from contract and requirements to final test results and delivery transfer, is necessary to fulfil regulatory concerns. For instance, all source code must be archived and stored for 10 years after the last production unit rolls off the assembly line. Chapter 2 contains the essence of the full documentation needed.

7.7 Requirements and Standards

7.7.1 Markets

The ultimate customer is the automobile buyer. Intermediate customers are the OEMs who contract or purchase ECMs for their cars. Each car model and combination of engine and drive train requires a specific ECM.

The markets for automobiles are all over the world—people drive cars everywhere. Automobiles must endure environments that range from artic cold to scorching desert heat to tropical moisture to ocean salt spray. Furthermore, ECMs must tolerate years of engine heat and vibration, as well. Surviving these assaults is a daunting task for ECM enclosures, let alone the circuit boards with sensitive electronic components within the enclosures.

7.7.2 Government Standards

In the United States, automobiles must meet emissions and fuel economy standards. These standards apply to the fleet averages of automobiles produced and sold. ECMs directly control the fuel economy and emissions produced by vehicles. Consequently, these government standards directly affect the requirements on the design of ECMs.

Emissions: The U.S. federal government is phasing in new federal emission regulations; they started in 2004 and extend to 2009. These new regulations are called Tier II emission regulations, and they apply to different vehicle classes, as shown in Table 7.1 [3].

TABLE 7.1

Emissions Specifications for the United States [3]

	Carbon Monoxide, CO	Hydro-Carbons	Nitrous Oxides, NOx	Particulates
Tier 1 gasoline	4.2	0.31	0.6	0
Tier 1 diesel	4.2	0.31	1.0	0.08
TLEV: transitional low-emission vehicles	4.2	0.156	0.6	0.08
LEV: low emission vehicles	4.2	0.090	0.3	0.08
ULEV: ultra-low emission vehicles	2.1	0.055	0.3	0.04
SULEV: super ultra-low emission vehicle	1.0	0.010	0.02	0.004
ZEV: zero emission vehicles	0.0	0.0	0.0	0.0

Units are in grams per mile, based on 100,000-mile durability.

The definitions in Table 7.1 have been somewhat stratified, as well. The California Air Research Board, or CARB, has required that light trucks and sport utility vehicles, among others, meet LEV II emissions reductions; they also extended the durability standards to 120,000 mi (192,000 km). CARB also specified that a fraction of vehicles must be zero emissions (ZEVs), but recognized ZEV as too stringent for immediate production. CARB has allowed partial ZEV (PZEV) credits to those vehicles with power plants and systems close to ZEVs—natural gas, fuel cell, and hybrid powered vehicles. Vehicles that are designated PZEV must meet SULEV emissions for 150,000 mi (240,000 km) [3–9].

The U.S. Environmental Protection Agency, or EPA, adopted many of the CARB requirements to develop the Tier II regulations.

Fuel economy: The U.S. government has also instituted the Corporate Average Fuel Economy, or CAFE, regulations to improve the fuel economy of cars and light trucks. The CAFE regulations apply to the sales weighted-average fuel economy to vehicles under 8500 lb (3856 kg). The fleet average for cars must exceed 27.5 mpg (11.62 km/L). Beginning in 2005, the average for light trucks must be better than 21.0 mpg (8.87 km/L), 21.6 mpg (9.13 km/L) in 2006, and 22.2 mpg (9.38 km/L) in 2007. The average for light trucks goes up to 23.5 mpg (9.93 kg/L) in 2010 [10].

Should a particular model of car fail to meet or exceed the fleet average, the manufacturer must pay a penalty at the rate of $5.50 per 0.1 mpg (0.0423 km/L) below the standard on each automobile produced [10]. Mileage credits are given for vehicles that exceed the fleet-average mileage required by the CAFE standards; consequently, manufacturers factor into their production plans the number of small cars to drive up the fleet-average mileage of the automobiles that they produce each year.

7.7.3 Preparing Requirements

Automotive OEMs place requirements on suppliers of ECMs. These suppliers also have internal guidelines and requirements to serve their current and future concerns. Their concern is to remain flexible and avoid repeat work in successive models of ECMs.

Specifications and high-level requirements are very important in the automotive world. Some automotive companies and suppliers are using model-based specifications to capture subsystem requirements. Chapter 3 lists some specific tools for capturing requirements.

7.8 Analysis

Suppliers of ECMs use a variety of analyses to develop products. University researchers are constantly developing new algorithms (e.g., fuzzy logic or neural networks) and demonstrating feasibility. The most successful concepts eventually find their way into ECMs over the course of years.

Specialists who use proprietary heuristics are called subject matter experts (SMEs). Calculations and approximations are used where the physical processes are well known.

Some analyses finish by putting an ECM prototype on the engine stand dynamometer. Eventually all development undergoes testing on the test track—and our public roadways—believe it or not!

7.9 Design Trade-Offs

7.9.1 System Architecture

The design cycle time for ECMs is 2 or 3 years, depending on the company, the team, and the particular product. During the first 3 years, only changes to the boot ROM are usually allowed (but this is not absolute—hardware changes can be made after a year if needed). Some hardware modifications begin filtering into the ECM somewhere between 3 and 6 years. The goal is to avoid hardware changes and only "tweak" the software in the boot ROM over the production life of the ECM.

Most companies are moving ECMs toward common platforms. They are trying to leverage their original efforts through reuse. It is a simple concept but challenging nonetheless. It requires predicting advances in engine control and allowing margin for upgrades while minimizing hardware costs.

7.9.2 Hardware

Most OEMs plan and design an automobile to last about 10 years. Currently, automobile warranties are moving from 3 years and 36,000 miles to 10 years and 100,000 miles; longer warranties require higher reliability from all components, including the ECMs.

Automobile ECM designers are more worried about the number of ignition cycles than miles driven; a cycle bounds the period when a car turns on and to when it turns off. It is defined by the ignition switch, not by the engine turning. Many designers plan ECMs to operate for 100,000 to 200,000 cycles. The problem here is that the EEPROM inside the ECM is written on power-down and EEPROMs have a limited number of write cycles before they no longer function properly.

As hybrid vehicles become more prevalent, a parallel set of problems exist for the energy-storage control module (ESCM) for batteries. ESCMs currently must run at least 50,000 execution cycles, although some OEMs and customers want 70,000, some even as high as 200,000.

ECMs need 32-bit processors for algorithmic sophistication and processing power. Designers also want full-featured controller chips with many peripherals on the integrated circuit (IC)—the higher the integration, the better—resulting in higher reliability, lower power consumption, and ultimately lower system cost. Peripherals include ADCs, timer processor units (TPUs), and flash memory.

The controller converts most analog signals with its internal ADCs. These conversions usually require resolutions of 10 or 12 bits. Some signals, such as pressure, require fairly high resolution, which means that separate 16-bit or 18-bit or even 20-bit converters are sometimes used.

TPUs can have different types of masks; two primary types in the Freescale TPUs are Mask A, for automotive, and Mask G, for general. Mask A, called Timing/Counting, has capture-and-compare timing circuits for automotive ECMs. Spark timing and injector controls need the resolution and speed of capture-and-compare timing circuits. Mask A also has circuitry to support pulse-width modulation (PWM) for electric motor control and stepper motor timing. Mask G, called General/Motion Control, has more circuits for driving electronic motor waveforms to control consumer appliances; these include commutation, Hall-effect decoding, and multichannel PWMs. Mask G also has frequency measurements and quadrature decode [11–13].

One way to speed up operations and decrease the time for calculations is to use a large LUT. A lookup in a LUT is much faster than performing a polynomial calculation. A large LUT gives higher resolution and more set points than a smaller LUT and can approximate the accuracy of a polynomial calculation. The trade-off is speed vs. resolution vs. memory size, all of which directly affect cost and reliability.

7.9.3 Power

Automotive power is always dirty; it is full of voltage spikes, sags, dropouts, ripple, and noise. Most automobiles are 12 VDC systems, but some may go to 24 or 42 VDC systems in the near future. Regardless, you still have to deal with unregulated, spiky, and noisy power. Appropriate filtering and circuit protection is necessary within the ECM; most ECMs have internal regulators.

7.9.4 Software

Software development processes vary greatly between different design groups and suppliers. Some groups have detailed processes with carefully prescribed code reviews, tests, and field tests. Others tend to be more "seat of the pants" and go to testing on the test track more quickly than their competitors.

Most software is written in C with some assembly routines for critical timing concerns, such as the bootstrap code and the low-level drivers. The selection and application of a real-time operating system (RTOS) varies widely for the 32-bit controllers within the ECMs. These vary from simple schedulers and round-robin kernels to pre-emptive and co-operative RTOSes. Some ECMs have a custom RTOS, while others use a commercial-off-the-shelf (COTS) package.

Some car companies are farming out the low-level drivers and RTOSs to vendors. Others keep all the software development in-house to retain control and corporate memory of the development effort.

7.9.5 Hardware vs. Software

The trade-off between hardware and software balances performance, cost, reliability, and the ability to upgrade. More specifically this balance reduces to speed vs. ROM size vs. the cost of components.

No surprise to a lot of you software developers, software sometimes makes up for the sins of the hardware design. Designers of ECMs often leave spare I/O pins for future use to fix things that inevitably had flawed specifications or design.

While cost is still prohibitive, suppliers of ECMs might be moving toward using field-programmable gate arrays (FPGAs). An FPGA can blend together the advantages of both software and hardware—flexibility and speed.

7.9.6 Buy vs. Build

Historically, buy vs. build has been a cyclical affair for ECMs. Car companies began developing ECMs in-house but found they did not always have enough resources to do the job properly. Then they contracted out the ECM design to vendors, but found that they were not always fast enough to meet their demand. Now that ECMs are so critical to the operation of automobile engines, more automobile companies are bringing or keeping ECM design in-house to maintain the design resources and capability. Companies have also worked to develop career paths for embedded engineers, making long-term employment more attractive to firmware engineers.

An interesting note is that car companies rely solely on outside vendors for certain integrated subsystems, such as antilock brake systems and airbags, which are mission-critical systems and components. This sort of outsourcing may portend a future swing back to outside vendors to supply ECMs.

7.9.7   Manufacturing

ECMs have aluminum or high-temperature plastic enclosures. Cost and durability dictate their design.

The PCBs within ECMs generally use FR-4 substrates; again cost, durability, and reliability all dictate the selection of materials. In the past, some PCBs were flex circuits to achieve packaging density in spite of cost! Today, however, surface-mount ICs and higher integration (microcontroller incorporating power electronic field effect transistors and external components such as flash memory and RAM) have reduced the need for flex circuits.

Nearly all PCBs are multilayer boards with vias and internal ground planes. The ground planes are the first line of defense in containing EMI and managing EMC compliance. The circuit boards have rigid attachments to the enclosure to reduce flexing during vibration.

Manufacturing of ECMs usually incorporates design-for-manufacture (DFM), design-for-assembly (DFA), and design-for-test (DFT). The goal is to maintain quality while reducing cost of rework or recall.

7.10   Tests

Testing, both formal and informal, varies from group to group and company to company. The variation in testing extends all the way through software unit tests, functional tests, system tests, and "feet in the car" tests on the test track. Interestingly, some groups go straight to the test track with each revision of the software.

I advocate a rigorous program for ECM development that includes laboratory tests, inspec-tion, peer review, subsystem tests of both hardware and software, and system tests on a dynamometer. Not everyone agrees with my approach; they see experimentation as quicker for finding problems, tuning the software, and understanding the issues of drivability and driver experience for noise, vibration, and harshness.

Another variation between designers is problem tracking. Some are much more rigorous about it than others.

How can such variation exist in a mission-critical industry? An important problem is that ECM software and its operation are hidden. Another reason for the variation in procedures is the attitude, prevalent in some circles, that "anyone can be taught software," which is akin to saying that anyone can be taught to draw. Sure, a guy can draw stick figures, but he is not Rembrandt! Combine the obscurity of operation with the attitude that anyone can do software and you have the current situation with development in many companies, "Hey, it's a great drawing in that box—trust me!"

7.11 Integration

Integration, particularly with numerous software builds, occurs frequently enough to approach a spiral form of development. Several steps exist for each revision before integration finishes. The software must perform and pass the following steps before going to dynamometer tests:

- Compile and link
- The "blinky light test"
- Bench tests and breakpoints
- Diagnostic trouble codes (only in later or more mature levels of software builds)

Interestingly, the "blinky light test" requires a lot of modules and operations to function correctly in the software to light simple light-emitting diodes (LEDs). This makes it an effective form of verification. Once the software development team blesses the results, then the ECM, with its new version of the software, heads to the test track for testing in a vehicle.

7.12 Manufacturing

Manufacturing of ECMs holds many of the same concerns that any other mass-produced item does. There are electrical and electronic concerns such as fabrication of circuit boards and interconnections and assembly of electronic components, fabrication of mechanical components and enclosures, inspections, and tests.

The automotive industry is peculiar in one respect—the philosophy of calibrations. Design and development teams use lots of "calibration" points in developing ECMs. "Calibrations" are parameters and parameter thresholds that are set in manufacturing, though some calibrations are managed as different ECM part numbers that have variations in content to serve different models of automobiles.

The company and the team just do not have the time to characterize everything. The software finishes long before the production parameters are fixed. These parameters concern the sensors and control system within the ECM and include thresholds for (this is not an exhaustive list):

- Timing
- Temperature
- Air pressure
- Gas chemical concentration
- Voltage

Some might view these "calibrations" as a crutch. The automotive industry, as mentioned already, is extremely conscious of schedule. Design teams just do not have time to fully analyze the automotive systems, the environments, and the interactions with ECMs. Moreover, flexibility, changes, and reuse dictate the need for calibrations. Designers often select or change sensors and actuators late in the development cycle, a practice that requires significant flexibility during development. Reuse of software, where a module or subsystem of software is used in multiple applications, uses "calibrations."

Example:

A calibration might indicate the number of engine cylinders, the number of cam shafts in the engine, whether an engine has electronic throttle control (ETC) or a manual throttle cable, whether air conditioning is included on the automobile, and whether the system supports dynamic cylinder deactivation (i.e. going from running on eight cylinders to four). A single software system configured by calibrations can support all these different situations.

7.13 Support

The automotive world, particularly ECMs, blurs the lines between product launch, logistics, and maintenance. The design team does continuous development of the software for an ECM, even after manufacturing begins. Calibrations that go beyond production release are called "service releases."

Onboard diagnosis, or OBD, for automobiles uses trouble trees and diagnostic trouble codes. These came about in the early 1990s. OBD helps mechanics determine where a problem is and what needs replacement. If an ECM is bad, then the mechanic replaces the whole module.

Sometimes a particularly difficult failure is brought in from the field. The development team will put the ECM on a bench and hook it up for the background debug mode (BDM) to get additional, nonstandard diagnostics from the ECM for laboratory analysis. These codes are meaningful to engineers but not useful to mechanics doing automotive repair. Furthermore, opening a sealed ECM is a somewhat delicate—and warranty voiding—operation.

Finally, revision management must be stable. The U.S. government regulations require automotive companies to keep all source code for 10 years after final production of a particular car model. This type of archiving provides the appropriate basis for upgrades and fixing recalls.

7.14   Disposal

The recycling of materials is big business in the automotive industry. Most of us are familiar with junkyards and automobile recyclers, those companies that crush cars, shred up the metals and plastics, and melt down the scraps.

The automotive industry is concerned with Restriction of the use of certain Hazardous Substances (RoHS), as well. European regulations and markets are driving both recycling and the reduction of hazardous materials. This will place a burden on automotive manufacturers, the extent of which is not yet fully understood.

7.15   Liability

The liabilities in the automotive industry are fairly obvious. We all are familiar with automobile recalls; no doubt many of us have had vehicles recalled for one thing or another. Recalls are a fact of life for automobile manufacturers; they all have their attendant economic impact, safety concerns, and potential legal problems. Some companies term "recalls" by the euphemism "campaigns" to soften the meaning for customers. Some campaigns are mandatory, dictated by the government—some are voluntary.

Recalls can be instructive; here are some examples taken from References 14 and 15:

- *"January 20, 2006* Certain vehicles equipped with V-6 engines may have a condition where fuel is no longer supplied to the engine. This condition occurs without the illumination of the fuel level low indicator light or the warning chime . . . . Dealers will reprogram the electronic control module (ECM) with new software free of charge. Owners were notified in December 2005 and asked to maintain at least 1/4 of a tank fuel level to avoid this condition until the parts are available. The recall is expected to begin during March 2006 . . .

- *January 11, 2006* On certain passenger vehicles equipped with all wheel drive and a 3.6 liter V-6 engine, the torque monitoring functions of the Electronic Throttle Control (ETC) are not enabled. These functions can limit engine speed and torque if unusual Engine Control Module (ECM) hardware or software failures occur . . . Dealers will reprogram the ECM on these vehicles free of charge. The recall is expected to begin on January 24, 2006 . . .

- *March 12, 2002* [Company E] recalled the [specific model] trailer hitch—circuitry in the converter is inadequate to properly manage voltage spikes that can lead to an electrical short or open circuit within the converter, causing a failure and an inoperative trailer light.

- *September 11, 2000* [Company F] recalled about 270,000 [cars]—air bags that may deploy unexpectedly because of corrosion in the inflator.

- *During 2000* [Company G] recalled ignition modules that could cause a car to stall. When the temperature of an ignition module rises above a certain value the chances of the module cutting out also increases.

- [Company H] recalled 263,000 1995–97 [vehicles] . . . The electronic control module for the airbag could corrode from water or road salt and then accidentally deploy the driver side airbag.

- [Company I] recalled 757,000 1992–97 [vehicles] because higher than specified electrical load through an accessory power feed circuit may cause a short circuit and allow current to flow through ground wiring. This could cause overheating and an electrical fire.

- [Company J] recalled 1995–97 [vehicles] because improperly routed wire harness for the air-conditioner may permit wires to rub together and short circuit, resulting in a blown fuse, dead battery, or fire.

- *December 11, 1998* [Company K] recalled 226 [electric vehicles] to reprogram the logic in the motor electronic control unit (ECU), which can mistakenly detect a failure of an electrical current sensor at speeds above 50 mph. It can cause the sudden loss of power and unexpected deceleration.

There are some common elements in these recalls.

- Passage of time—these were all fielded units
- Nonobvious or obscure causes
- Environmental interactions, i.e., corrosion, overheating
- Failure modes with significant effects, i.e., fire or injury." [14,15]

ECMs have the mitigating factor of "calibrations," which reduce the potential for costly changes. Flash memory also allows field upgrades to operational software if appropriate and necessary. The word is "reduce"—it is not "eliminate." A software fix or upgrade can solve many problems, but not all.

7.16 Summary

7.16.1 Emphases

Automobile ECMs control internal combustion engines and balance their performance, fuel economy, and emissions. Designing and building ECMs is a challenging business. It is a mission-critical design; each automobile must run correctly and well.

The automotive industry lives and dies according to schedule. The three main concerns are schedule, cost, and quality; of the three, "schedule always wins."

One way to do continuous development and still meet production schedules is to incorporate "calibrations" in the software. "Calibrations" are parameters and parameter thresholds that are set in manufacturing; they allow the developers to finish the algorithms and test the vast majority of software operations without knowing some of the final parameter values.

7.16.2 Gotcha's

There are two primary problems in developing ECMs for the automotive market: schedule and variability in rigor. Some development teams do treat the design of ECMs with diligence and care; some, unfortunately, are more haphazard. ECM development is mission-critical and needs appropriate consideration for good processes and procedures.

Acknowledgments

My genuine thanks to a colleague who has worked in the automotive sector for over a decade supporting the development of control firmware for power trains. Unfortunately, I cannot acknowledge him here by name so as to maintain the confidentiality of his clients. My friend, your help is greatly appreciated!

I also thank Richard Bishop at IVSource for helping me with this field; I recommend his book, *Intelligent Vehicle Technology and Trends*, Artech House, 2005, for a look into the future of embedded systems in vehicles. His Web site is www.ivsource.net.

References

1. http://www.oica.net/htdocs/statistics/tableaux2005/worldproduction_cars2005. pdf
2. http://www.oica.net/htdocs/statistics/tableaux2005/worldproduction_lightCV2005. pdf
3. http://www.dupontelastomers.com/apps/autofocus/regulations.asp
4. IANGV (International Association for Natural Gas Vehicles) Emission Report 31 March 2000.
5. www.engva.org (European Association for Natural Gas Vehicles).
6. www.arb.ca.gov (California Air Resources Board).
7. U.S. Environmental Protection Agency, Office of Transportation and Air Quality, publication EAP420-B-00-001, February 2000.
8. www.autofieldguide.com/columns/article (Automotive Design and Production).
9. www.aecc.be/en/european_legislation.htm (European Association for Emissions Control).
10. http://www.nhtsa.dot.gov/cars/rules/cafe/overview.htm
11. http://busy.lab.free.fr/download/tpu-functions-list.txt
12. http://www.eslave.net/tpu/source/source.shtml
13. http://www.cloudcaptech.com/MPC555%20Resources/TPU3/tpurm.rev3. pdf#search=%22TPU%20masks%22
14. Fowler, K., Class 307: Fantastic Failures, *Embedded Systems Conference*, San Jose, April 5, 2006, pp. 9–10.
15. http://autorepair.about.com/library/recalls/

8

Case Study 5—Industrial Flowmeter

8.1 Concept and Market

8.1.1 Who, What, Why, How, Where, and When

Agar Corporation in Houston, TX, designs and builds multiphase flowmeters. The primary application for these flowmeters is monitoring raw material flow (oil, water, and gas) from wellheads in oil fields. Actually, they may be used anywhere crude oil is pumped—the oil field, into the ship, and into or out of a tank farm.

Agar's flowmeters measure the unseparated fluid from the wellhead and give the percentage of flow contributed by oil, water, and gas. These sensors use the venturi principle to measure flow without separating the oil, water, and gas into individual streams. Figure 8.1 shows a multiphase flowmeter in use.

Monitoring the flows and measuring the volume of constituent materials is important to oil companies. They want to know how much water vs. oil that they are transporting; it affects their business decisions and revenue. The water has to be removed and treated so that it does not pollute the environment with crude oil byproducts.

Agar sells its products directly to end users such as petroleum companies, refineries, and loading facilities. It has sales offices around the world.

8.1.2 Economics

Equipment in oil fields can last for decades. Agar Corporation designs multiphase flowmeters to last for a long time—decades—and supports their installation and maintenance.

The flowmeter business follows the oil industry. The recent upturn, during 2005 and 2006, in petroleum prices worldwide has caused oil companies to plunge profits into research, development, and capital assets. This means that Agar is currently very busy supplying flowmeters to the industry.

FIGURE 8.1
A model MPFM-408 multiphase flow meter. (Used with permission from Agar Corporation.)

8.2 People and Disciplines

While Agar Corporation is a small company with about 50 employees at headquarters, it is truly a multinational company with offices and people from all over the world. Multiethnicity in the workforce helps communications between Agar and customers in other countries; customer relationships are further aided by the corporate understanding of the various cultures.

People in the Houston office include

- Mechanical engineers
- Electrical engineers
- PhD physicists
- Fluid dynamicists

- Software engineers
- Manufacturing professionals
- Purchasing and procurement personnel

Engineers comprise the sales force, which means they can better understand and communicate with their technical customers.

8.3 Architecting and Architecture

8.3.1 Process

Agar Corporation does not have a traditional process for product development. Agar uses a form of waterfall development. The team documents carefully and tests for certification, as product safety is of paramount importance in their industries. Agar Corporation maintains ISO 9001 certification, which means that they use prescribed and documented procedures.

8.3.2 Parameters

Safety is the parameter of paramount importance in design and development. Of critical importance to Agar's customers is avoiding sparks that might ignite volatile vapors and gases. The designers of flowmeters have to consider energy that can be stored and delivered from electrical circuits to a potentially flammable environment.

Agar Corporation designs to standards that define and categorize environments. These standards are quite similar between different sets of organizations: ATEX, Underwriter Laboratories (UL), International Electrotechnical Commission (IEC), and EN. Agar designs to IEC 60079-11 and its various gas groups for petrochemical above-ground environment. Gas group IIC designates a hydrogen environment for the most flammable gases and has the lowest threshold-to-energy delivery that can ignite such a gas. Gas group IIB designates an ethylene environment, which is less dangerous and can tolerate greater energy delivered. Gas group IIA designates a propane environment, which is the even less dangerous and tolerates even more energy.

The standards also define various operating zones. IEC Zone 0 is a place in which an explosive atmosphere is continually present. Zone 1 is a place in which an explosive atmosphere is likely to occur in normal operating conditions. And Zone 2 is a place in which an explosive atmosphere is not likely to occur in normal operation, but if it does, it only occurs for short periods.

Dependability is another important parameter for products from Agar Corporation. Their flowmeters often are "out in the middle of nowhere" and reliability is critically important. Robust operation is a part of dependability; if any component fails and the flowmeter stops functioning, then the valves go into a safe mode.

8.3.3   Analysis

Engineering design at Agar uses various forms of fault tree analysis (FTA), and failure modes and effects analysis (FMEA), and risk analysis to ensure safe operation of their flowmeters. They carefully design for temperature extremes, which can range from Alaska's North Slope to Middle East deserts; the same design and equipment operates in any and all environments.

Agar Corporation also does testing to prove designs too. More on this later.

8.3.4   Architecture

These flowmeters are the essence of embedded systems. They have digital and analog electronics sealed in a weatherproof, and sometimes explosion-proof, enclosures. Higher power circuits, such as processor boards, often reside in enclosures remote from the flowmeter and potentially volatile environments. This architecture keeps power density low in the vicinity of flammable gases to reduce the possibility of ignition.

Agar engineers carefully follow previous designs, if it is an upgrade design. The reason is that safety considerations and certifications drive the designs and previous designs have already surpassed hurdles for safe design.

8.3.5   Interfaces

The primary interface into the electronics is a serial interface. Agar provides RS-232, RS-422, RS-485, and a current loop for communications. These flowmeters can use Modbus or HART industrial network protocols. Agar is planning to incorporate Fieldbus and Profibus network protocols in the future.

The equipment also can provide relay closures. Agar dedicates one relay to provide an alarm, which can drive a light, buzzer, or an input to a status system.

The control box often has an liquid crystal display (LCD) of two lines by 40 characters. This simple display supplies basic status and alarms. No keypads are needed for this equipment.

8.4   Phases

Agar Corporation follows a course of prototype development and testing, as opposed to specific phases. To begin, engineering and production personnel meet to discuss a concept for a new product. Once they agree to an approach to an idea, the designers begin development of several prototypes.

The first prototype is a breadboard that proves feasibility of the concept in the lab. Next Agar develops an initial prototype; at this time they also contact the appropriate certification body to begin the certification process. The concern is for the very tight restrictions on safety, and so the initial prototype and working with the certification body early in development is necessary.

Example

A sensor immersed in a fluid is one example of concern for certification that an initial prototype might address. The sensor must incorporate DC-blocking capacitors; these capacitors must be connected in series to prevent problems from a short circuit failure in one capacitor. Also, capacitors must be small enough to avoid storing too much charge for the rated environment and having the potential to ignite the fluid. Finally, the capacitors must have a high voltage rating to accommodate human mistakes from connecting wrong voltages to the sensor.

The final stage is the production prototype; it is as close to the final product as possible to find any problems in manufacturing.

The development work on all three prototypes has feedback loops to previous stages. Revision can go all the way back to the original concept.

8.5 Scheduling

Agar Corporation uses both top-down and bottom-up planning. For a small project that a customer requests, they use top-down planning to meet deadlines. For the bigger projects that begin as a good idea, they use bottom-up planning; development can take twists and turns and present challenges that defy deadlines. For those projects that use bottom-up planning, sales pressure can exert influence to set deadlines and eventually enforce some top-down schedules.

8.6 Documentation

Agar Corporation is an ISO 9001 certified organization and undergoes audits from any number of certifying bodies (see the next section). This forces them to follow their procedures and produce the specified documentation for each product.

8.7   Requirements and Standards

Agar certifies through government-sanctioned bodies to examine and test designs for flowmeters. Some of these include:

- ATEX (Europe)
- UL (US)
- Canadian Standards Administration (CSA) (Canada)
- GOST (Russia)

They pay for audits from these certifying bodies. These audits help ensure that Agar Corporation is following and meeting the necessary standards.

Agar Corporation initiates requirements several different ways. One way is through the marketing and sales staff, which brings in requests from the field for a new product or upgrade. Another way is having the engineering design team generate a new idea. After the new idea is brought forward, the design team considers the environment and the required safety standards. The standards define the limitations in all new designs.

8.8   Analysis

Agar Corporation begins examining feasibility of a new product by discussing it among the team. If the new product or upgrade appears to be a software change to a current model, they inspect the software and discuss the implications of change. If significant questions remain about the feasibility, they will build and test a prototype. They have facilities such as flow loop and temperature environmental chambers to test prototypes.

Agar Corporation sometimes relies on heuristics or numerical simulations to assess a new idea.

8.9   Design Trade-Offs

8.9.1   Architecture

Agar Corporation maintains designs of its flowmeters and installations for many years. There are some old designs still in the field using computers running DOS® (Microsoft's disk operating system). Newer designs need great computational power; they are migrating to microcontroller and DSP-based architectures and away from general-purpose microprocessors. The newer processors reduce power consumption and physical size.

Agar is moving the system design of their flowmeters toward more distributed architectures. New, low-power microcontrollers make preprocessing at the sensors quite straightforward. These processors lower the overall power consumption from that of the older single-board computers.

Agar separates higher-power circuits in remote "safe" areas away from lower-power circuits that provide signal conditioning on the flowmeters. Usually this means that the processor board, sometimes a single-board computer, is in the remote, safe area.

8.9.2 Hardware

One of the biggest problems for Agar Corporation is the inventory of components and the long-term stability of the supply of components. This problem affects the design decisions for the electronics. One way this problem plays out is that Agar engineers will measure parameters of components and fix their exact values so that Agar's flowmeters make stable and precise measurements. Unfortunately, after the parameters are measured, a vendor might later change a manufacturing parameter that affects the design.

Example

In one situation, Agar Corporation originally designed a circuit with a crystal oscillator that needed a 100 pF capacitor to start and maintain resonance. A test in manufacturing checked the start-up of the oscillator by repeatedly turning the circuit on and off. Over time, manufacturing technicians found that they had to place a capacitor in parallel with another. If the manufacturing test found that the oscillator did not start, assembly personnel then followed a procedure to remove the parallel capacitor so the oscillator would start reliably. After some time, engineers investigated and found that the vendor had changed the crystal fabrication and no longer specified the 100 pF capacitors for the crystal oscillator; instead the specification had moved to a 33 pF capacitor and the manufacturer had not informed them!

As Agar Corporation designs systems with more distributed architectures, they are moving away from single-board computers and desktop operating systems, such as DOS, and are using more microcontrollers and DSP chips. The microcontrollers tend to be 16- and 32-bit processors, though 8-bit controllers are finding their way into sensor heads. Agar has used or has considered using a variety of different microcontrollers; these range from the venerable 8051 series from Intel to the PIC family from Microchip and AVRs from Atmel.

While Agar migrates to microcontrollers and DSP chips for greater computational processing, a problematic interaction arises between size and power density. Agar must make sure that the physical area of each component is large enough to reduce power dissipation and avoid igniting a volatile environment. This means that components cannot be "crammed together" as you might find in consumer appliances, such as cell phones. This power density concern, which even focuses down to individual resistors, forces the need for physical space to dissipate power appropriately.

Agar also must design and build the circuit boards for diverse and severe environments. They use conformal coatings on the assembled circuit boards and potting compounds around larger components to protect against dust, sand, and condensation.

8.9.3   Power

The power input for Agar's flowmeters is either 120 or 240 VAC for most systems; it must also be 50 or 60 Hz capable. Generally, the flowmeters have a 12 VAC-to-DC converter in a remote, safe area.

The power circuits usually have two types of safety "barriers" to reduce the potential of ignition from high-energy discharges. One is a Zener barrier; the other is a galvanic isolation barrier.

The Zener barrier (Figure 8.2) clamps the voltage with a fuse and Zener diode. If the input voltage goes too high, the fuse will open and prevent the delivery of energy that exceeds a design threshold. The resistor limits instantaneous current, for the very same reason—it prevents the delivery of energy that is above the limit. One caution here is that capacitive or inductive energy storage in the supplied circuit must also be understood and designed to avoid storing too much energy that could be delivered in an accidental discharge.

The galvanic isolator is a DC–DC converter designed to prevent any charge from moving from safe area into the hazardous area. There is no DC electrical connection between the two areas. This provides a higher level of safety and isolation for the components in the hazardous area.

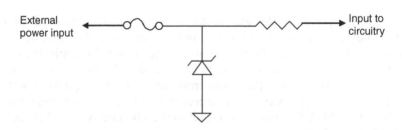

FIGURE 8.2
Schematic diagram for a Zener safety barrier. The fuse and zener diode clamp the voltage to an appropriate level; should it go too high, the fuse will open. The resistor limits instantaneous current.

8.9.4 Cable Harnesses

Cables are an important part of the multiphase flowmeters. The designers work to keep cables and wires to a minimum in the flowmeters. They also keep cables short and compact. The cables may be tied down within enclosures to keep them from resonating with potential sources of vibration. The cables are also sheathed or placed in conduits to meet the standards for the specified gas-group environment.

8.9.5 Cooling

Agar Corporation avoids active cooling in its flowmeters because of the additional complexity and certification required. The design team tried a thermoelectric cooler inside an enclosure once but they had problems with positive thermal feedback; it doubled the power consumption and ultimately did not cool the enclosure.

Enclosures must remain sealed from volatile environments. No vents are allowed, which excludes fans and forced-convection cooling from the ambient external atmosphere through the enclosures. Furthermore, the concern for mechanical reliability of the fan bearings rules out fans from even circulating air internal to the enclosure for cooling. Flowmeters often operate for years without interruption—a difficult requirement for fans.

8.9.6 Mechanical Structure

Flowmeters potentially must withstand years of constant vibration; this particularly applies to the cables and mechanical enclosures. If a flowmeter sits next to a big pump within a refinery, the vibration can be unrelenting. Agar Corporation uses shock mounts on some boxes in the flowmeters to dampen vibration. They also use cable glands to seal cable penetrations and mitigate the effects of vibration on the cables entering an enclosure.

8.9.7 Software

The engineers at Agar Corporation use careful and thorough processes to develop software. They write code in C, although some subroutines are written in assembly language to achieve performance. Engineers use code reviews, tests, and field tests to assure software functionality. They archive developed software and version control it. Like certain other applications in the case studies in this book, they cannot easily upgrade software once it is fielded; it has to be right the first time.

Historically, Agar engineers have used single-board computers that either run DOS or use a custom operating system. With newer systems using newer versions of single-board computers, they have considered several commercial or open-source operating systems, such as Embedded Windows™,

QNX™, and Linux. Their biggest concern is the stability of the software platforms; flowmeters may run for 15–20 years, Agar must maintain their software and software tools for a long time.

As mentioned previously, Agar engineers are using microcontrollers in distributed architectures. They are considering commercial RTOSes for the 16- and 32-bit processors in the core, while they are writing their own custom RTOSes for the 8-bit microcontrollers doing the preprocessing in the sensor heads. Generally, the operations of the sensor heads are quite simple, which means that the real-time operating system (RTOS) does not need to be complex.

8.9.8   Buy vs. Build

Agar Corporation has used commercial off-the-shelf (COTS) single-board computers in their flowmeters. Newer products, however, have a more distributed architecture and Agar is moving to all custom design. Agar is doing a mix of things with software RTOSes; some are custom-made while others are commercial products.

8.9.9   Manufacturing

Agar Corporation does not have an explicit program for DFx (design–for-manufacture [DFM], design-for-assembly [DFA], or design-for-test [DFT]), but the engineering team does work closely with the production staff during design and development to produce flowmeters amenable to manufacturing. They work together to develop test jigs. An example of working together on test jigs would be those used to test circuit boards arriving from a fabrication vendor.

8.9.10   Test and Maintenance

The engineering team also works with the field sales staff and affiliates from around the world. During design, this type of teamwork helps them prepare better ways to maintain flowmeters in the field.

An example of speeding maintenance is to provide a test head outside the electronics enclosure; it is potted with the appropriate glands to seal the cable penetration into the box. The test head speeds diagnostics by allowing maintenance personnel to connect to the test head, which provides the signals and controls to the electronics, without opening the main enclosure. This is important because opening the enclosure is a lengthy process that requires obtaining a special permit, potentially shutting down adjacent or related equipment, testing of the environment, and then a tedious effort to undo all the bolts sealing the enclosure.

Agar continues to research ways to speed maintenance. They are considering radio frequency (RF) links, such as the Bluetooth standard, to interrogate the enclosures without opening them or even physically connecting to them. Another advantage to a wireless approach is that in very cold climates, such as northern Canada, the maintenance personnel would not have to leave the warm confines of their vehicle to test the flowmeter.

8.10   Tests

Agar Corporation performs a variety of different types of tests to assure design quality. They test prototypes; they do subsystem functionality tests on the hardware; and they hold periodic reviews of the design and software.

The team performs thermal testing on prototypes to understand and demonstrate stability in the measurements and calibration. They use a flow loop to confirm operation. They also use ESD equipment to test and confirm electromagnetic compatibility (EMC) for static charge.

The team has periodic reviews of the design and software. These are set according to stages of both initial and production prototypes. They review the software and hardware subsystems separately and then together as a system.

8.11   Integration

Integration is considered a part of testing. Agar Corporation has a large flow loop to test and demonstrate flowmeters. Agar has a detailed factory acceptance test (FAT) that customers will often sign off before acceptance.

8.12   Manufacturing

8.12.1   Outsourcing

Agar Corporation outsources the fabrication and assembly of most subsystems: circuit boards, metal structures and components, and assembly. Agar does not have the volume of products to justify maintaining basic manufacturing capability in-house.

Agar has qualified certain vendors to fabricate the circuit boards and to solder the components onto them. These electronic "board houses" are qualified to UL standards to ensure that the circuit boards meet UL ratings.

Agar also has qualified vendors to weld spool pieces and metal frames for the flowmeters. They also outsource some assembly of the frames, such as the shock mounts and cabling.

8.12.2   Assembly

Agar Corporation performs the final integration of the subsystems on their premises. For some of the smaller meters, such as the water-cut meters, they will do more of the assembly in-house than would be done on the larger multiphase flowmeters.

8.12.3   Tests

Agar Corporation does all the final testing in-house. ISO standards and certification means that they must do many different types of tests to maintain quality. One example might be the purchase of 100 enclosures; Agar must check the dimensions of each enclosure to confirm the size even though they are catalog items and Agar has ordered them for years. The production personnel follow forms and checklists. Another example of proper procedure is pressure tests of the sensor seal housing; they do strip-chart recordings of the pressures and send copies to the customer on delivery after the sale.

8.13   Support

8.13.1   Logistics

Agar Corporation maintains an inventory of components for replacement. They will replace modules and subsystems in the field but not individual components. Agar plans for minimal software upgrades in the field, and new circuit boards or enclosures might be installed to replace an older unit.

8.13.2   Maintenance

Periodic maintenance is an involved procedure. First the technician must obtain the appropriate permits and then set up an explosive-gas analyzer to "sniff" for volatile gas for 1 hour. Then the technical personnel go through a shutdown procedure, possibly involving adjacent or related equipment, and then remove the bolts on the enclosure. Finally, they tag the flowmeter as hazardous before running the diagnostics. Afterwards, they have to follow proper procedure to "button-up" the enclosure, bolt it, and then restart all the associated equipment.

Trucks with appropriate equipment will go into the field to test and calibrate Agar's flowmeters.

Field service personnel use a notebook computer to test the embedded system through one of the serial interfaces on the electronics box.

8.13.3 Technical Support

Agar Corporation maintains a large, international network of field sales and service staff and affiliates who are highly technical. They can answer questions and speak directly to customer staff, in their language and technical vocation, to address concerns.

8.14 Disposal

Agar relies on their outsourcing vendors to handled concerns with Restriction of use of certain Hazardous Substances (RoHS) properly. They also advise their customers on the proper procedures to dismantle and recycle old equipment.

8.15 Liability

8.15.1 Economics

Economics is a large liability for these multiphase flowmeters. If one shuts down and stops the flow of crude for even a short time, it causes financial repercussions throughout the industry. Reliability and accuracy are directly tied to the economics of oil field production.

8.15.2 Safety

Safety in a potentially volatile environment is probably the number-one concern in developing multiphase flowmeters. Agar engineers expend much effort to understand all the aspects of safety and to maintain low-energy discharge limits. A spark that could ignite an explosion affects human life and safety. This makes safety a very large liability for Agar Corporation in designing their products.

8.15.3 Legalities

With economics, safety, and international markets, the legal liability is also large. Careful design and attention to customer concerns help reduce legal liability.

8.16   Summary

Multiphase flowmeters used in the petrochemical industry place very particular demands on design and development. First, they must be safe; energy distribution, charge storage, and discharge receive much attention to avoid igniting volatile environments. Second, reliability is very important; these flowmeters must operate in extreme environments—from blazing desert heat to Artic cold—for years. They also must endure constant vibration. Finally, they must be accurate and stable in their measurements. A lot of money rides on the exact percentage of oil, water, and gas transported through the flowmeters.

Acknowledgment

My thanks to Steven Bates at Agar Corporation for providing the information for this chapter.

9

Case Study 6—Military Support Equipment

9.1   Concept and Market

9.1.1   Who, What, Why, How, Where, and When

The Support Systems business unit of Boeing Integrated Defense Systems makes support equipment for military aircraft; the equipment is sold to all four major U.S. service branches (Army, Air Force, Navy, Marines) and to foreign militaries. Boeing Support Systems started in this market to support its own aircraft. Boeing Support Systems has since expanded and tries to compete for equipment that supports other military systems, as well.

Boeing Support Systems makes and sells the support equipment, both mechanical and electrical, to fit one of three different levels of maintenance: Organizational Level (O-Level), Intermediate Level (I-Level), and Depot Level (D-Level).

- O-Level indicates that the equipment tests aircraft on the flight ramp. The equipment must endure dust, rain, snow, very cold and very hot temperatures, and humidity. Figure 9.1 gives an example of O-Level equipment.

- I-Level indicates a controlled environment with the appropriate heating, cooling, and humidity control for the tested units or "black boxes." These black boxes are called weapons-replaceable assemblies (WRAs) by the Navy and line-replaceable units (LRUs) by the Air Force; they are the repairable elements within a system that can be changed at the flight line. Figure 9.2 gives an example of I-Level equipment.

- D-Level indicates maintenance performed at either a military facility or by the manufacturer in a controlled environment on the next lower assemblies from the I-Level, typically at the circuit card level. These circuit cards are called shop-replaceable assemblies (SRAs) by the Navy and shop-replaceable units (SRUs) by the Air Force. Figure 9.3 gives an example of D-Level equipment.

FIGURE 9.1
An example of O-Level equipment, a radar simulator for the F-15 fighter jet. (Used with permission from the Support Systems business unit of Boeing Integrated Defense Systems.)

FIGURE 9.2
An example of I-Level equipment, the F-15 Avionic Maintenance Support System (AMSS). (Used with permission from the Support Systems business unit of Boeing Integrated Defense Systems.)

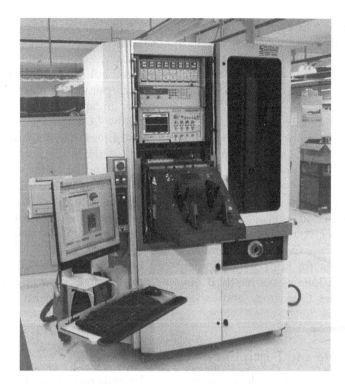

FIGURE 9.3
An example of D-Level equipment, the F-15 Digital Test Station (DTS). (Used with permission from the Support Systems business unit of Boeing Integrated Defense Systems.)

Most of the material in this case study is in the O-Level arena. Consequently, the decisions and trade-offs in this case study will slant in that direction. I-Level and D-Level equipment are often called automatic test equipment (ATE) in other industries.

9.1.2 Economics

Boeing Support Systems invests in both research and internal process improvement to remain competitive and stay abreast of market trends. Boeing evaluates individual opportunities on a case-by-case basis and either responds to request for proposals (RFPs) or talks with customers during the proposal phase for a particular aircraft program, such as an upgrade to an existing capability.

For very small numbers Boeing Support Systems generally makes the equipment in-house. Once production gets underway they will subcontract most of the parts and act as the system integrator. There are some products that Boeing Support Systems subcontracts entirely to a vendor, who may ship the product-directly to the customer. Boeing Support Systems sells equipment to first-time customers via a System Compatibility Test (SCT) for the Air Force or First Article Test (FAT) for the Navy. Subsequent sales usually only require passing an Acceptance Test Procedure (ATP).

One size does not fit all in the world of military support equipment; it is not off-the-shelf in any sense. Boeing designs and builds support equipment for specific needs and purposes. Boeing will incorporate commercial off-the-shelf (COTS) into system solutions primarily in applications where the environmental operating requirements will allow COTS products.

Boeing generally has short production runs of less than 100 and usually less than 50. For O-Level support equipment, volumes range from one or two items per squadron up to one per aircraft, depending on the system that is supported. The smaller jobs tend to be updates that add new capabilities to existing fielded equipment. The larger jobs tend to be new designs or large production runs of existing designs. The I-Levels and D-Levels tend to have smaller volumes because the support equipment is located in fewer places than the O-Level equipment.

Life cycles for support equipment can be fairly long; 20 years is not atypical. As long as people fly Boeing aircraft, Boeing Support Systems has opportunities to upgrade and enhance their fielded equipment.

9.2   People and Disciplines

The composition of the development team depends on the requirements for the design. Customers generally dictate what they want in terms of reliability, maintainability, documentation, and special needs. Boeing Support Systems might add a discipline to meet unusual requirements. In general, Boeing uses the following types of people during the design phase:

- Team leader
- Electronic hardware engineers
- Mechanical engineers
- Software/firmware engineers
- Technicians
- Quality assurance (both hardware and software)
- Reliability and maintainability engineers
- System safety engineer (depending on the design requirements)
- Supplier management and procurement
- Contracts
- Technical publications
- Training

The actual design effort usually only includes the engineers and the team leader, but the others come and go as necessary.

9.3 Architecting and Architecture

9.3.1 Process

Boeing Support Systems follows a waterfall process in development. They perform an analysis for the requirements by looking at the inputs and outputs from the unit-under-test (UUT) and then they specify test approaches that cover all the signals. The team designs the equipment to meet those requirements.

Boeing Support Systems is an ISO9001 company and as such has many processes and procedures; a description of them would be far beyond the scope of this book.

9.3.2 Parameters

The main parameters that affect design and development are typical for mission-critical equipment, for example, schedule, requirements, weight, power, size, cost, reliability, fault tolerance, testability, and availability. Boeing also considers environmental conditions, including electromagnetic interference (EMI) and electromagnetic compatibility (EMC) for O-Level equipment. Ultimately, there are different build requirements based on the environment that the equipment must survive.

9.3.3 Analysis

Boeing Support Systems uses failure modes and effects analysis (FMEA) to ensure that any failure of their test sets do not cause any single point failures in the UUT. They also screen out components that might get too hot or might have a weakness of some sort as revealed by a reliability analysis.

9.3.4 Architecture

O-Level support equipment is portable, so the keypads are integrated and simple; they might be numeric only, for instance. The displays tend to be simple, such as an liquid crystal display (LCD) with four or five lines of text. There are circuit cards and a central processing unit (CPU) tailored to the functions needed in a particular design, such as programmable loads, memory, and digital and analog stimuli. The equipment also has a power supply. The programs generally run in firmware. Updates to the program might load externally or require a change of programmable read-only memory (PROM).

Equipment for the I- and D-Levels comprise an entire test station, anywhere from one to five racks and up to seven feet tall, full of spectrum analyzers, waveform digitizers, counter-timers, digital multimeters, and high-power loads. Typically, this type of equipment has circuit cards that plug into a backplane, for example, VME or VXI. Since most of these systems support multiple UUTs, the programs are stored on some form of removable media, and, depending on the system, might actually compile on the station itself. A patch panel (e.g., Virginia Panel or MAC Panel) generally handles all of the interfaces (with the possible exception of high power loads and stimulus) via an interface adapter. The interface adapter takes the hundreds of pins available on the patch panel and routes them to cable interfaces so that cables can be connected from the interface adapter to the UUT. The test station must sometimes supply cooling air or liquid cooling. A test station can have multiple displays; one might be used as the primary interface for the operator while another might display pictures or schematics or wiring diagrams that the operator can use for guidance to perform a procedure. Full keyboards or touch-panel displays are used to input information and run programs on the test station. There may be multiple CPUs to provide all the necessary processing.

9.3.5 Interfaces

Environmental concerns are critical for O-Level equipment. Not only must the equipment survive the environment but also the operator must be able to use it in harsh environments, so operation while wearing winter gloves or chemical/biological protection gear is also necessary. Consequently, it is imperative to keep user interfaces as simple as possible.

9.4 Phases

9.4.1 Concept

Boeing Support Systems can take anywhere from 3 to 12 months depending on the complexity of the project in this phase. Boeing moves through the Concept phase that takes them to the preliminary design review (PDR). The design engineers map the requirements into a workable design and generate block diagrams. The customer will come in for the PDR and the team will present the overall design concept. If the customer agrees that Boeing Support Systems is on the right path, then Boeing moves on to the next phase; if not, Boeing makes corrections.

9.4.2 Detailed Design

This phase can take from 6 to 12 months depending on complexity. During this time, Boeing Support Systems moves through the detailed design phase to the critical design review (CDR). The design team develops the details of the design, such as schematics and wiring diagrams. Boeing develops a full drawing package during this phase. They conduct analyses for reliability and maintainability and vet the overall design. The customer will come in for the CDR and the team will present the final design. If the customer agrees that Boeing has met the requirements, then Boeing moves on to the next phase; if not, Boeing makes more corrections. Once the CDR is complete, Boeing starts buying parts for either the engineering model or for the integration phase.

9.4.3 Engineering Model

Sometimes Boeing builds an engineering model (prototype) to prove out designs and makes sure everything works the way Boeing thinks it should. These prototypes are not always pretty, that is, there are cuts and jumpers, incorrect or nonexistent metal finishes, and so forth. The design continues to evolve, but the basic concept generally stays intact. This is done in a lab environment with UUTs that Boeing Support Systems owns or has borrowed from the customer. Sometimes it requires the team to operate test benches. As Boeing builds the engineering model, the team refines the design and moves toward the first production models. These production models are often used for environmental qualification testing. Once Boeing completes environmental testing, the team might have to refine the design further, but often they are fully into the integration phase by then.

9.4.4 Integration

Usually the designers and the requirements people carry out the integration. The effort is intended to validate the design concepts and make sure that all the requirements are still being met.

9.4.5 Sell-Off

Once integration is complete, Boeing Support Systems does a final design update, if necessary, and goes to "sell-off", which is conducted with the customer. They perform a go-path test with the test equipment in a real-life environment, for example, on the aircraft or with the real UUTs. At the same time, they verify the procedures. If the equipment passes this test, Boeing is off to production.

9.4.6 Timing and Acceptance

From CDR to sell-off can take from 6 to 12 months depending on complexity. A general rule of thumb is that O-Level development takes from 18 to 24 months, and I-Level and D-Level development takes from 24 to 36 months.

Production copies are sold via an ATP, which may or may not be witnessed by the customer. Depending on the size of the production run the design may be built in-house, divided up between in-house and out-of-house, or contracted out as a build-to-print. Depending on the contract with the customer there may or may not be follow-on support for fielded items. On large complex systems there is often some form of sustaining contract in place.

9.5   Scheduling

Boeing Support Systems uses both top-down and bottom-up scheduling. Often customers set the end date and Boeing builds a schedule to try and support that date from the bottom up. If it is not possible to meet their desired end date, Boeing works with them to either decrease the requirements or move out the end date. If no end date is specified, Boeing tends to use bottom-up-type planning.

The deployment date is usually the constraint on the schedule for the item Boeing Support Systems supports. The customer usually requires that the support be available, either prior to the first deployment or in conjunction with it.

9.6   Documentation

9.6.1   Purposes

Documentation serves a number of purposes. It can instruct an operator on the proper procedures to follow when using the equipment, or it can identify parts, such as with a drawing or a parts list, or it can be used in the repair of equipment, for example, schematics.

The documentation goes to various branches of the government for review. High-level description type documents might go to the program office, while lower-level design documents might go to the project engineer. Documentation is prepared to either government standards when specified by a data item description (DID), or to company standards when DIDs are not called for. Documentation can be required during all phases of the program as called for in the contractor data requirements list (CDRL). The CDRL dictates what data are prepared, what format they should follow, when they are submitted, and to which parts of the government they are submitted.

9.6.2   Types

The government has a lengthy list of data items from which to choose [1]. Boeing Support Systems generally deals with drawings, schematics, wiring diagrams, block diagrams, reliability predictions, software source code, testability analysis, acceptance test reports, Acceptance Test Procedures, master program schedules, technical publications, computer program identification number (CPIN) requests, requests for nomenclature, and support-equipment requirements documents (SERDs).

9.6.3   General Formats for Documents

Details for documentation descriptions can found at the Web site in Reference 1.

9.7   Requirements and Standards

9.7.1   Military Standards

There are many military standards that this equipment must follow. Some examples include

- MIL-STD-461 Requirements for the control of electromagnetic interference characteristics of subsystems and equipment
- MIL-STD-810 Environmental test methods and engineering guidelines
- MIL-HDBK-2165 Testability program for systems and equipments
- MIL-PRF-28800 Test equipment for use with electrical and electronic equipment, general specification for
- MIL-DTL-31000 Technical data packages

The U.S. government has been moving away from MIL-STDs over the past decade. Now they tend to put exactly what they want in a statement of work (SOW) or a Procurement Specification (PS) instead of calling out a military specification.

9.7.2   Preparing Requirements

In some cases the customer writes the requirements document for Boeing; in others Boeing writes a SOW or PS or both for the customer. All affected disciplines have input to the documents, which are generally included in the customers request for proposal. Depending on the complexity of the product, this process can take from days to weeks.

9.8   Analysis

Since Boeing's support equipment is used for maintenance, the system being tested either exists or is in the design phase already. Boeing generally does not attempt to do something that has never been done before, so the feasibility of a new project is based on their experience with the system Boeing hopes to maintain. Consequently, the Boeing team determines a design approach based on their knowledge and experience with similar systems that Boeing already supports.

In general, Boeing does not apply heuristics during the design process. They do not maintain a central library of calculations or approximations. They generally do not simulate a design.

In some cases, Boeing will test in conjunction with a development effort. For instance, Boeing may develop a lab unit or prototype for use in environmental testing or for early integration efforts. These assets are used by the engineering community and are sometimes kept around for future developmental or field support needs.

9.9   Design Trade-Offs

9.9.1   Architecture

Boeing Support Systems considers many aspects of architecture issues for support equipment. One consideration is that Boeing's overall goal tends to be higher reliability rather than cool features. Here are a few of the factors they consider:

- Fault isolation and tolerance
- Integration risks
- Power conversion efficiency
- Power consumed
- Cooling
- Data throughput
- Design complexity
- Mass
- Cost

As Boeing's hardware is used for maintenance, the importance each of these factors varies by the level of maintenance. O-Level equipment, for example, must be portable and must run off aircraft power or batteries because facility power is not generally available. It must operate in very hot and very cold environments and must be cooled with ambient air, that is, no external cooling. Reliability and maintainability of the equipment is important because the user may take the equipment to a remote location for a long period of time and provide little or no external support. Boeing usually custom designs the support equipment for each application. Common bus structures (VME, VXI, etc.) are used in some applications but others are custom architectures.

I-Level and D-Level equipment pose a different set of requirements. The equipment usually resides in a facility with cooling air, power, and controlled temperature and humidity. It tends to be large and is not often moved. It can be custom designed or built from existing technology such as a Teradyne 9100 mainframe. The architecture often uses common backplanes (VME, VXI, PXI, etc.). The designs tend to be more complex and costly and take much more time to develop than O-Level equipment.

9.9.2   Hardware

Temperature range is the biggest concern for Boeing for selecting integrated circuit (IC) components. Boeing tries to buy components with the longest life possible. They design all of their support equipment for a 20-year life. Some of it has been in the field longer than that. If necessary, Boeing will buy critical components and inventory them for the life of the project.

Boeing assures EMC through proper circuit board layout and enclosure shielding and gaskets. As in other industries, shrinking design rules within IC fabrication cause concerns for Boeing's support equipment. Boeing hermetically seals switches. They also put an EMI rubber boot on top of them. Lamps and displays have EMI grids.

Vibration is a bigger concern for Boeing Support Systems than shock. They use proper mounting and control of dimensions to minimize damage from vibration. Boeing Support Systems has few problems with temperature since their designs are based on components and subassemblies that are intended for the temperature extremes Boeing expects to encounter.

9.9.3   Power

The input power for O-Level is typically either 28 VDC or 115 VAC, 400 Hz. Equipment for the I- and D-Levels usually requires 115/240 VAC 60/50 Hz power, but it can use 400 Hz or 28 VDC power. All equipment has an input filter on incoming power.

Some O-Level equipment uses standard disposable batteries. I- and D-Level equipment generally does not use batteries.

9.9.4   Cable Harnesses

Boeing Support Systems keys connectors and labels them at both ends of each cable. They also put a label in the middle of the cable. They put a reference for every leg of a cable with corresponding markings on the receptacles. Boeing generally uses scoop-proof MS connectors. Unused holes in the connector are all filled to protect the connectors. The cables are conduit style for both physical and EMI protection. They test each cable for continuity, discontinuity and insulation resistance. Furthermore, each cable is tested during self-test of the equipment.

9.9.5   Cooling

Boeing Support Systems prefers heat sinks over fans in O-Level equipment. Fans are a second choice if necessary. Equipment for both I- and D-Levels generally uses heat sinks and cooling fans. Boeing considers liquid cooling a last resort. Liquid cooling and refrigeration are avoided because of environmental hazards and the mess associated with their use. Fans can be noisy, and hence the preference for heat sinks.

9.9.6   Mechanical Structure

Most support equipment uses one of the standard form factors and backplanes with rigid circuit boards. Rigidity resists problems from vibration. The design often depends on the application and its environment.

9.9.7   Software

Boeing Support Systems develops software according to a software development plan. Key steps in this activity include requirements definition, detailed design specification/review, coding, coding integration reviews, test readiness reviews, final testing, and delivery. They conduct code reviews to verify compliance with coding standards. They also conduct design and requirements peer reviews.

The level of software sophistication and difficulty depends again on the level of maintenance. Software for support equipment at the O-Level tends to be less complex and can be prepared in some of the more simple languages like HP Basic or possibly in assembly language.

I-Level and D-Level equipment tend to use higher level languages. Typically the software is written in ATLAS or C. Boeing Support Systems does not use Ada in their support equipment; Ada is considered an avionics language. The software for I-Level and D-Level equipment tends to be much more complex to develop and time-consuming, but it is still no more complex than the hardware of the support equipment.

Boeing Support Systems develops custom RTOSes because the equipment does not usually run in a standard environment like Windows. Boeing upgrades the software in the field with EEPROMs or some sort of removable media to support the end user in a timely manner. There are no planned maintenance actions unless the customer contracts with Boeing to provide updates. Updates are usually accomplished via an engineering change proposal (ECP) with the customer.

9.9.8   Hardware vs. Software

Flexibility of upgrades is the prime driver for Boeing Support Systems. It is much easier to upgrade fielded software than to upgrade fielded hardware.

Boeing Support Systems tries to make the hardware as robust as necessary to handle future upgrades and do the rest in software.

9.9.9 Buy vs. Build

Boeing Support Systems uses both COTS and custom design. They buy standard COTS equipment like meters, scopes, and instruments on a card. Their custom-built designs are specialized to the application and may contain some COTS hardware. The advantages of COTS equipment is that it readily available and less expensive to buy than building custom equipment. However, COTS equipment is not always capable of satisfying the environmental extremes that may be required. While custom equipment is more expensive and takes longer to develop, its operating characteristics and configuration are more readily controlled. In general, Boeing tries to use COTS when it will meet their requirements without modification. The system architect usually makes those decisions.

9.9.10 Manufacturing

Since Boeing Support Systems makes limited quantities of most items, the fabrication and assembly process is not highly automated. Boeing uses design-for-manufacture (DFM) practices where practical, but since the quantities are low, it is not the most important factor. ATE is particularly used to check cables.

9.9.11 Test and Maintenance

The goal for testing and maintenance is to provide the smallest footprint possible at the O-Level because military personnel transport it to places with ongoing military operations. I-Level and D-Level support and test equipment must provide fast and accurate testing (the faster it can be done, the less equipment and manpower is needed). To address these concerns, Boeing Support Systems incorporates built-in-test (BIT) and self-test provisions in their support equipment for both maintenance and calibration of the test equipment itself.

9.10 Tests

9.10.1 Formal and Informal

Formal testing is only required at the end of the effort. Boeing performs informal testing throughout the process with engineering models and prototypes. All tests are performed by the engineers on the design team. The tests are run from the early design phase throughout the end of the program.

9.10.2   Laboratory Tests

Boeing sometimes prepares lab units or prototypes for use in environmental testing during the development and early integration efforts. There is no set rule (unless required by the customer) for when lab tests are required or not. New or more difficult design issues are generally tested prior to moving from bench testing and integration to aircraft testing and integration due to the potential damage to the aircraft and the relative value of aircraft integration time.

9.10.3   Inspection

Boeing engineers perform inspection during the development phase. Once Boeing is ready to formally demonstrate a product inspectors for hardware quality-assurance get involved. Software and hardware quality personnel take part in the peer reviews during the build process.

A quality check occurs following the completion of each major sub-component (circuit cards, cables, etc.) once a product is moved to production. A quality check takes place again once the subcomponents are assembled into the final product. The customer's quality-assurance representatives perform the final inspections during buy-off.

9.10.4   Peer Review

Most modern military equipment contains both hardware and software. Hardware development is generally done in compliance to an international quality standard such as AS9100. This standard requires design verification and validation. Preliminary and CDRs are held to assure customer requirements are addressed by the design.

Boeing develops software for support equipment in accordance with a software development plan, as mentioned above. The development plan defines participation in the reviews; it usually includes members of the engineering team, the team lead, and sometimes quality assurance, and supplier quality. Peer reviews are performed from requirements definition through final testing, again according to the software development plan. Entry and exit criteria are defined for each phase. In most aerospace companies today, these reviews are in line with the SEI CMMI principles.

9.10.5   Environmental

Environmental tests are part of the requirements dictated by the customer. The environmental tests can include temperature, humidity, EMI, salt/fog, fungus, altitude, vibration, shock, and splash proofing. Some of these tests can be performed in-house, but some must be subcontracted out to vendors. The engineers generally participate in putting the test plan together, and in some cases they participate in the tests themselves. These tests are generally performed on a preproduction or prototype unit; they take place either in serial sequence as part of the production program or in parallel with the production program, depending on the schedule and customer requirements.

9.10.6 Manufacturing

There are no specific types of manufacturing tests required other than inspections and successful completion of the ATP (see earlier).

9.10.7 BIT, BITE, and Simulators

Boeing Support Systems uses equipment to simulate aircraft equipment to reduce use of the aircraft time. There are really no other types of simulators or simulations required. Boeing designs BIT into the product during the design phase. The design engineers prepare the BIT.

9.11   Integration

Boeing usually integrates hardware and software together for O-Level equipment. Integration activity generally occurs in three phases.

The first phase is when the hardware and the software are first married together. Lab tests are run to see that the proper inputs and outputs are generated when required.

Once the equipment seems to be operating properly, Boeing moves the equipment to bench integration. This phase occurs on a test bench that simulates the various aircraft systems exercised by the support equipment. These tests represent a realistic environment for a high percentage of the aircraft systems.

Finally, once the bench integration is complete, aircraft integration begins. Boeing connects the equipment to the aircraft during this phase. Boeing performs environmental testing either in parallel or in conjunction with the integration phase. Field testing is not generally required.

9.12   Manufacturing

Boeing Support Systems can assemble circuit cards, make cables, produce simple metal work, and other similar type activities. Boeing Support Systems does not manufacture components, connectors, or printed circuit boards (PCBs). Most of the basic building materials are purchased from outside suppliers. Contracting with outside suppliers is done in two primary ways: build to print and unique design.

Boeing Support Systems does not use a dedicated in-house manufacturing facility for its military test equipment. Boeing relies on outside manufacturing sources for larger production quantities rather than attempting to manufacture those within the engineering labs. During the initial design phase, Boeing prepares a procurement package, which then forms the basis for a competitive procurement among potential suppliers. The engineering team works with the selected supplier during the manufacturing phase.

9.13   Support

Support equipment itself is not generally large enough to warrant an Opeval or a Techeval on its own. It can often be part of a larger program that does have those phases of evaluation. If Boeing introduces the support equipment on its own, it can be an On-Site Verification (OSV) or some other type of field introduction. Both engineers and technicians support an OSV, especially if it is a retrofit to existing equipment. Once an OSV is completed, authorities at the operational site generally sign off on the installation. The introduction is almost always completed in the field at the customer's site.

Engineers, technicians, or a field support organization can provide support of the delivered equipment. Boeing Support Systems does not maintain a website or a 24-h helpline. The users call Boeing directly when they need assistance. Boeing also supplies extensive manuals so that the user can provide his own technical support to the greatest extent possible.

9.14   Disposal

The military takes care of the disposal of the equipment when they are done using it; so there is no burden to Boeing Support Systems.

9.15   Liability

In dealing with military equipment, both personnel and equipment safety are primary concerns. Boeing Support Systems designs all test equipment to rigid safety requirements; the system safety group performs system safety analyses at all levels during the design and qualification process.

9.16   Summary

Boeing Support Systems builds support equipment with embedded real-time systems that test military equipment—particularly aircraft systems. The volumes produced are quite small, 50–100 units for any particular piece of equipment. The emphases are dependable operation (long periods of uninterrupted functioning without failure) and extreme environments. The equipment must provide simple and robust operation.

Acknowledgment

My thanks to Tim Murphy at Boeing Support Systems, a division of Boeing Integrated Defense Systems, for providing the information for this chapter.

Reference

1. U.S. Government requirements for documents: http://dodssp.daps.dla.mil/assist.htm

10

Case Study 7—Designing Instruments for Space Flight

10.1 Concept and Market

10.1.1 Who, What, Why, How, Where, and When

This chapter provides a general format for designing instruments for spacecraft. The two chapters that follow will provide specific instantiations of this format. I have drawn on several decades of experience from different people for this chapter and case study.

Often a team of 20 or 30 or more work, at least part time, on any instrument destined for space flight. The team comprises scientists, engineers, technicians, fabrication personnel, and administrative staff. The effort to develop the instrument can take anywhere from 2 to 4 years. The fabrication requires a number of resources including circuit board fabrication and assembly, machine work, and test facilities; these activities can take place in many different locations.

10.1.2 Economics

Instruments that fly on spacecraft take time and money to design, fabricate, and operate. Many instruments that orbit the Earth on a satellite or fly into the Solar System on a spacecraft cost between US$5MM and US$50MM. Most spacecraft have multiple instruments to maximize the opportunity to gather unique science data. This drives the cost of the satellite up to 100s of millions of U.S. dollars. Furthermore, a launch vehicle can easily cost between US$100MM and US$200MM.

Obviously, low-altitude missiles and sounding rockets can host very-short-term missions for considerably lower cost. They do not have the same constraints for radiation hardness or low-power consumption that longer-term satellite missions do.

The final cost of a space instrument depends heavily on the set of features and specifications. If features and capability are minimized and the time stretched out so that fewer people need to work on the instrument, then its final cost can be lower. Full features and capability and a short duration for development drive the instrument costs much higher.

10.2 People and Disciplines

You will need a variety of people and expertise to staff a team to design, develop, and fabricate a space instrument. The project usually begins years before as a proposal from a scientist or a group of scientists. Once the project is funded, these folks work in concert with the design and fabrication teams and communicate often with your teams. They also participate in all the design reviews.

First, you need a program manager, a lead project engineer, and a systems engineer. These folks may have trained as aerospace, mechanical, electrical, or systems engineers; they should, however, have experience with designing and building space instruments. The program manager handles the administrative side of things: the scheduling, estimation, and resource management. The lead project engineer is responsible for the architecture of the instrument and preparing the specifications and plans and monitoring the test results. The systems engineer should know the spacecraft and its interface with the instrument.

Next, you will need a design team to work together and to take responsibility for the instrument:

- Software engineers to develop the code for the instrument
- Electrical engineers to design the hardware
- Component engineer to find, test, and inventory space-qualified components
- Hardware or software engineers to develop the ground support equipment (GSE)
- Mechanical engineers to design the chassis and mechanical mechanisms
- Administrative staff to support them

Test engineers, technicians, and a fabrication team also work closely with the design team.

For fabricating a space instrument, you will need some or all of the following people:

- Computer aided design (CAD) designers
- Technicians to fabricate circuit boards
- Machinists for the fabricating the mechanisms and chassis
- Assembly personnel to solder the circuit boards
- Assembly personnel to fabricate the cable harnesses
- Technicians and assembly personnel to integrate the instrument on the spacecraft

- Quality workmanship inspectors
- Administrative support

For testing a space instrument, you will need some or all of the following staff:

- Technicians to perform the thermal vacuum tests
- Technicians to perform the shock and vibration tests
- Technicians to run the end-to-end systems tests
- Administrative support

10.3 Architecting and Architecture

10.3.1 Process

You generally only get one opportunity to get a space instrument right, although reprogrammable logic, field-programmable gate arrays (FPGAs), and computer-based controls are removing these limitations; on-orbit corrections can now be performed. The development is mission-critical and it must reflect that reality. A V-model development process is best for a system that is programmed once. A spiral model might work for on-orbit reprogramming.

10.3.2 Parameters

A space instrument can have many different parameter types, including:

- Environment—temperature extremes and cycles, shock and vibration, radiation
- Mechanical—size (volume), configuration, mass
- Cooling—thermal conduction and dissipation
- Power—consumption, fault tolerance
- Electromagnetic compatibility (EMC) and electromagnetic interference (EMI) margins and tolerances
- Data—memory size, throughput and channel bandwidth, fault tolerance to improper communications
- Command and data handling—telemetry to the GSE on Earth, control to the processor
- Mechanical operations—stepper motors or brushless direct current (DC) motors to move filter wheels, mirrors, scan the instrument, spin momentum wheels, etc.

10.3.3   Architecture

There are two basic configurations for a space instrument. One configuration is a fairly "dumb" instrument (or sensor) that is tightly coupled to a central processor; this is the centralized approach. The other is a distributed approach, which uses "smart" instruments (or sensors), each with its own embedded processor, all networked together. Both configurations have their advantages and disadvantages.

A centralized approach can be the simpler and lower mass configuration if there are only a few instruments (sensors) on the spacecraft, usually fewer than five. The cables may be bulky, but they are limited in number. This approach can optimize the size and number of components (fewer and smaller than with a distributed approach if only a few sensors are used) and can be more efficient in power consumption. Its downfall comes when sensors increase in number beyond a handful. Then it becomes unwieldy and heavy; moreover, fault tolerance becomes far more difficult to achieve.

A distributed approach is better for larger configurations with many instruments (sensors). This configuration has smaller and lighter cables and makes fault tolerance somewhat easier by isolating failures. A network cable for data transfer and command handling between each instrument (sensor) can also be significantly lighter than all the cables needed for the point-to-point scheme of a centralized configuration. A distributed approach can ease the integration of sensors through a "plug-and-play" philosophy.

10.3.4   Interfaces

A space instrument has a number of specific interfaces:

- Mechanical—to the spacecraft
- Electrical signaling—command and control signals, connectors
- Power—raw or regulated VDC from the spacecraft
- Thermal—thermal paths from the instrument frame to the spacecraft
- Data—formats of commands, signals, and data transfers

10.4   Phases

10.4.1   Concept

During the Concept Phase, the design team should establish the mission goals, objectives, and constraints. The team needs to understand the requirements of the project and demonstrate that the proposed architecture will meet these requirements.

The conceptual design review (CoDR) concludes the concept phase. During the CoDR, the design team should present the following:

- Program organizational structure
  - Organizational interfaces
  - Schedule
  - Cost
  - Policy
- Review mission objectives and science goals
- Requirements
  - Mission: environment, host resources, science requirements
  - Performance: technical characteristics
  - Major instrument function and interfaces
- Research—literature, patent searches
- Design constraints and major trade studies performed
- Requirements process and management
- System architecture
  - Concept
  - Hardware components
  - Software components
  - Operations concept including the GSE
  - Support systems and logistics
- Planned test program
- Planned integration
- Development drivers
- Risk assessment

The output of the CoDR will constrain the baseline design following the closure of any action items resulting from the review. Long-lead items, development-support equipment, breadboard parts, and materials can be purchased following the successful completion of the CoDR.

10.4.2 Preliminary Design

During the preliminary design phase, the design team should prepare the design and interfaces with block diagrams, signal flow diagrams, schematics showing logic diagrams, first interface circuits, packaging plans, configuration and layout sketches, preliminary analyses, and modeling. The design team should have established the estimates of weight, power, volume, and the basis for the estimates. The design team should also have prepared the mechanical, power, thermal, and electronic designs with load, stress, margins, and reliability assessments. The software engineers should specify the software requirements, design, structure, logic flow diagrams, computational loading, design language, and development systems.

The preliminary design review (PDR) concludes the preliminary design phase. It is the first major review of the detailed design and will be held prior to the preparation of most of the formal design drawings and software code development. The PDR is held when the design advances sufficiently to begin some breadboard testing or the fabrication of engineering models.

During the PDR, the design team should present the following:

- Technical objectives, requirements, general specification
- Closure of action items from CoDR
- Completion of research, trade-offs, and feasibility
- Requirements—function, performance, interface
- Analyses
    - Mechanical/structural design, and analyses
    - Weight
    - Power
    - Electrical, EMI/EMC
    - Thermal paths
    - Radiation design and analyses
    - Data rates, throughput, bandwidth, and commands
- Software requirements and design
- GSE design
- System performance budgets
- Design verification, test flow, and test plans
- Host interfaces and drivers
- Parts selection and qualification
- Event tree analysis (ETA), failure modes effects analysis (FMEA), and fault tree analysis (FTA)
- Contamination requirements and control plan
- Quality control, reliability
- Materials and processes

The completion of the PDR and the closure of any action items generated by the review provide the basis for the start of the detailed design effort and the purchase of parts, materials, and equipment.

10.4.3   Critical Design

During the critical design phase, the design team should complete all the parts of the design and interfaces: mechanical, power, thermal, and electronic designs with load, stress, margins, and reliability assessments. The software engineers should have written and tested all code; if the development is spiral, then the software engineers should have completed the first cycle and the software should be in a stable, functional state. The GSE should be specified, designed, and well on its way to being coded.

The critical design review (CDR) concludes the critical design phase. It will be held near the completion of engineering evaluation using the breadboard model of the project. It should be held prior to any design freeze and before any significant fabrication activity begins.

The CDR should present all the same basic subjects as the PDR, but in final form. During the CDR, the design team should present the following:

- Closure of action items from the PDR review
- Changes from the PDR review
- Final parts list
- Final implementation plans including: engineering models, prototypes, flight units, and spares
- Final software design and process implementation
- Final GSE design and process implementation
- Engineering model and breadboard test results
- Design margins
- Completed design analyses
  - Mechanical/structural design, and analyses
  - Weight
  - Power
  - Electrical, EMI/EMC
  - Thermal paths
  - Radiation design and analyses
  - Data rates, throughput, bandwidth, and commands
- Safety Requirements
- Operations Plan
- Updated ETA, FMEA, and FTA
- Qualification
- Test
  - Plans
  - Status of procedures and verification plans

- Test flow
- Schedule
- Documentation status
- Test history of the hardware
- Product assurance
- Previous anomalies, deviations, waivers, and their resolution
- Identification of residual risk items
- Plans for shipping containers, environmental control, transportation

Completion of the CDR and resolution of all the action items generated by it constitutes the baseline design.

10.4.4   Fabrication

During the fabrication phase, the design team should confirm that all components are tested and that the results are acceptable. The team should also complete the integration plans. The GSE should be completed with sufficient capability to aid the integration of the instrument to the spacecraft. The team should ensure that the design of the instrument has been validated through the environmental qualification and the acceptance test program, that all deviations, waivers, and open items have been satisfactorily closed, and that the project, along with all the required support equipment, documentation, and operating procedures, is ready for integration.

A fabrication review can conclude the fabrication phase or can be combined with the pre-environmental review (PER) after integration. If a Fabrication Review is held, here are some items to address:

- Rework/replacement of hardware, regression testing, or test plan changes
- Compliance with the test-verification matrix
- Project assessment of any residual risk
- GSE status
- System integration support plans

10.4.5   Integration

During the integration phase, the design team should confirm that the instrument integrates into the spacecraft and that the results of both the end-to-end system tests and the environmental tests are acceptable. The PER occurs before the end-to-end system tests and after the environmental tests. The pre-ship review (PSR), which signals the preparation to move the spacecraft to the launch facility, ends this phase. Some items that should considered in the PSR:

- Measured test margins versus design estimates
- Demonstrate qualification/acceptance temperature margins
- Trend data
- Total failure-free operating time of the item
- Could-not-duplicate failures should be presented along with assessment of the problem and the residual risk that may be inherent in the item
- Project assessment of any residual risk
- GSE status
- Review shipping containers, monitoring/transportation/control plans, postshipment plans

The GSE team will train the operators of the GSE to access data from the instrument and send commands to the instrument. The remainder of the integration phase has to do with the spacecraft and its attachment (integration) to the launch vehicle or missile.

10.4.6   Launch and Mission

Most of this phase is taken up with spacecraft concerns and not the instrument. During this time only the GSE operators are active. Once the mission begins, the science team will collect and analyze the data from the instrument. These data pass through the GSE to the science team; the scientists often communicate frequently with the GSE operators, if not actually operate the GSE to control the instrument themselves.

If possible, try to debrief the project to record lessons learned. Few people or organizations do this, but a record of lessons learned is the single most important tool to improve effort and quality on future projects.

10.5   Scheduling and Estimating

The project always has one important deadline, the launch. In some projects, the launch window is only open for a few weeks once a year or once a decade. Delaying the launch to complete your instrument development or testing is not an option.

Planning for a space mission requires top-down planning because everything flows backwards from the mission objectives and the launch date. Then National Aeronautics and Space Administration (NASA; or ESA in Europe) will constrain the schedule further with more deadlines, such as spacecraft integration, environmental tests, and availability of the launch pad and its facilities.

Once you have the mandated deadlines, carefully plan for each phase and leave margin for contingencies (fight for these—you will need them and sponsor organizations seldom recognize the need). Use management planning software, such as Microsoft Project™ to perform bottom-up planning to meet these deadlines. I suggest that you avoid loading anyone on the project more than 50% of their time; going over this limit often overloads people. This is a general heuristic that full time effort is really about 50% of a person's time at work; so much of what we do on the job is not directly productive for the project at hand.

Never use more than two digits of precision in high-level analyses and presentations. Few people, if any, ever do any better than one digit of precision in any estimate of effort and cost and time.

Here's another problem: engineers will rush a design into fabrication under pressure from management to demonstrate progress on the schedule to the sponsor, but then end up changing the design repeatedly. This forces the fabrication team to start the entire process over each time—printed circuit board (PCB) fabrication and inspection, assembly and inspection, and finally circuit test and verification.

10.6   Documentation

10.6.1   Purposes

As mentioned in Chapter 1, documentation for a space instrument serves three purposes:

- To record the specifics of development
- To account for progress,
- To instruct the use of the instrument.

Hopefully the record will be useful for future projects, but it is also necessary to survive quality audits should one be leveled on your organization. Documentation certainly is needed to account for progress in developing a space instrument—otherwise, design reviews would go out of fashion. Finally, documentation of the space instrument can reveal the extent of its utility and capability.

10.6.2   Types

Table 10.1 lists the major documents. In addition, you will have the standard stock-in-trade documents, for example, engineering notebooks,

TABLE 10.1

Example List of Documents for Developing a Space Instrument

	CoDR	PDR	CDR	PER	PSR	Designated Author
Mission-Level documents						
Concept of operations document	D	F	△			Sponsor organization
Mission requirements document	F	△				Sponsor organization
Host spacecraft documents						
Spacecraft Interface Control Document (ICD)	D	F	△			Systems engineer
Safety data package			D	F	△	Systems engineer
Instrument documents/databases						
Project plan	F					Systems engineer
Configuration management plan	F	△	△			Systems engineer
Problem resolution plan	F	△	△			Systems engineer
Infrastructure plan	F	△	△			Systems engineer
Product acceptance plan	F	△	△			Systems engineer
Risk management plan	F					Systems engineer
• Risk management plan	D	F	△			Systems engineer
• Risk management database	△	△	△	△	△	Systems engineer
• Risk watch list	△	△	△	△	△	Systems engineer
• Fault tree analysis (FTA)	I	D	F	△	△	Systems engineer
• Failure modes effects analysis (FMEA)		I	D	△	△	Lead engineer
Development plans	D	F				Systems engineer
• Architecture development plan	D	F	△			Lead engineer
• Software development plan	D	F	△			Software lead
• Electronics development plan	D	F	△			Hardware lead
• Mechanical Packaging	D	F	△			Mechanical lead
Development plan						
• GSE development plan	D	F	△			GSE engineer

(Continued)

TABLE 10.1

Continued

	CoDR	PDR	CDR	PER	PSR	Designated Author
Quality assurance plan	D	F				Component engineer
• Parts control plan	D	F				Component engineer
• Parts inventory list		D	F	△		Component engineer
Requirements document	D	F	△			Systems engineer
Compliance matrix	I	D	F	△	△	Systems engineer
EMC/EMI test plan	I	D	F			Systems engineer
Test plan	I	D	F			Systems engineer
• Test procedures			I	F	△	Lead engineer
• Test results					F	Test technicians
• Integration procedures			I	F	△	Lead engineer
• System verification & validation (V&V)					F	Test engineer
Documentation plan	D	F	△			Systems engineer
Documentation release schedule		D	F	△	△	Systems engineer
Software users manual/maintenance documents		I	D	F		Software lead
Training manual		I	D	F		Systems engineer
Electronic design documents and schematics	D	F	△			Hardware engineers
Mechanical design documents and schematics	D	F	△			Mechanical engineers
Software design documents and source code	D	F	△	△		Software engineers
Fabrication plan and databases	D	F	△			Lead engineer
• Vendor data	D	F	△			Lead engineer
• Bill of materials (BOM)	D	F	△			Lead engineer
• Fabrication and assembly instructions	D	F	△			Hardware engineers
• Inspection reports			△			Inspectors
Design reviews and reports	△	△	△	△	△	Systems engineer
Signature list	I	F				Systems engineer
Action item database	△	△	△	△	△	Systems engineer

Key: I: Initial development—A full outline of the document has been established. Writing of some sections has begun. D: Complete draft—The document is completely written and is undergoing review. A very small number of TBDs can remain, but these are limited to specific pieces of information, not entire sections or subsections. F: Released final version—Completed initial release. △: Updates to released version—Re-released with changes. CoDR: Conceptual Design Review. PDR: Preliminary Design Review. CDR: Critical Design Review. PER: Pre-Environmental Review. PSR: Preship Review.

e-mail messages, memos, letters, project documents, and manuals. As always, engineering and management presentations will be numerous, and they will become a part of your documentation load.

10.6.3 General Formats for Documents

The appendices have sample documents and plans. The project plan can combine all the system engineering issues and project concerns into one document, as follows:

Project Plan

1. Introduction
   1.1. Purpose
   1.2. Scope
   1.3. Definitions, Acronyms, and Abbreviations
   1.4. References
   1.5. Overview
2. Project Overview
   2.1. Project Purpose, Scope, and Objectives
   2.2. Assumptions and Constraints
   2.3. Project Deliverables
   2.4. Evolution of the Project Plan
3. Project Organization
   3.1. Program Structure
   3.2. Organizational Structures
   3.3. External Interfaces and Organizations
   3.4. Roles and Responsibilities
      3.4.1. Program Manager
      3.4.2. Project Lead Engineer
      3.4.3. Systems Engineer
      3.4.4. Hardware Engineering
      3.4.5. Software Engineering
      3.4.6. Mechanical, Packaging, and Thermal Engineering
      3.4.7. Fabrication Engineering
      3.4.8. Parts Quality Assurance
4. Management Process
   4.1. Project Estimates
   4.2. Project Plan
      4.2.1. Phase Plan

10.7 Requirements and Standards

10.7.1 NASA and Military Standards

Several different sets of regulations and standards can apply to instrument design for spacecraft. Table 10.2 lists a few examples of those standards.

10.7.2 Preparing Requirements

You need to work with your sponsor organization, whether it is NASA, ESA, government, or a commercial space contractor, to develop the

TABLE 10.2

Examples of Some Standards That Might Apply to Developing a Space Instrument

Category or Source	Standard	Description
NASA	NASA RP-1124	Outgassing data for selecting spacecraft materials
	311-INST-001	Instructions for EEE parts selection, screening and qualification
	PPL 21	GSFC preferred parts list
	GEVS	Appendix A: General Environmental Verification Specification
US Military	MIL-STD-461E	Electromagnetic emission and susceptibility requirements for the control of electromagnetic interference, Part 3 for class A2 equipment
	MIL-STD-462 (Notice 2)	Measurement of electromagnetic interference
	MIL-STD-1540C	Test requirements for launch, upper stage, and space vehicles, September 15, 1994
	MIL-STD-883	Test method standard, microcircuits
	MIL-B-5087B	Bonding, electrical, and lightning protection for aerospace systems
ISO	AS9001	Quality Systems Aerospace standard (aerospace equivalent of ISO9001)

requirements. The science team, who probably originated the proposal for the instrument, will be intimately involved as well.

10.8 Analysis

10.8.1 Feasibility

The science team poses a scientific problem that an instrument might solve. The engineering design team determines if it can be done or if only reduced capabilities are possible. The team determines feasibility through experience, calculations, and trade studies.

The primary concerns for feasibility are resolution, accuracy, speed, power consumption, memory capacity, data throughput, weight, and volume. The engineering team typically comprises project lead engineer, the systems engineer, and an instrument engineer. Feasibility is finalized during the concept phase.

10.8.2 Heuristics

Heuristics or rules-of-thumb are one of the most valuable means to analyze feasibility, lay out the design concept, and estimate effort and schedule. Here are samples of heuristics that might be used.

Management issues:

- Expect to spend about US$1MM/kg to develop an instrument for space [1]
- Expect an instrument to consume power at a rate of about 1 W/kg [1]
- Simple things like cables and connectors and alignment cause many problems.
- Have a checklist for everything.

Antennas and measurements:

- Sensitivity is proportional to weight and size; higher sensitivity requires more weight and greater size.
- Light sensitivity in optical instruments is proportional to the size of the aperture.
- Measurement of electrical fields is proportional to the size of the antenna.

Power supplies are always a problem:

- Low-voltage power supplies regularly have anomalies in power management, DC–DC conversion, and distribution.
- High-voltage power supplies always have to deal with contamination; outgassing can sustain a plasma, which allows corona discharge and arcing.

Optics:

- Optical contamination is always an issue—the whole spacecraft has to be cleaned to protect lenses and mirrors.
- Molecular contamination due to outgassing condenses on every cold surface including mirrors and lenses, which results in smears in images.
- Particulate contamination causes glint in optics.
- If optics can point at the sun, they will. (Can the instrument do it safely? Sensitive optics may not even be able to view the Earth.)
- Optical detectors usually need cooling. (Do you passively couple to the cold side of the spacecraft? Or use thermoelectric coolers? Or fly cryogens to cool the detector?)

EMC:

- When measuring particles or electric fields, then external biases or magnetic fields (either static or dynamic) from the spacecraft will require various fixes.

- Material surface properties will cause the spacecraft to accumulate charge and thereby distort measurements of plasma or electric fields. Low Earth orbit has less of this problem (the residual atmosphere helps). Geosynchronous orbit is a bigger problem. Finally, light or sha-dow affects the charge buildup due to photoemissions   and photoelectrons.

- Internal EMC—battery currents and switching noise will couple into sensitive analog circuits and detectors.

10.8.3 Calculations

Calculations are the center of most designs, for example, weight distribution, center of mass, power consumption, focal length (for optical instruments), and level of effort, to name just a few issues. Calculations for mean time between failures (MTBF) can approximate reliability; MTBF is really only good for comparing design approaches. FTA and FMEA can provide an indication of failure modes and fault tolerance.

The systems engineer, project lead, and hardware engineers may all be involved in preparing these calculations and analyses. Most of these calculations are done in the concept and preliminary design phases.

10.8.4 Numerical Simulations

Some aspects of spacecraft instrument design cannot be verified on Earth, so you can only model and simulate. Mission profiles and orbital mechanics directly affect instrument design; they are extraordinarily challenging and sometimes receive less attention then they should.

Certain other aspects require huge efforts to verify; if so, simulation becomes a tool of expediency. Simulating the thermal conduction paths and EMC/EMI effects of an instrument are two examples.

Specialists on the development team handle these numerical simulations. Most simulations occur during the concept and preliminary design phases. They often serve to clarify feasibility.

10.8.5 Testing

Many different types of tests can reveal design concerns. One example of testing for analysis is the building of an engineering model of the instru-ment before space-qualified fabrication. An engineering model can help find problems, such as a power-up glitch in specific models of FPGAs that drew too much current from the power supply of a recent spacecraft. Tests with engineering models can also find incompatibilities between the in-strument and the spacecraft of the mechanical attachment, thermal con-duction, and data protocols.

Primarily the hardware engineers on the development team handle these engineering model tests. Most such tests occur during the preliminary design phase.

10.9 Design Trade-Offs

10.9.1 Architecture

Every mission is different, which means the instrument design is nearly always custom. In balancing scientific objectives, the architecture of the spacecraft is always a compromise and will drive the architecture of the instrument. In the Chapter 12, I compare architectures for a subsystem's power distribution and data networking between a traditional, centralized "star" configuration and a distributed configuration. Many smaller satellites benefit from the optimizations brought by a centralized "star" configuration. Larger satellites and spacecraft tend to benefit more from a distributed approach.

A centralized "star" configuration can allow for a less "smart" but more optimized and tightly integrated instrument. Its disadvantages are that it is often less fault-tolerant, more complex to isolate failures, and more difficult to integrate.

A distributed configuration is more easily developed to isolate failures, tolerate faults, and integrate. Its disadvantages are that it is often more complex.

Regardless of configuration topology, the design team will have to calculate:

- Mass
- Power-conversion efficiency
- Power consumed
- Cooling needed
- Cost
- Data throughput
- Design complexity
- Integration risks
- Fault isolation and tolerance

There are design references for space instruments and spacecraft that will help you [2, 3].

10.9.2 Electronic Hardware

Space flight places special demands on circuit design. The circuit components, particularly transistors, and integrated circuits (ICs), must be radiation-hard, and they have to minimize power consumption. Weight and power constraints of space flight keep memories small. Circuit boards and components should not outgas either—this can lead to coating critical subsystems, such as optics, with an undesirable material that alters their operations. Finally, all space-qualified components and subsystems are long-lead items—typically 6 months to one year to deliver.

Radiation hardness: Radiation has three primary effects on electronics: total ionizing dose, displacement damage, and single-event effect (SEE). The total ionizing dose is the accumulated exposure to cosmic rays and energetic particles that eventually degrade a component drive to outside its design range. Displacement damage results from prolonged exposure to low-energy particles; they tend to create crystalline defects that increase the resistance of the device. An SEE results when a single energetic particle encounters the IC; there are a variety of SEEs [4]:

- Single-event upset (SEU)—the passing of the particle causes a change in the logic state.
- Single-event latch-up (SEL)—the passing of the particle not only causes a change in the logic state, it activates a parasitic circuit between power and ground that can destroy the IC if the parasitic circuit is not current limited. Sometimes, turning off the power and then back on can clear the SEL.
- Single-event burnout (SEB)—an SEL does not clear and then destroys the device. Such events can happen in power metal-oxide semiconductor field-effect transistors (MOSFETs).

Reference 5 is a good source concerning radiation effects and analysis. ICs fabricated as silicon-on-sapphire or silicon-on-insulator tend to resist SEE problems.

Outgassing: Choosing the appropriate materials can significantly reduce outgassing. Avoiding certain thermal greases and lubricants is important. Space-qualified polymeric encapsulants for the circuit boards, connectors, and cables will also help.

Conduction cooling: Almost all circuit design relies on conduction cooling. This means that heat travels through the materials, casing, and enclosure to dissipate away from the circuit source. Components must have packaging with low thermal resistance to allow conduction cooling.

Processor trade-offs: Besides being radiation-hardnened, the chosen processor needs to have sufficient but not excessive computational power; this usually helps maintain low power consumption. Another important aspect in choosing a processor is whether there is corporate familiarity for working with its software development tools; components may have long-lead times, but learning new tools can take even longer.

Support peripherals tradeoffs: Space components and ICs are unlike those in terrestrial applications and markets, where many different models of the same processor incorporate a variety of peripheral functions. Most space-qualified components have older architectures that have a long-established history, and they tend to not have high levels of integration. You will need to find peripheral components to support them—things like direct memory access (DMA), memory, timers, analog-to-digital converters (ADCs), and digital-to-analog converters (DACs).

Radiation-hard peripheral components are out there but they won't be the lowest power, fastest speed, or highest resolution. The variety in support components is small. You may need to design functions in rad-hard FPGAs to support the main processor.

Memory trade-offs: As is true with the support peripheral ICs, so it is with memory. Space-qualified memory is not the densest, largest, or fastest available. You will need to do smart system and circuit design to get the necessary functions to reside in the memory available.

10.9.3 Power

Power systems in instruments and spacecraft tend to be the source of a surprising number of problems [6]. A well-designed instrument can help avoid some of these problems through early recognition of them and an effort to design carefully. Focusing on the DC–DC converters and their interactions with the power distribution system should help.

DC–DC converter trade-offs: Several small companies build space-qualified DC–DC converters. Different models have different levels of radiation tolerance. Sometimes they are available off the shelf, but like most space-qualified components or subsystems, they usually have long lead times for delivery. Their biggest weakness is the type of power-switching transistors used, which determines the switching frequency and hence the efficiency. Power transistors are susceptible to both total-dose and SEE.

10.9.4 Electromechanical Hardware

Sometimes a space instrument needs mechanical movement, maybe to scan a mirror, rotate a filter wheel, or open a door. An electric motor can generate many of these motions. In selecting a motor, you need to be aware of several concerns:

- The insulating epoxies used on the motor windings must not outgas.
- Thermal conduction paths must be short, e.g., through the motor's case to a cold plate. There is no convection cooling in space.

- Bearings must be sized for the driven mass and for the vibration of launch.
- Lubrication must not outgas and the separators between the balls and the race within the bearings must be compatible with space applications.

You have the choice in motor types between step (or stepper) and brushless DC. Brushless DC motors have two concerns over step motors for space applications:

- The integrated Hall sensors for the commutation may not be radiation-hard.
- Cogging is always a problem for smooth rotation.

Mechanical design for space needs to be appropriately rugged to survive the vibration of launch. The design must also account for temperature extremes and swings to avoid binding or fracture due to expansion and contraction.

10.9.5   Cable Harnesses

Cables and connectors are another perennial source of problems in spacecraft and instruments. Low-tech problems such as connector keying, alignment, and attempted male-to-male connector matings do happen far too frequently. Conscientious and clear instructions and proper oversight can reduce these problems significantly.

Routing cable harnesses on a spacecraft to instruments always remains a challenge. A physical mock-up, made of wood and cardboard, is invaluable to help with orientation, alignment, and tie-downs.

Space-qualified connectors are only allowed a small number of mate and de-mate cycles, often 5 or 10 cycles. Connector savers, which are two connectors—one male, one female—soldered back-to-back, attach to the flight hardware and remain there until final system integration. This allows simulators, other instruments, test equipment, and engineering boards to attach to the flight hardware for testing.

10.9.6   Cooling

Removing heat from circuitry in a spacecraft has very few options. Realistically all you can do is conduct the heat through a base plate into the spacecraft. If you have a lot of heat to dissipate, heat pipes or a thermoelectric cooler may be needed to drive a large radiator. In the extreme, you may have to resort to cryogenic cooling, but spacecraft only have limited capacity for storing cryogens.

Most circuit designs rely on conduction cooling. Hot components or those that dissipate more power need special attention—they might need metal conduction fingers that extend to the enclosure casing, or even a heat pipe. Even a few watts of dissipation are a major concern for a spacecraft.

10.9.7 Mechanical Structure

Most instruments use either a backplane with rigid circuit boards or a stacked "sandwich" chassis (described in Chapter 12). Often these structures are custom machined from blocks of aluminum alloy. Some experimentation with carbon fiber is being done for larger structures where stiffness and size with low weight are needed, for example, booms, solar cell panels, and sun shades.

10.9.8 Software

Wars are fought over software languages, so I won't even go there. What I do know is that good processes for developing software are a must: careful design, regular design reviews, regular code inspections, and defined, rigorous tests to verify functionality. Software design should be modular and well-documented.

More and more often, a real-time operating system (RTOS) is becoming necessary, both in the development of the software and to assure proper operations. A commercial RTOS can take less time to install and run, and it can sometimes have fairly decent technical support from the vendor. Fault tolerance is a major concern for space instruments, and the RTOS must handle many different situations and anomalies gracefully. Unfortunately, many commercial RTOSs have either too many features or take too much memory to be used in a space instrument.

10.9.9 Ground Support Equipment

The GSE is specific equipment that tests and then supports your instrument during the mission. Often GSE is a high-end desktop computer that can receive, display, and store the data sent down from the spacecraft. Occasionally, you might have to design a piece of hardware to swallow large amounts or specific types of data. Generally, most of the effort in developing GSE lies in programming the system to handle your data.

10.9.10 Buy vs. Build

Most space instruments must be radiation-hard or -tolerant, have low mass, and consume low power, all of which indicate custom design. Typically, the only commercial off-the-shelf (COTS) components might be the RTOS software, the GSE, and maybe some of the GSE software.

10.10 Tests

10.10.1 Laboratory Tests

Hardware engineers usually prototype circuits and modules on the laboratory bench. Prototypes help prove concepts and confirm designs. GSE engineers might occasionally need to build prototypes. Most prototyping occurs during the conceptual design phase.

During the preliminary and critical design phases, design engineers can use engineering models, which are circuit boards and chassis fabricated to space-qualified standards but using non-space-qualified components with the same electrical properties as space-qualified components, but without all the process and inspection steps. Such models are particularly useful to study interactions between boards and subsystems in a similar configuration to the flight version. Software engineers can implement and test functions that might not otherwise be easy to simulate.

Engineering models also provide a good approximation to the flight instrument for GSE engineers. These models can generate data that replicates the operation of the flight instrument.

10.10.2 Peer Review

The software engineers and the GSE engineer should perform regular code reviews to confirm the functionality and quality of the software. The entire system should undergo regular design reviews (CoDR, PDR, CDR, etc.) as should reviews for fabrication. The entire team should be involved in most design reviews. Design reviews occur at the end of each phase; sometimes targeted reviews to examine important issues will occur during the middle of a phase.

10.10.3 Subsystem Tests—Hardware

These tests should be developed as part of the design process. They should test power-up operations, simulated fault conditions, the power distribution system, the command and data communications, and built-in-test (BIT) or diagnostics.

Hardware engineers and technicians design and run these tests. They should design the tests during preliminary design and refine them during critical design. They should run the tests on the engineering units during the critical design phase and on the flight units during the fabrication and integration phase.

10.10.4 Subsystem Tests—Software

These tests of the software should be developed as part of design process. They would test functionality and operations during all conditions: power up, power down, dormancy, data transfer and communications, command,

and control. They should monitor the instrument's response to simulated fault conditions, simulated commands, and simulated data transfers.

The software engineers and technicians should design and run these tests. A GSE engineer might help by providing the GSE as part of the test system and simulated commands. They should develop these tests during preliminary design and refine them during critical design. They would run the tests on the engineering units during the critical design phase and on the flight units during the fabrication and integration phase.

10.10.5 Simulators

Clearly the subsystem tests need simulators to support the tests. The simulators can represent both the host spacecraft and other instruments. The entire design team, i.e., hardware, software, and GSE engineers and technicians, might contribute to the design of these simulators during preliminary design and refine them during critical design.

Simulators represent and exercise data and power configurations of the satellite, command-and-control streams, and other instruments. They can also exercise the cable harnesses. They are used all the way through from the preliminary design phase through the fabrication and integration phase.

10.10.6 Ground Support Equipment

The GSE's primary purpose is to support the mission after launch by receiving, analyzing, displaying, distributing, and storing data. It has an important secondary purpose—that of supporting testing, system tests, and integration. The GSE can simulate the telemetry to the host spacecraft by generating appropriate commands and data formats. It can also receive simulated data downloads from your instrument during system tests and integration, process the data, and then display the processed data.

GSE engineers design these tests and technicians run them. They should design the tests during preliminary design and refine them during critical design. They run the tests on the engineering units during the critical design phase and on the flight units during fabrication and integration phase. The GSE also supports integration and system tests when the host spacecraft mates to the launch vehicle and on the launch pad.

10.11   Integration

10.11.1   System

Integration begins in the critical design phase with connecting the engineering models together and driving the mock-up with both spacecraft and other instrument simulators. This activity helps assure that the interactions between modules are understood. Insertion of the flight units occurs during the fabrication and integration phase. An orderly sequence of tests verifies operation and functionality as each module is added to the system. Finally, when all components and instruments of the spacecraft are in place, then a full suite of system tests can be run both to verify the operation against the specifications and to validate the design.

System tests exercise the functionality and operations of the spacecraft and its instruments during all conditions: power up, power down, dormancy, communications with the instruments, command and control, and data transfer to Earth. The test team monitors the instrument's response to simulated fault conditions, simulated commands, simulated data streams from other instruments, and simulated host communications.

The entire design team, consisting of hardware, software, and GSE engineers and technicians, contribute to designing and running these system tests. The GSE engineer is integral to the process by providing both the GSE as part of the test system and simulated commands from Earth to the spacecraft.

The team develops these tests during preliminary design and refines them during critical design. They run the tests on the engineering units during the critical design phase and on the flight units during fabrication and integration phase.

10.11.2 Environmental

Environmental tests assure operation during launch and space flight. These environmental tests primarily exercise the instrument to check it for survival during the shock and vibration of missile launch and the vacuum of space. These tests almost exclusively exercise the instrument hardware, but the software needs to be running while undergoing these tests. Environmental tests occur during the fabrication and integration phase. Sometimes the entire spacecraft with all instruments and subsystems in place are subjected to environmental tests.

Thermal vacuum: These tests occur within steel chambers that are large enough to accommodate a large instrument, subsystem, or sometimes an entire satellite. See Figure 10.1 for an example of a chamber. These chambers can pump down the air pressure to simulate space. Heating cord wrapped around the instrument or system or high wattage lamps on the walls of the chamber can produce a hot thermal environment. Cooling coils around the walls of the chamber can produce a cold thermal environment.

Typically, a thermal-vacuum test runs for days or weeks. After installing the instrument inside the chamber and connecting the signal lines to monitor its operation, technicians pump out the air. Then they run a prescribed series of tests that cycle the temperature from low to high and back down again. Figure 10.2 illustrates one potential set of temperature profiles. At each extreme, very cold or very hot, they let the instrument dwell, sometimes for hours, to reflect the particular mission.

Thermal-vacuum tests tend to open fractures, particularly in circuit boards. The thermal expansion and contraction work the materials to extend any fractures present.

FIGURE 10.1
An environmental chamber for running thermal vacuum tests. (© 2006 The Johns Hopkins University Applied Physics Laboratory. All rights reserved. Used with permission.)

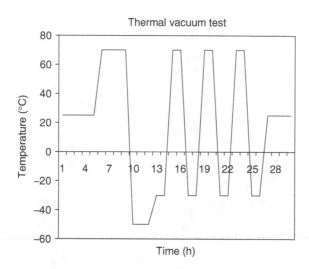

FIGURE 10.2
An example profile for temperature swings in a thermal vacuum test. The instrument might "soak" at the high temperature of +70°C for several hours before being run in a functional test of its operation. The instrument might then "soak" at the survival temperature of –50°C for several hours before the test chamber raises the low operating temperature to –30°C and then the instrument runs in a functional test of its operation. (© 2007 by Kim Fowler, used with permission. All rights reserved.)

(a)

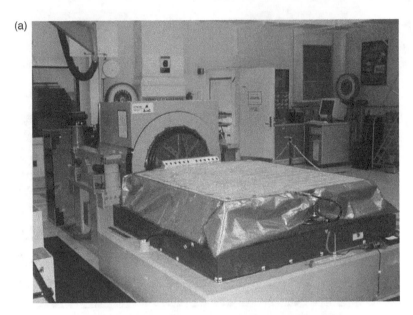

(b)

FIGURE 10.3
Vibration tables for running shock and vibration tests. (a) A vibration table before attaching a test article. (b) A large vibration table with a satellite positioned for vibration testing. (© 2006 The Johns Hopkins University Applied Physics Laboratory. All rights reserved. Used with permission.)

During these tests, engineers continuously monitor and record the operation of the instrument, as well as the temperature profile. They note anomalies in operations, which later must be rectified in the laboratory or constrained by the flight-operation rules of the mission.

Thermal-vacuum facilities are maintained by technicians who run the tests. Usually a mechanical engineer oversees the facility and its operations.

Shock and vibration: Shock and vibration tests simulate the mechanical environment during missile launch. Large tables with powerful voice coils provide the shock and vibration tests. These tables can accommodate an instrument, subsystem, or sometimes an entire satellite. See Figure 10.3 for examples of a vibration table and a satellite on another table. They can produce a mechanical impulse (shock) and a variety of vibration waveforms, such as sinusoidal, square, and triangular. Furthermore, the control system for a "shock-and-vibe" table can drive swept or random frequencies and amplitudes. Figure 10.4 gives an example of a vibration profile.

Typically, a "shock-and-vibe" test runs for less than an hour. Often a series of tests are run in different axes of orientation to fully test the instrument.

During these tests, engineers continuously monitor and record the operation of the instrument, as well as the vibration profile. Resonant peaks in the mechanical structures are noted and compared to those predicted in the mechanical models. Engineers note anomalies in operations, which later must be rectified in the laboratory or constrained by the mission.

Shock and vibration facilities are maintained by technicians who run the tests. Usually a mechanical engineer oversees the facility and its operations.

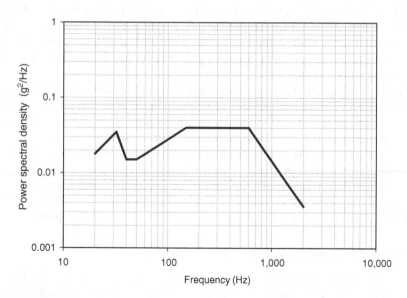

FIGURE 10.4
Example of a vibration profile using random vibration excitation.

10.12   Manufacturing and Fabrication

There are a number of considerations in fabricating an instrument for a spacecraft. Nearly all subsystems for a space instrument are custom-built. Procuring components and modules for fabrication always requires long lead deliveries.

My own experience in fabricating space instruments indicates that the CAD design, circuit board fabrication and assembly, and the subsystem or instrument assembly takes 10 months on average. The absolute minimum is 4 months, if nothing else is in fabrication and the engineers do not make any revision to the design. Tables 10.3 through 10.5 illustrate minimum, average, and backlog times for fabricating circuit boards and instruments.

New software tools are arriving on the scene to provide product lifecycle management. They tie a lot of things together, such as CAD programs and terminals with scheduling, inventory, and resource management.

10.12.1   Electrical and Electronic Fabrication

Hardware engineers either draw up schematics or collaborate with a designer in the CAD team. The CAD designer lays out the boards, back-planes, connectors, and cable harnesses (Figure 10.5). This goes on during the preliminary design and critical design phases.

The CAD designer transfers the completed drawings to the fabrication group who builds multilayer circuit boards (Figure 10.6). An inspector visually analyzes the circuit boards and then performs a microscopic inspection of the test coupons from the circuit boards (Figure 10.7). The inspector is looking at cross-sections of vias in the test coupon for attached voids that plated in fabrication or for delamination of the layers or for metal plating that is too thin or too thick. These circuit-board fabrication activities begin in the critical design phase but primarily occur during the fabrication and integration phase.

Here's a big problem; engineers often will rush a design into fabrication under pressure from management to demonstrate progress on the schedule to the sponsor; they want to show someone a significant fraction of boards in fabrication at the next design review. Unfortunately, the design is not complete and usually not correct. They then find problems and end up revising the design repeatedly. This forces the fabrication team to start over each time—PCB fabrication and inspection, assembly and inspection, and finally circuit test and verification. Tables 10.3 through 10.5 illustrate how starting over in the middle of the process loses days if not months.

10.12.2   Mechanical Machining and Fabrication

The chassis or enclosure or sandwich-stack of circuit boards for the instrument must be custom-designed and built. A mechanical engineer

FIGURE 10.5
CAD designer laying out a circuit board. (© 2006 The Johns Hopkins University Applied Physics Laboratory. All rights reserved. Used with permission.)

FIGURE 10.6
One example of a multilayer circuit board. (© 2006 The Johns Hopkins University Applied Physics Laboratory. All rights reserved. Used with permission.)

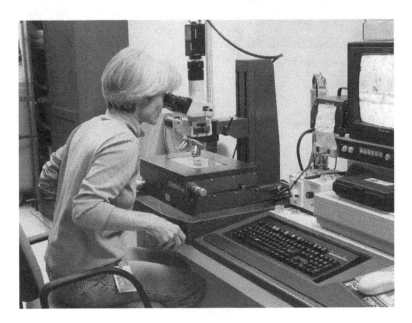

FIGURE 10.7
Microscopic inspection of a circuit board test coupon. (© 2006 The Johns Hopkins University Applied Physics Laboratory. All rights reserved. Used with permission.)

prepares schematics that are transferred to a designer in the CAD team. The designer then lays out the chassis and cable harnesses. This CAD-layout process happens during the preliminary design and critical design phases.

The CAD designer then transfers the completed drawings to the fabrication group, where a machinist mills the chassis enclosure or the metal chassis bands for the circuit boards from blocks of aluminum alloy. Later another technician might anodize or paint the aluminum structure to protect it from corrosion. These fabrication activities begin in the critical design phase but primarily occur during the fabrication and integration phase.

10.12.3 Assembly

Assembly includes the circuit boards, the cable harnesses, and the mechanical chassis components. Either personnel hand-solder the ICs to the circuit boards (Figure 10.8) or a pick-and-place machine is programmed, and the circuit boards are fed through it to a solder reflow oven. One or more technicians fabricate the cable harnesses. Other technicians assemble the circuit boards into either the aluminum chassis or into their metal bands (for a sandwich stack). After each step, inspectors check the quality of the workmanship of each assembly (Figure 10.9). Finally, the circuit boards and cable assemblies are potted with a durable encapsulating polymer to protect the components from condensation or contamination before launch and thus prevent the components from outgassing in space. All these steps take place during the fabrication and integration phase.

TABLE 10.3

Minimum and Average Times in Design and Fabrication for a Simple Circuit Board (© 2007 by Kim Fowler, used with permission. All rights reserved)

Category	Description	Minimum Days	Average Days	Backlog Days
Design (in CAD)	Received into design	0.25	1	5
	Part placement and board routing (after all revisions)	5	10	10
	Engineering design review	5	10	0
	Breadboard release	0.25	1	0
	Informal flight fabrication review	5	10	0
	Engineering model release	0.25	5	0
	Flight fabrication review	5	10	15
	On-table for sign-off	5	10	0
	Flight release to PDM	0.25	1	0
	Design subtotal	**26**	**58**	**30**
Board fabrication	Computer aided manufacturing—design rule check	0.375	3	8
	Request quotes from external vendor	1	3	5
	Place requisition and get signatures	2	4	0
	Contract released to external vendor	0.125	1	2
	Kitting—requests parts kits from inventory	10	10	5
	Fabrication—not complex	5	15	15
	Bare board—in receiving	1	3	0
	Bare board—inspection and coupon tests	3	5	5
	Prepare task control card	0.25	1	2
	Board fabrication subtotal	**23**	**45**	**42**
Board assembly	Release from kitting	1	2	0
	Assembly of passive components	1	5	10
	Inspection passive components assembly	0.5	2	0
	Test—contingent upon engineer to test	0	5	0
	Assembly of active components	1	7	10
	Inspection active components assembly	0.5	2	0
	Test—contingent upon engineer to test	0	5	0

(Continued)

TABLE 10.3
Continued

Category	Description	Minimum Days	Average Days	Backlog Days
Board assembly (Continued)				
	Tailor process	1	5	0
	Inspection	1	3	0
	Coating and encapsulation	2	5	5
	Inspection after coating	0.5	2	0
	Coating touchup	1	3	0
	Inspection after coating touchup	0.5	2	0
	Final inspection	0.5	2	0
	Release to program	1	3	0
	Assembly subtotal	**12**	**53**	**25**
System assembly	Harness assembly	1	5	5
	Module plug-in	1	2	10
	Stake jack screws	1	2	0
	Environmental stress screen	5	15	0
	Repair of boards	5	10	0
	Sign-off	5	10	0
	Release to program	1	5	0
	Design subtotal	**19**	**49**	**15**
Calendar—minimum, average, backlog (days)		79	205	112
Calendar—minimum, average, backlog (months)		3.8	9.8	5.3

Most designs experience a significant amount of revision. While average CAD design time is 360 hours or a little over 2 months, most flight boards take 8 to 12 months to complete just the CAD design alone. The minimum time is the best that can be expected in any one category. *DO NOT use minimum time as an estimation tool.* Use the average time and add extra time for design revision. In essence, the average time total is a bare minimum for scheduling!

TABLE 10.4

Minimum and Average Times in Design and Fabrication for a Complex, Multilayer (8 to 24 Layers with Hidden and Blind Vias) Circuit Board (© 2007 by Kim Fowler, used with permission. All rights reserved)

Category	Description	Minimum Days	Average Days	Backlog Days
Design (in CAD)	Received into design	0.25	1	5
	Part placement and board routing (after all revisions)	5	10	10
	Engineering design review	5	10	0
	Breadboard release	0.25	1	0
	Informal flight fabrication review	5	10	0
	Engineering model release	0.25	5	0
	Flight fabrication review	5	10	15
	On-table for sign-off	5	10	0
	Flight release to PDM	0.25	1	0
	Design subtotal	**26**	**58**	**30**
Board fabrication	Computer aided manufacturing design rule check	0.375	3	8
	Request quotes from external vendor	1	3	5
	Place requisition and get signatures	2	4	0
	Contract released to external vendor	0.125	1	2
	Kitting—requests parts kits from inventory	10	10	5
	Fabrication—large, many holes, many layers	10	20	0
	Bare board in receiving	1	3	0
	Bare board—inspection and coupon tests	3	5	5
	Prepare task control card	0.25	1	2
	Board fabrication subtotal	**28**	**50**	**27**
Board assembly	Release from kitting	1	2	0
	Assembly of passive components	1	5	10
	Inspection passive components assembly	0.5	2	0
	Test—contingent upon engineer to test	0	5	0
	Assembly of active components	1	7	10
	Inspection active components assembly	0.5	2	0

(Continued)

TABLE 10.4
Continued

Category	Description	Minimum Days	Average Days	Backlog Days
Board Assembly (Continued)				
	Test—contingent upon engineer to test	0	5	0
	Tailor process	1	5	0
	Inspection	1	3	0
	Coating and encapsulation	2	5	5
	Inspection after coating	0.5	2	0
	Coating touchup	1	3	0
	Inspection after coating touchup	0.5	2	0
	Final inspection	0.5	2	0
	Release to program	1	3	0
Assembly subtotal		**12**	**53**	**25**
System assembly	Harness assembly	1	5	5
	Module plugin	1	2	10
	Stake jack screws	1	2	0
	Environmental stress screen	5	15	0
	Repair of boards	5	10	0
	Signoff	5	10	0
	Release to program	1	5	0
	Design subtotal	**19**	**49**	**15**
Calendar—minimum, average, backlog (days)		84	210	97
Calendar—minimum, average, backlog (months)		4.0	10.0	4.6

Most designs experience a significant amount of revision. While average CAD design time is 360 hours or a little over 2 months, most flight boards take 8 to 12 months to complete just the CAD design alone. The minimum time is the best that can be expected in any one category. *DO NOT use minimum time as an estimation tool.* Use the average time and add extra time for design revision. In essence, the average time total is a bare minimum for scheduling!

TABLE 10.5

Minimum and Average Times in Design and Fabrication for a Complex, Rigid-Flex Circuit Board. (© 2007 by Kim Fowler, used with permission. All rights reserved)

Category	Description	Minimum Days	Average Days	Backlog Days
Design (in CAD)	Received into design	0.25	1	5
	Part placement and board routing (after all revisions)	5	10	10
	Engineering Design Review	5	10	0
	Breadboard release	0.25	1	0
	informal Flight Fabrication Review	5	10	0
	Engineering Model release	0.25	5	0
	Flight Fabrication Review	5	10	15
	On-table for sign-off	5	10	0
	Flight release to PDM	0.25	1	0
	Design subtotal	**26**	**58**	**30**
Board fabrication	Computer aided manufacturing—design rule check	0.375	3	8
	Request quotes from external vendor	1	3	5
	Place requisition and get signatures	2	4	0
	Contract released to external vendor	0.125	1	2
	Kitting—requests parts kits from inventory	10	10	5
	Fabrication—rigid/flex multilayer	30	40	0
	Bare board in receiving	1	3	0
	Bare board—inspection and coupon tests	3	5	5
	Prepare task control card	0.25	1	2
	Board fabrication subtotal	**48**	**70**	**27**
Board assembly	Release from kitting	1	2	0
	Assembly of passive components	1	5	10
	Inspection passive components assembly	0.5	2	0
	Test—contingent upon engineer to test	0	5	0
	Assembly of active components	1	7	10

(Continued)

TABLE 10.5
Continued

Category	Description	Minimum Days	Average Days	Backlog Days
Board Assembly (Continued)				
	Inspection active components assembly	0.5	2	0
	Test—contingent upon engineer to test	0	5	0
	Tailor process	1	5	0
	Inspection	1	3	0
	Coating and encapsulation	2	5	5
	Inspection after coating	0.5	2	0
	Coating touchup	1	3	0
	Inspection after coating touchup	0.5	2	0
	Final inspection	0.5	2	0
	Release to program	1	3	0
	Assembly subtotal	**12**	**53**	**25**
System assembly	Harness assembly	1	5	5
	Module plug-in	1	2	10
	Stake jack screws	1	2	0
	Environmental Stress Screen	5	15	0
	Repair of boards	5	10	0
	Sign-off	5	10	0
	Release to program	1	5	0
	Design subtotal	**19**	**49**	**15**
Calendar—minimum, average, backlog (days)		104	230	97
Calendar—minimum, average, backlog (months)		5.0	11.0	4.6

Most designs experience a significant amount of revision. While average CAD design time is 360 h or a little over 2 months, most flight boards take 8–12 months to complete just the CAD design alone. The minimum time is the best that can be expected in any one category. *DO NOT use minimum time as an estimation tool.* Use the average time and add extra time for design revision. In essence, the average time total is a bare minimum for scheduling!

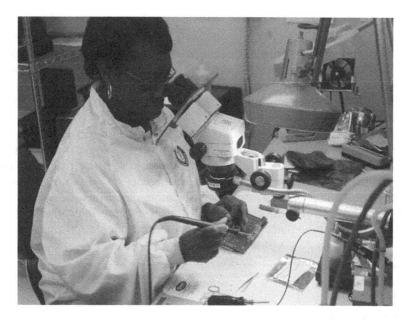

FIGURE 10.8
Assembling a circuit board. (© 2006 The Johns Hopkins University Applied Physics Laboratory. All rights reserved. Used with permission.)

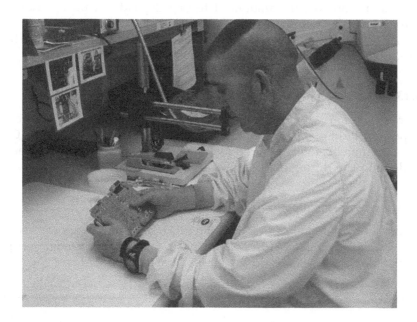

FIGURE 10.9
Inspecting a circuit board. (© 2006 The Johns Hopkins University Applied Physics Laboratory. All rights reserved. Used with permission.)

10.12.4 Tests

All fabrication steps undergo inspections—circuit boards and test coupons, cable assembly, potting, and mechanical assembly. Each flight circuit board and each cable are connected, in turn, to the engineering model and verified for functionality. The design team then verifies the entire flight instrument with an end-to-end system test.

After a design review, the PER, the instrument undergoes shock and vibration tests and then thermal-vacuum testing. Assuming the tests reveal no problems or anomalies, the instrument would then be ready for integration with the spacecraft. Otherwise, contingency plans require remedial action to solve the problems or remove the anomalies. The design team then holds a PSR to confirm the completion of all testing of the instrument.

All these steps take place in the fabrication and integration phase.

10.13   Support

10.13.2 Spacecraft Integration

This is the final set of actions to attach the instrument to the host spacecraft. It includes mechanical bolting, cabling, and alignment. The entire spacecraft might then undergo environmental system tests, such as thermal vacuum, shock, and vibration; the amount of testing depends on the sponsor and contractor for the host spacecraft. During this integration to the spacecraft, the GSE engineer and technicians support all testing with the GSE. These tests are the final steps in the fabrication and integration phase.

10.13.2 Launch

After integration with the spacecraft, the GSE engineer and technicians support all testing and operations with the instrument through the GSE. They also train operators to run the GSE. This is the primary involvement of the design team during the Launch and Mission phase.

10.13.3 Technical Support

The design team is "on call" should anything go wrong or appear anomalous with the instrument. They would have prepared some standard scripts, prior to launch, for actions during various potential scenarios.

10.14   Disposal

There is very little to dispose. The spacecraft falls out of orbit and burns up in the atmosphere or disappears into the Solar System or beyond. The fabrication facility must handle and dispose of all its materials according to

NASA and OSHA (in the United States that is the Occupational Safety Hazards Administration) regulations.

10.15 Liability

The primary liability is the risk of a mission failing to obtain any scientific data and losing the return on investment for any commercial investors. This risk is borne equally by the instrument developers, the contractor for the host spacecraft, and the contractor for the launch vehicle (NASA or ESA or commercial contractor). For commercial satellites, insurers might or might not risk covering the launch and deployment of the satellite.

The only other legal risk is if the spacecraft has a radioisotope thermionic generator (RTG) to provide power. Sometimes environmental groups will use legal means to prevent its launch; their concern is the risk of a failure during launch that could spread man-made radiation over populated areas. RTGs are used only on deep-space missions; so you probably will seldom encounter one.

10.16 Summary

10.16.1 Emphases

The architecture of an instrument must balance weight, power, fault tolerance, and integration ease. All subsystems, ICs, and components in space flight must be radiation tolerant. The system must also survive the shock and vibration of launch and wide temperature swings on orbit.

10.16.2 Gotcha's

Beware of and remove the "low-tech" problems of misalignment, incorrect connector mating, and glitches in power conversion and distribution.

Acknowledgments

My thanks to both Brian Alvarez and Larry Frank at The Johns Hopkins University Applied Physics Laboratory for providing some of the information for this chapter.

References

1. According to personal communications with Mr. Larry Frank in September 2006, these heuristics are remarkably good in many situations even though there are always some major exceptions.
2. Brown, C.D., Elements of Spacecraft Design (AIAA Educational Series), *American Institute of Aeronautics and Astrophysics*, April 2003.
3. Griffin, M.D. and French, J.R., Space Vehicle Design (AIAA Educational Series), *American Institute of Aeronautics and Astrophysics*, March 2004.
4. http://creme96.nrl.navy.mil/cm/RadEffects.htm
5. NASA/GSFC Radiation Effects & Analysis Home Page, http://radhome.gsfc.nasa.gov/top.htm
6. Personal communications with colleagues at JHU/APL, 2002–2005.

11

Case Study 8—Aerospace Video Processor

11.1 Concept and Market

11.1.1 Who, What, Why, How, Where, and When

This chapter illustrates some specific issues outlined in Chapter 10. I had a firsthand view of some of the effort in developing the system because the customer called me in as a consultant.

Ecliptic Enterprises Corporation in Pasadena, CA, developed a video system for data acquisition. It is a space version of a commercial off-the-shelf (COTS) video system; it is designed for shorter-term missions that approach low-earth orbits. The system uses commercial components to speed development and reduce cost. In many ways, the system is simpler and cheaper than most instruments designed for space travel because it does not have some of the requirements that long-lived missions might have.

The system captures, processes, transmits, and displays images in real time for a mission-critical application. The sensors were very similar to those in video cameras. The system compresses and multiplexes the data from the sensors into a single data stream and then sends the data via a telemetry link to a remote location. At the remote location, the support equipment decommutates (demultiplexes), decompresses, displays, and stores the data stream on disk.

Figure 11.1 outlines the design of the system. The flight hardware has four circuit boards: a video-compression board, a multiplexer board, a power supply board, and an analog housekeeping board. A field-programmable gate array (FPGA) on the multiplexer board buffers up the data streams from each sensor. Digital signal processor (DSP) chips on both the compression board and the multiplexer board multiplex data and to compress the images. The analog housekeeping collects temperature data and analog signals from various places on the sensors, converts it to digital format, and then multiplexes the housekeeping data into the data stream for telemetry.

Source software code for the DSP chips is written in C. The display software on the support equipment is a purchased COTS package.

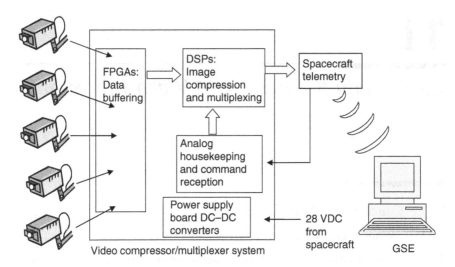

FIGURE 11.1
Block diagram of the video data acquisition system for aerospace. (© 2006 by Kim Fowler, used with permission. All rights reserved.)

The entire project took about a year to complete at Ecliptic Enterprises in California. It took another 8 months to integrate into the launch vehicle in Arizona and then to fly on a mission over the Pacific Ocean. It worked as expected.

11.1.2 Economics

This type of system sells in low volume (5 to 10 per year) and at a high margin. Each one is hand made—machined, fabricated, and assembled. Each undergoes thorough environmental testing. None of these fabrication processes makes for a cheap product, but the final product is still much cheaper than a custom space-qualified instrument.

The system is not solely tied to making sales in the space business. Commercial aerospace and military experiments also buy the system. Consequently, it does not need all the qualification testing for space flight.

11.1.3 Project Background

Until development of this product, Ecliptic had primarily built analog systems. They had completed a digital system once before, but it was a much less ambitious system—it only had two cameras and no image compression.

This project had several complications: the contract with the first customer for the new system was devoid of necessary requirements, and a key person left partway through the project. The effort had begun with legacy code (from the previous digital system with two cameras), but that code was poorly designed and implemented, and it did not work for this new product.

Moreover, they designed the system with a new and unfamiliar DSP processor and associated development system; the nuances of the new processor were something of a stumbling block.

The folks at Ecliptic had to regroup, completely revise the design, and institute new development processes. They buckled down and worked 70 and 80 hours per week for 6 months. It was a terrible grind for them, but they completed the project and had a working product at the end of it.

What follows is a hybrid case study—it combines some of what happened to develop this video compressor and some of what happens now at Ecliptic in developing other products.

11.2 People and Disciplines

Ecliptic started out with a team of four engineers and a mechanical designer; as mentioned earlier, one engineer left about 5 months into the project.

They relied on outside vendors to supply some of the subsystems and services. They purchased a commercial software package that handles data from space flight and worked with that vendor to tailor it for their purposes. They also contracted another firm to do shock and vibration tests on the flight portions of the equipment.

They also worked closely with their customer to develop the product.

11.3 Architecting and Architecture

11.3.1 Process

This sort of product development requires processes that address mission-critical applications. Ecliptic used a combination of waterfall and spiral development to complete this project.

In the beginning, while trying to use the legacy code, they did not follow defined processes. Partway through, they had no development plans of any kind: design, schedule, or test; they had no code reviews to assure quality and functionality; a single code developer, with no accountability, was working with the legacy code.

The software was a mess! It was what some people call "spaghetti code." The style guide had not been followed; the software it had no consistent format. The software did not use a common set of event handlers; each sensor was handled differently, even though the sensors were very similar in data output. Orphan code segments floated throughout the software. The whole program was a set of nested interrupts—a major no-no!

New world order: Ecliptic recognized the problems and completely changed things around. They defined and instituted new software processes that required code reviews and metrics for anomalies and production. They held monthly reviews of their progress with the customer. They staged releases of the software (using a spiral development model). They also prepared a complete set of documents to cover the products design, development, and test.

The new processes included record keeping, metrics, and production guidelines. Records of production metrics are important to quantify aspects of quality; they include bug rates, such as lines of code per hour or per day (LOC/h or LOC/day), bugs found, bug severity, and status of fixes for bugs.

11.3.2 Parameters

The parameters covered a wide variety of concerns: power, weight, size, environment, and the characteristics of the compression algorithms. The environment had to survive the shock and vibration of a missile launch. The compression algorithms were both lossless and lossly, depending on the application.

11.3.3 Analysis

There was no major effort to analyze the system. It could have been done but only portions had been completed. The main concern for the revised system was computational robustness—could the system gracefully recover from a sensor outage or a failure in the data stream? Most analysis was performed by laboratory bench tests on prototype circuit boards.

11.3.4 Architecture

The video compressor/multiplexer has a centralized architecture to save weight and power (Figure 11.1). The mechanical configuration has circuit boards that plug into a backplane. The sensors feed raw video data to the video compressor/multiplexer. Figure 11.2 shows a system on a laboratory bench undergoing software tests. Figure 11.3 shows two types of circuit boards that fit into the chassis; note the commercial components that comprise the circuit boards; also note the channel locks on the sides of the boards to clamp them and reduce vibration resonances.

11.3.5 Interfaces

The main interfaces for the video compressor/multiplexer were with the host vehicle and the sensors—primarily video imagers. The interfaces each

FIGURE 11.2
The video compressor/multiplexer in lab tests with two attached cameras. (© 2005 by Kim
Fowler, used with permission. All rights reserved.)

had defined electrical signal levels, signaling types, and data formats with
frames and specified sequences.

11.4 Phases

For a small, agile company using a spiral type of development, defining phases is not as useful as it is with a larger product or company. Ecliptic
has a general design phase for preparing schematics of circuit boards and
chassis or enclosure. Then they fabricate and assemble several "turns" of
circuit boards and test them for functionality. Finally, they build the space-
qualified product, which is tested to environmental specifications for temperature swing, vibration, and shock, then it is shipped to the customer. The
software goes through staged releases that implement subsets of the requirements.

11.4.1 Design

During this "design phase," early conceptual design leads to requirements,
iterating until the requirements are understood and complete. For the video
compressor/multiplexer, two engineers designed the circuit boards and
drew up the schematics; of these two, one developed the FPGA design and

(a)

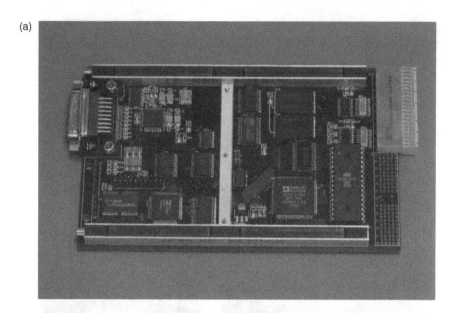

(b)

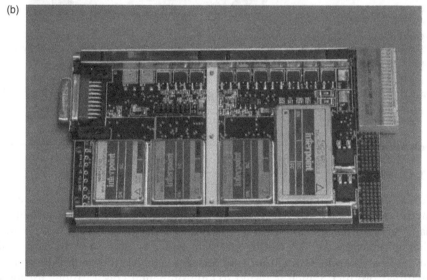

FIGURE 11.3
Two types of circuit boards used in the video system: (a) a video compressor board and (b) a power supply board. (Ecliptic Enterprises Corporation, used with permission.)

code; the other developed the code for one of the DSPs. The mechanical designer drew up the schematics for the chassis and sensor attachments. The VP of engineering managed the process and wrote the majority of the DSP software.

For other products and projects, the three engineers, including the VP, traded software tasks. The VP usually has the task of understanding the customer's requirements and then cleaning up specifications to a sensible form.

11.4.2 Fabrication and Delivery

The engineers made minor corrections to the schematics and then had outside firms fabricate the chassis and fabricate and assemble the circuit boards. The software went through three major iterations; the company held monthly design reviews with the customer to discuss progress.

11.4.3 Commercial Production

After fabrication and assembly, an outside firm performed the shock and vibration tests on the system. The methods and results of the tests were uncertain, so Ecliptic then took the system to the customer, who had a shock and vibration table, to complete the testing. After running the system and its software for many hours without failure, the system was delivered to the customer.

This video compressor/multiplexer is now available as a COTS product. Most customers, though, require some changes to its software or structure to fit their applications.

11.5 Scheduling

As in most projects, the scheduling began with the end in mind, and then they did a top-down timeline. There was not much bottom-up planning because there are so few people involved. They simply worked the hours needed (70–80 h/week) to complete the project. Hopefully this pace will not continue for future projects.

11.6 Documentation

11.6.1 Purposes

Documentation serves several purposes in aerospace electronics. It aids quality assurance by providing a basis for design reviews. It confirms that requirements are met and acts like a checklist for all activities. It serves as part of the subsystem delivery for customers who will integrate it into larger systems and provides necessary information for the integration. Finally, some of it can serve as part of the advertising and support literature for marketing the product.

11.6.2 Types

Figure 11.4 outlines the general types of documents found in similar aerospace products. While Ecliptic Enterprises did not produce all of the documents shown in Figure 11.4 for this project, it covered the major ones.

Several of these documents can be standard company documents that do not change from project to project—or they only require a minor addendum to tailor the document to the project. Standard company documents might include

- Infrastructure plan
- Problem resolution plan
- Documentation plan
- Risk assessment and management plan
- Configuration management
- Software style guide

Most documents can follow templates, in spite of variations between projects.

A software style guide, for instance, is a necessary component of good software processes. It helps the code developers standardize source code with headers, comments, and acceptable formats. It should not change from project to project.

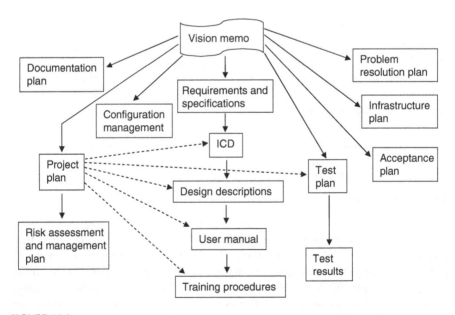

FIGURE 11.4

Example outline of documents needed for aerospace electronics such as Ecliptic's video compressor/multiplexer. (© 2005 by Kim Fowler, used with permission. All rights reserved.)

11.7   Requirements and Standards

11.7.1   Markets

There are two markets: aerospace flights to low-earth orbit and lunar exploration. This means that the video compressor/multiplexer addresses commercial, government, and military aerospace experiments. It does not address long-term missions, high-altitude orbits, or earth-bound consumer applications. Most experiments that use the video compressor/multiplexer are attached to missile boosters. Generally, the video compressor/multiplexer does not need all the qualifications for space flight, but Ecliptic could revise the design to include components with higher levels of radiation tolerance and could perform more environmental tests should the mission require.

11.7.2   Standards

There are no standards that the video compressor/multiplexer must follow. Those outlined in the previous chapter are usually sufficient. Otherwise, customers may require adherence to specific standards, which usually relate to the environmental tests or to levels of radiation tolerance.

11.7.3   Preparing Requirements

The requirements for the video compressor/multiplexer had some of the following characteristics:

- Compress data streams from five imaging sensors
- Use lossless compression for radiometric observations from several of the sensors
- Use lossly compression for video streams from several of the other sensors
- Record housekeeping data—primarily temperatures from thermo-couples
- Have GPS time and one pulse per second to synchronize data
- Multiplex data into single, serial data stream for telemetry
- Select and sequence the sensors during the mission or have all five going at once
- Demultiplex data, decompress images, store on disk, and display video images
- Survive a harsh environment until the end of the mission.

One of Ecliptic Enterprises' biggest and ongoing problems is deciphering the customer's requirements. Typically, it is because the written specifications from the customer are unclear or ambiguous or nonsensical.

11.8   Analysis

Most products built by Ecliptic, including the video compressor/multiplexer, do not require either numerical analysis or simulation for feasibility. Nearly all analyses on the video compressor/multiplexer were laboratory bench tests performed on prototype circuit boards and systems. These are effective forms of confirming functionality. All three engineers, developing both hardware and software, tested prototype designs in Ecliptic's laboratory. These tests went on throughout development and fabrication.

11.9   Design Trade-Offs

11.9.1   Hardware

The main set of decisions for the electronic hardware centered on the number of FPGAs vs. DSPs and on the number of circuit boards in the enclosure. To keep weight and size down, the chassis could hold no more than four circuit boards. This drove the design to use a combination of FPGAs and DSPs for data manipulation, compression, and multiplexing. They also upgraded a small DSP to a more recent model to handle the housekeeping and receiving the commands from the spacecraft.

Ecliptic relied on legacy software and a former board design—in hope of reducing the effort through reuse—which forced some of the data to circulate between several different boards. The system had a video compression board, a multiplexer board, a power supply board, and an analog housekeeping board (Figure 11.1). In hindsight, the system might have been reduced to three optimized boards with a complete redesign.

The VP and two design engineers prepared the system design. The two design engineers then each took two boards and designed or redesigned them.

11.9.2   Power

Most spacecraft supply raw +28 VDC to instrumentation. The dedicated power supply board had DC–DC converter modules to change the +28 VDC to 3.3, 5, and ±15 VDC.

Many aerospace applications have some sort of problem with the power system. This project was not spared either; power-up glitches turned on some field-effect transistors, which then caused lens covers over the sensors to open. These covers were supposed to prevent contamination during launch, which was a short time after the power-up and glitches occurred. A hardware redesign eliminated the problem. This situation pointed out the value of bench testing to reveal problems—all of which you know, but it bears repeating.

11.9.3 Cooling

This product, like so many other spacecraft instruments, relies on conduction cooling. The low power consumption of the DSPs is important. A short mission time also helps to reduce the concern over cooling, but that is not a given for longer missions such as found in orbiting satellites.

Both the hardware design engineers and the mechanical design were involved in considering the cooling design.

11.9.4 Software

The software had to receive both data from the sensors and uplink commands from the spacecraft (initiated by the ground support equipment and sent via telemetry to the spacecraft), compress the data, and then multiplex the desired data streams onto the telemetry output. The compression could take one of two different algorithms: lossless compression for ratiometric, scientific data and lossly compression for television imaging.

The software development was the most difficult part of this project. It involved between 15,000 and 18,000 lines of code (LOC). With the 70- and 80-hour weeks put in over 6 months, the production rate averaged 100 LOC/day or about 10 LOC/hour. This rate of production for the software is unusually high when compared with most teams developing code for mission-critical instruments. (The principal developer was extremely skilled and highly motivated.) Typical production rates for most companies range from 5 to 25 LOC/day or between 1 and 3 LOC/hour.

Exacerbating the situation, Ecliptic took on the joys of learning a new DSP processor and its development system for this particular project. Some of the stumbling blocks of the new processor included problems (undocumented features) with the serial ports, the cache memory, and the reset lines. After solving these problems, they then had to struggle with deciding whether to implement either queues or a system of buffers to receive the data from the sensors.

11.9.5 Hardware vs. Software

Clearly, the complexity of processing all the instrument functions required implementation in software (15,000–18,000 LOC). To implement these functions in hardware, such as FPGAs, would have required far too many components, which would have added circuit boards with their attendant size and weight.

Concerns over potential fixes or patches that might be needed at the last minute necessitated a Joint Test Association Group (JTAG) port, requiring both hardware and software design to connect the JTAG line through all the circuit boards. The JTAG port on the end panel of the chassis allows software upgrades without opening the enclosure to replace or program the EEPROM.

11.9.6 Buy vs. Build

Originally, Ecliptic had planned to write custom software to run the ground support equipment (GSE). Because of devoting all their efforts to the software in the video compressor/multiplexer, they ran out of time to write a custom software package. Instead, they contracted with a developer in a small company to modify his product to run on their GSE. While the commercial software was a fine product, it still took time to work out the kinks with this particular application. Handling high-bandwidth data is always challenging, regardless of how good the software support is.

11.10 Tests

11.10.1 Laboratory Tests

I have already mentioned that bench testing was important to this project. Not only did it confirm the functionality of the hardware, it also helped chase down bugs in the software. On several occasions, Ecliptic needed to run the equipment through extremes in temperature to exercise the sensors and to locate a persistent problem that showed up at low temperatures.

One of the hardware engineers and the mechanical designer handled Ecliptic's environmental chamber for the laboratory tests. The entire team participated in planning some of the bench tests. Most of this work took place during the design and fabrication phases.

11.10.2 Peer Review

Ecliptic held monthly design reviews with the customer. The primary focus in the reviews was the software development. A list of concerns and software bugs were collected leading up to each review, then Ecliptic distributed the list just before the review. During the course of the review, they recorded action items to correct and dispose the concerns. Leading up to the next review, they incorporated the activities that resolved the action items.

The entire team at Ecliptic participated in each review with the customer, who often had four or five engineers attending. Each design review took a full day of about 10 hours of work.

11.10.3 Subsystem Tests—Hardware and Software Integration

As part of the spiral development practice, Ecliptic performed iterations of integrating subsets of the software into prototype versions of the hardware. For each sensor, they had simple but effective tests to confirm functionality. They exercised the system for each integration, beginning in the design phase and continuing through the fabrication phase. Eventually, the fully-functional version of the software ran on the final, flight-quality hardware. The entire team participated in these integration tests.

11.10.4 Environmental

Once Ecliptic finished fabrication of the chassis and circuit boards, they subjected the video compressor/multiplexer system to both thermal tests and to shock and vibration tests. These were in accordance with the final requirements of the customer. The mechanical designer ran the thermal tests or worked with an outside firm to do the initial shock and vibration testing. One of the hardware engineers eventually had to accompany the system to the customer site for a final set of shock and vibration tests.

The thermal tests took about 3 days to complete. The shock and vibration tests took much longer, mainly because they had to be repeated. The final set of shock and vibration tests took about 4 days at the customer's facilities. These tests took place at the end of the fabrication and delivery phase.

11.11 Integration

Obviously in a spiral model of development, integration occurs repeatedly. This mode of integrating subsystems of hardware with subsets of software is described above.

The customer integrated the video compressor/multiplexer and the sensors onto the launch vehicle at a facility in Arizona for the final system integration. The contractor for the launch booster reviewed the results from the subsystem environmental tests; they also performed the actual integration of all the final subsystems onto the booster.

11.12 Manufacturing

11.12.1 Electrical and Electronic

Two engineers designed the circuit boards; their layouts had between 8 and 12 layers. Ecliptic sent the schematics out to a board fabrication house who then fabricated the boards. Typical turn time to fabricate new boards is a week, but paying a premium can turn around new boards within 2 or 3 days.

11.12.2 Mechanical

The mechanical designer designed the chassis, enclosures, and attachment fixtures. Ecliptic sent the schematics out to a machine shop to mill the sides of the enclosures and the attachment fixtures from aluminum alloy. Fabrication time was between 1 and 2 weeks.

11.12.3 Assembly

Once the circuit boards had been fabricated and returned to Ecliptic, they sent the boards out with a parts kit and instructions to skilled contract help to assemble the boards. Assembly can take from 1 to 5 days, depending on complexity of the board and the work load in the electronic assembly firm.

11.12.4 Tests

Ecliptic does not have standard manufacturing tests because so many projects are nearly custom designed. The laboratory bench tests and environmental tests, described earlier, serve as their manufacturing tests, as well.

11.13 Support

Ecliptic Enterprises provides technical support on an as-needed basis. Most customers have one time missions, at the end of which, the equipment is destroyed upon re-entry to the atmosphere. If the mission is long term and if technical support is extensive then Ecliptic will contract to support the customer.

11.14 Disposal

Like most spacecraft, there is very little to dispose of. This system for the first customer fell into the ocean. The final customer, the U.S. government, will dispose of the GSE and its computer when the project finishes.

Other Ecliptic customers with other missions allow the equipment to burn up in the atmosphere upon re-entry. Beyond the actual product, Ecliptic Enterprises must handle and dispose of all its scrap materials according to NASA and OSHA regulations.

11.15 Liability

There are no major safety issues with this product; it does not have high voltages or currents; it does not have hazardous materials; it is not large or heavy. The only legal liability is in the contracts with customers and it is small; they are standard provisions found in most contracts, such as failure to perform.

11.16   Summary

11.16.1   Do It Right

These folks recognized the problems with bad processes midway through the project. They reevaluated, regrouped, instituted good processes, and pushed through the problems to a successful end result.

This project had enormous potential for severely damaging the customer's reputation and its relationship with its final customer. Both sides took tough measures to complete the work; the customer insisted on recovery and good processes and then worked as a team member to accomplish those goals; Ecliptic agreed to change its processes. These two companies ended up working together well and hopefully have a new level of trust.

11.16.2   Emphases

This subsystem was simpler than many spacecraft projects because it did not have the level of documentation and qualification required for those space projects. It represents a good middle ground for space flight experiments and instruments that do not need the standard space guidelines or bureaucracy.

This case study ends well—which is unusual for projects that encounter serious difficulties midway. The development of this video compressor/multiplexer system had complex components, interactions, and processes. It only succeeded because good people with good attitudes did good work with the right processes.

11.16.3   Gotcha's

The customer, who called me in to help solve the problem, was part of the problem in the beginning. Their contract was weak; it did not provide any requirements for custom development. This meant that they did not have recourse if the product did not perform as they had expected. Expectations must be clearly communicated—the requirements must be clear and well-thought-out! Contracts must state the requirements and development processes expected if any custom work is to be done.

Acknowledgment

My thanks to Douglas Caldwell at Ecliptic Enterprises Corporation for helping to provide some of the information for this chapter.

12

Case Study 9—Satellite Subsystem

12.1 Concept and Market

12.1.1 Who, What, Why, How, Where, and When

This case study is a subsystem that is somewhat more complex than a single instrument. It illustrates some specific concerns that Chapter 10 outlined.

The U.S. National Aeronautics and Space Administration (NASA) has a program called Living With a Star, which is seeking ways to study the sun and space environments around the Earth through university collaboration. A particular component of that program is Space Environment Testbeds (SET). NASA began planning back in 2001 to orbit university experiments on a variety of satellites with SET.

NASA collaborated with The Johns Hopkins University Applied Physics Laboratory (JHU/APL) to build a platform for the SET experiments. The platform was to host 6–12 experiments for each satellite mission and different groups of experiments for different satellite missions. JHU/APL was to study, design, and build the SET platform and then to work with NASA to integrate it onto various launch vehicles. A major obstacle to the SET platform, though, was that each satellite was different from other satellites, and each experiment was different from the other experiments.

NASA and JHU/APL developed the concept of a "carrier" to overcome the differences between satellites and between experiments. The carrier would be a complete subsystem to support the experiments while providing minimal, unobtrusive interference with the host spacecraft; it would be a mechanical platform with standard interfaces for data, power, and thermal dissipation (cooling) for the experiments. The SET carrier would also isolate faults in power and data in its interface with the different experiments and prevent them from propagating to the host spacecraft.

The benefit of the carrier concept is to save money over multiple missions and to increase the number of experiments that can fly. Doing one design for the carrier would provide a standard interface for data, signaling, and power to many different experiments. Universities could easily use a simulated interface for the carrier to develop their unique experiments and have good assurance that their equipment would work immediately upon integration with the actual carrier before attaching the entire subsystem to a spacecraft.

The carrier concept generated a number of requirements. First, it had to "piggyback" modules mechanically onto the host satellites. Next, it needed to be small in both size and volume and have low mass. Third, it had to keep power consumption low and have low heat dissipation. Finally, it needed to isolate from the host spacecraft faults or failures both among the experiments and within itself.

As a program manager, I worked on the carrier at JHU/APL together with a lead project engineer and a systems engineer from late winter 2002 to the fall of 2002—about 8 months. A change in the collaborative effort moved the project to NASA in October 2002 for the remainder of the effort. JHU/APL no longer had any involvement. Though we did not finish the SET carrier project, we did make a number of useful trade-offs that illustrate some basic principles for designing subsystems for spacecraft.

12.1.2   Economics

One of the important determinations that we made was that the effort to design and build the first SET carrier subsystem and then integrate it with a spacecraft within three years would cost between US $5MM and US $8MM. The final cost would depend heavily on the final set of features and specifications, which NASA and JHU/APL had not completed by the time the project switched to full NASA oversight. If features had been minimized and the time stretched out to 3.5 years, the final cost would have been closer to US$5MM. Full features and an accelerated schedule of 3 years would have driven costs closer to the US$8MM.

The benefit of doing the design once and then replicating it for successive missions would reduce cost of the carrier to less than US$1MM per mission. This compared very favorably with spending more than US$10MM per mission in a more typical effort that would use a custom design for every mission.

Another benefit of the SET carrier concept is that it gives opportunity to universities to fly experiments in space at a low cost. Generally, they could get by with between US$50K and US$100K to spend on an experiment, which is within the range of most monetary grants to universities.

12.2   People and Disciplines

The initial design team at JHU/APL had one program manager, one lead project engineer, and one systems engineer who consulted part time for the team. All were trained as electrical engineers, and two of us had significant software experience. We also worked closely with a group of colleagues from NASA—Goddard Space Flight Center in Greenbelt, MD, to develop the requirements.

If JHU/APL had continued the design, development, fabrication, and integration of the SET carrier (and the project had not moved to NASA), the team would have been expanded significantly for the remaining phases of development. The design team would have added two electrical engineers to design the hardware and develop the ground support equipment (GSE), one to two software engineers, and one mechanical engineer to the original design team of three.

For fabricating the SET carrier platform, JHU/APL would have employed the following staff in a part time role:

- Two computer aided design (CAD) designers
- Six fabrication and assembly staff
- Two inspectors
- Two machinists
- One component engineer
- Administration support

12.3 Architecting and Architecture

12.3.1 Process

A spacecraft subsystem is mission-critical. We planned to use the V-model process to develop the SET carrier.

12.3.2 Parameters

We had a number of parameter types to define and consider in designing the SET carrier. They include

- Mechanical—size (volume), configuration, mass, and thermal conduction and dissipation
- Power—consumption, fault tolerance
- Component selection—low power and radiation hardness
- Data—memory size, throughput and channel bandwidth, fault tolerance to improper communications
- Command and data handling—telemetry to the GSE on Earth, control to the processor

12.3.3 Architecture

We studied two different configurations for the SET carrier, particularly the cable harnessing. This was important because cables can contribute significant weight to a subsystem; they also affect the ease of integration and fault tolerance [1].

We settled on a distributed approach to power distribution and data networking. By distributing the raw 28 VDC and providing a local DC–DC converter at each experiment, we could reduce the size and weight of the power cable. It also made fault tolerance easier; if any experiment failed, its converter would prevent it from dragging down the remaining experiments. Using a network cable between each experiment in a ring configuration would also reduce the size of the cable over a centralized "star" configuration.

12.3.4   Interfaces

A spacecraft has a number of specific, physical interfaces:

- Mechanical—to the host spacecraft and to the experiments
- Electrical signaling—command and control signals, connectors
- Power—raw 28 VDC from the host spacecraft, regulated DC voltages to the experiments
- Thermal—thermal paths from the experiments through the SET carrier to the host spacecraft
- Data—formats of commands, signals, and data transfers

12.4   Phases

This section of the case study is thin. A better example of a set of development phases with many more details is found in Chapter 10. We never got to complete the Concept Phase before the project moved out of our hands. If it had stayed and we finished developing it, the SET carrier would have followed this course:

- Concept—complete all specifications for features and development timeline, define the architecture, and select basic components (duration ∼10 months)
- Preliminary design—finish breadboarding all prototype circuits and modules, define all the fabrication processes, and have all design processes running with initial drafts of all documents (duration ∼10 months)
- Critical design—prepare engineering models, begin system tests, begin fabrication of flight components and circuit boards, and have an initial GSE system working (duration ∼12 months)

- Fabrication and integration—complete the fabrication, assembly, inspection of all modules and the SET carrier platform, perform all environmental tests, and integrate the SET carrier on the launch vehicle (duration ~ 14 months)
- Launch and mission (duration ~ 6 months)

Please recognize that these durations would overlap to give a shorter calendar time then a simple summation of these phases.

12.5 Scheduling and Estimating

The project had one important deadline, a launch opportunity in December 2005. This constraint indicated top-down planning because all scheduling choices flowed backwards from that launch date. The integration of the SET carrier to the host spacecraft had to complete about 6 months before launch, and delivery of the SET carrier to integration had to occur about 5–6 months before that, or about 1 year before launch.

I used Microsoft Project™ to perform bottom-up planning to meet these deadlines. For the people and projected team mention in Section 12.2, I planned for dividing the tasks to take advantage of parallel effort and to avoid sequential bottlenecks. The most important thing that I did was to avoid loading anyone with more than 50% of their time (per the reasons given in Chapter 10).

Microsoft Project™ allowed us to calculate a person's total effort quite easily so that we could estimate cost fairly well—to about two digits of precision. As mentioned in Chapter 10, no one really does any better than two digits of precision in any estimate of effort and cost and time.

12.6 Documentation

12.6.1 Types

We had plans for a full slate of documents to cover the project. Table 12.1 lists the major documents we planned to prepare had we been allowed to finish the project.

12.6.2 General Formats for Documents

We did develop separate plans: a systems engineering plan and a project plan. Normally, I would have combined these two documents into a single project plan, as done in Chapter 10; JHU/APL already had a systems

TABLE 12.1

List of Documents for the Set Carrier Subsystem [2]

	CoDR	PDR	CDR	PER	PSR	Designated Author
Mission-level documents						
Concept of operations document	D	F	△			Sponsor organization
Mission requirements document	F	△				Sponsor organization
Mission specific requirements	F	△				Sponsor organization
Carrier-level documents/databases						
Systems engineering plan (SEP)	F					Systems engineer
Risk management plan (included in SEP)	D	F				Systems engineer
Risk management database	△	△	△	△	△	Systems engineer
Risk watch list	△	△	△	△	△	Systems engineer
Fault tree analysis (FTA)	I	D	F	△	△	Systems engineer
Failure modes effects analysis (FMEA)		I	D	△	△	Lead engineer
Development plans	D	F				Systems engineer
Architecture development plan	D	F	△			Lead engineer
Software development plan	D	F	△			Software lead
Electronics development plan	D	F	△			Hardware lead
Mechanical packaging development plan	D	F	△			Packaging engineer
GSE development plan	D	F	△			GSE engineer
Product assurance implementation plan (PAIP)	D	F				Product assurance engineer
Parts control plan	D	F				Product assurance engineer
Parts inventory list		D	F	△		Product assurance engineer
Requirements document	D	F	△			Systems engineer
Compliance matrix	I	D	F	△	△	Systems engineer
System test plan	I	D	F			Systems engineer
Test procedure(s)			I	F	△	Lead engineer
Test results					F	Lead engineer

Document						Responsible
Signature list	I			F		Systems engineer
Action item database	Δ			Δ	Δ	Systems engineer
Documentation release schedule (document list)	D		F	D	Δ	Systems engineer
Software users manual/maintenance document	I	D	F			Software lead
Training manual	I	D	F			Systems engineer
Experiment documents						
Experiment Interface Control Document (ICD)	D	F	Δ			Systems engineer
Host spacecraft documents						
Host to payload ICD	D	F			Δ	Systems engineer
Safety data package	D	F		Δ		Systems engineer

Key: I, Initial development—a full outline of the document has been established. Writing of some sections has begun; D, Complete draft—the document is completely written and is undergoing review. A very small number of TBDs can remain, but these are limited to specific pieces of information, not entire sections or subsections; F, Released final version—completed initial release; Δ, Updates to released version—re-released with changes; SEP, system engineering plan; CoDR, conceptual design review; PDR, preliminary design review; CDR, critical design review; PER, pre-environmental review; PSR, pre-ship review.

engineering plan in place, so I wrote the project plan to fill in the specifics it left out. Both plans are outlined as follows [2].

Systems Engineering Plan

1. Objective
2. Scope
3. Roles and Responsibilities
4. System Specification and Performance Verification
    4.1. Requirements and Requirement Flow Down
    4.2. Technical Performance Standards
    4.3. Interface Definition and Control
    4.4. Configuration Management and Change Tracking
    4.5. System Validation
    4.6. Performance Verification
    4.7. Technical Performance Trending
    4.8. System-Level Design Guidelines
5. Risk Management
    5.1. Project Risk Management
    5.2. Fault Tree Analysis (FTA)
    5.3. Failure Modes and Effects Analysis (FMEA)
    5.4. Margin Management
6. Independent Reviews
    6.1. Peer Reviews Requirements
    6.2. Formal Reviews
    6.3. Action Item Management
7. Systems Engineering Documentation

Project Plan

1. Introduction
    1.1. Purpose
    1.2. Scope
    1.3. Definitions, Acronyms, and Abbreviations
    1.4. References
    1.5. Overview
2. Project Overview
    2.1. Project Purpose, Scope, and Objectives
    2.2. Assumptions and Constraints
    2.3. Project Deliverables
    2.4. Evolution of the Project Plan

3. Project Organization

3.1. Program Structure

3.2. Organizational Structures

3.3. External Interfaces and Organizations

3.4. Roles and Responsibilities

    3.4.1. Program Manager

    3.4.2. Systems Engineer

    3.4.3. Hardware Engineering

    3.4.4. Software Engineering

    3.4.5. Mechanical, Packaging, and Thermal Engineering

    3.4.6. Fabrication Engineering

    3.4.7. Parts Quality Assurance

4. Management Process

4.1. Project Estimates

4.2. Project Plan

    4.2.1. Phase Plan

    4.2.2. Iteration Objectives

    4.2.3. Releases

    4.2.4. Project Schedule

    4.2.5. Project Resources

    4.2.6. Budget

4.3. Iteration Plans

4.4. Project Monitoring and Control

    4.4.1. Requirements Management Plan

    4.4.2. Schedule Control Plan

    4.4.3. Budget Control Plan

    4.4.4. Quality Control Plan

    4.4.5. Approval, Distribution, and Archiving Plan

4.5. Risk Management Plan

4.6. Close-out Plan

5. System Architecture Development Process

5.1. Overview

5.2. Management and Staffing

5.3. Schedule and Iteration Plans

5.4. Design Inputs, Design Outputs, and Documents Required

5.5. Standards and Practices

12.7   Requirements and Standards

12.7.1   NASA Standards

Several sets of regulations and standards would have applied to SET carrier. Table 10.2 lists some of those standards.

12.7.2   Preparing Requirements

NASA and JHU/APL worked together at weekly meetings through the spring and summer of 2002 to refine the requirements. This may seem a bit unusual in its frequency but the NASA Greenbelt facility was only about 25 mi away from JHU/APL and travel was fairly easy for the team from NASA.

12.8   Analysis

We only just began the analyses that are so important to designing space subsystems before the project moved over to NASA. We performed trade-off studies for feasibility, architecture, and planning. We did not have much opportunity to do further analyses before the project moved over to NASA. Had we continued the project at JHU/APL, we would have simulated operations of the system including the ground support equipment (GSE). We eventually would have performed both FTA and FMEA on the architecture and design.

The GSE engineer would have worked with both software engineers and the hardware engineer to simulate operations and to exercise aspects of the system. Both the lead project engineer and the hardware engineer would have performed the fault tree analysis (FTA) and failure modes effects analysis (FMEA). All of these analyses would have occurred primarily during the preliminary design phase and finished early in the critical design phase.

12.9   Design Trade-Offs

12.9.1   Architecture

We compared several distributed approaches to power distribution and data networking with a traditional, centralized "star" configuration, which provided the baseline [1]. We calculated mass, DC–DC converter efficiency, power consumed, cost, and harness weight, on a per-experiment basis. With these values and assuming a payload of eight experiments, we then calculated the differentials for mass, power, and cost between the baseline configuration, which was a traditional, centralized "star" and other approaches. Finally, we examined and compared the following concerns between the different approaches:

- Design complexity
- Integration risks
- Fault isolation
- Experiment flexibility
- Mission adaptability

Table 12.2 lists these comparisons.

The distributed approaches (one is illustrated in Figure 12.1) were better than the centralized "star" configuration, illustrated in Figure 12.2. The centralized, "star" configuration would produce the regulated 15 and 5 VDC

TABLE 12.2

Comparing the Relative Merits of Centralized "star" vs. Distributed Approaches to Subsystem Architecture

Description	Mass per Experiment (grams)	Converter Efficiency (%)	Power per Experiment (W)	Cost per Experiment ($)	Harness Weight (grams)	Mass Differential (%)	Power Differential (%)	Cost Differential (%)	Design Complexity	Reduced Integration Risks	Fault Isolation	Experiment Flexibility	Mission Adaptability
Baseline centralized power distribution	231	75	4.02	13,407	296	—	—	—	Mid	Mid	Mid	Mid	Mid
Distributed—separate power converter #1 and network communications per experiment	299	68	5.22	28,719	199	38	17	114	Less	More	Much more	Much more	More
Distributed—shared power converter #1 and network communications between two experiments	129	74	4.08	6,113	153	−44	2	−54	More	Slightly more	Less	Mid	Mid
Distributed—separate power converter #2 and network communications per experiment	217	68	4.45	13,044	151	−6	11	−3	More	Much more	Much more	Mid	Mid
Distributed—separate power converter #3 and network communications per experiment	236	62	4.86	7,778	151	2	21	−42	More	Much more	Much more	Mid	Mid
Distributed—separate power converter #4 and network communications per experiment	266	72	4.21	11,215	151	15	5	−16	More	Much more	Much more	Mid	Mid

Source: Modified from Fowler, K. R., Frank, L. J., and Williams, R. L., *IEEE Transactions on Instrumentation and Measurement*, Vol. 53, No. 4, August 2004. pp. 1065–1070, © 2004 IEEE. Used with permission from IEEE.

NOTE: Boxed line was the selected design approach.

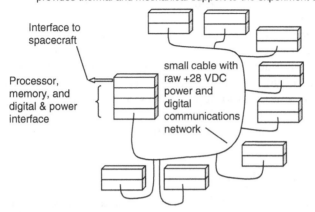

Individual experiments (on top) with local power converters and analog-to-digital converters in bottom boxes, each of which also provides thermal and mechanical support to the experiment card

Interface to spacecraft

Processor, memory, and digital & power interface

small cable with raw +28 VDC power and digital communications network

FIGURE 12.1
Distributed approach to the architecture of the SET carrier. This configuration had the distinct advantages of lower weight, better fault tolerance, and easier integration. (*Source:* Fowler, K. R., Frank, L. J., and Williams, R. L., *IEEE Transactions on Instrumentation and Measurement*, Vol. 53, No. 4, August 2004, pp. 1065–1070, © 2004 IEEE. Used with permission from IEEE.)

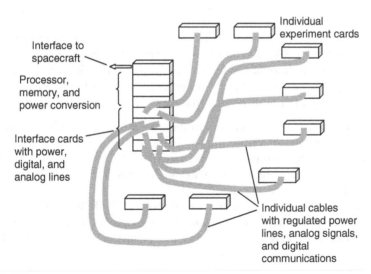

Individual experiment cards

Interface to spacecraft

Processor, memory, and power conversion

Interface cards with power, digital, and analog lines

Individual cables with regulated power lines, analog signals, and digital communications

FIGURE 12.2
Centralized "star" approach to the architecture of the SET carrier. While this configuration was slightly cheaper in the cost components, it was outweighed by the complexity of assembly, sheer mass, and more complicated integration. (*Source:* Fowler, K. R., Frank, L. J., and Williams, R. L., *IEEE Transactions on Instrumentation and Measurement*, Vol. 53, No. 4, August 2004, pp. 1065–1070, © 2004 IEEE. Used with permission from IEEE.)

in the central processor, which would require heavier and more conductors in a point-to-point scheme than the distributed approaches, which route 28 VDC to the experiments, and provide a local DC–DC converter at or near each experiment.

The centralized "star" configuration could minimize the number of active components at the expense of more cabling and less fault tolerance. A distributed approach required more active components, DC–DC converters and analog-to-digital converters (ADCs) sited next to each experiment, but with smaller and fewer cables and ultimately lower weight. A distributed approach not only reduces weight, but also eases integration and reduces the risk of attaching the wrong cables [1].

12.9.2 Electronic Hardware

Space flight places special demands on integrated circuits (ICs) and circuit design. ICs must be radiation-hardened. Both components and the circuit design must have low power consumption. Finally, weight and power constraints keep memories small.

Processor trade-offs: We studied a number of different processors for the SET carrier. Besides radiation hardness and low power consumption—which were required of any IC considered—we focused on ease of use and corporate experience with their development tools. We also considered computational power. See Table 12.3 for the details of our trade-offs.

We selected the UT80CRH196KDS from United Technologies, a radiation-hard version of the Intel 80196 microprocessor, because it had the necessary qualities we desired. JHU/APL had corporate knowledge working with its software development tools and its price and availability were acceptable [1].

ADC trade-offs: We studied a number of different ADCs for the SET carrier. We focused on radiation hardness and low power consumption; resolution and speed were not big factors. The SET carrier ADCs measured temperature; only 8-bit resolution was needed. See Table 12.4 for the details of our trade-offs.

We selected the APL Temperature Remote I/O (TRIO) chip because it appeared to have the best combination of qualities that we desired. JHU/APL had corporate knowledge working with this particular ADC, as well [1].

12.9.3 Power

A surprising number of problems are associated with power systems in satellites [3]. Consequently, we spent quite a bit of time considering power conversion and distribution schemes. The primary issue was whether to distribute the DC–DC converters or to use a set of centralized DC–DC converters. A centralized configuration would be slightly more efficient and probably use fewer components than a distributed approach, but the

TABLE 12.3

Comparing the Relative Merits of Different Processors

	UT69RH051	UT80CRH196KDS	Mongoose V	RAD 6000	RAD 750	RTX2010	UT69R000	Honeywell ESN
Selection criteria								
Radiation hardness	3	3	5	5	5	5	4	1
Low operating power	4	4	2	1	1	3	3	3
Performance	1	2	4	5	5	3	4	5
Small package size	5	5	3	2	2	5	4	1
Low part cost	5	5	2	0	0	4	4	2
Development tools	4	4	3	5	5	1	1	1
Availability	5	5	4	1	5	1	1	0
Future expansion	0	2	4	5	5	2	2	2
Features								
Architecture (bits)	8	16	32	32	32	16	16	16
Instruction/data space	64K/64K	64K/64K	4G	4G	4G	1M	1M/64K	1M/64K
Clock speed (MHz)	20	20	15	33	166	16	16	16
MIPS	<1	1.5	12.5	35	300	3.5	8	8

Rating: 0 = lowest, 5 = highest. High rating is better. Availability: 0 = no longer available, 1 = end of life.
Source: Modified from Fowler, K. R., Frank, L. J., and Williams, R. L., *IEEE Transactions on Instrumentation and Measurement*, Vol. 53, No. 4, August 2004, pp. 1065–1070. © 2004 IEEE. Used with permission from IEEE.
NOTE: Boxed column was the selected processor

TABLE 12.4
Comparing the Relative Merits of Different ADCs

	AD7572	AD7672	AD1672	Maxwell 7872RFP	APL TRIO Chip	Intersil 9008RH	AD571S	AD574
Selection criteria								
Radiation hardness	2	2	3	2	5	5	2	4
Low operating power	2	3	2	4	5	1	2	0
Accuracy	5	5	5	5	3	5	4	5
Small package size	1	1	3	4	5	3	2	1
Acquisition rate	1	1	5	4	3	5	1	1
Resolution	5	5	5	5	4	2	4	5
Availability	5	1	4	5	3	0	5	5
Features								
Resolution (bits)	12	12	12	14	10	8	10	12
Required voltages	+5,−12	+5,−15	+5	5	+3.3	15	+5,−15	+5,15
Maximum power dissipation (mW)	215	179	363	95	11	>400	275	725

Rating: 0 = lowest, 5 = highest. High rating is better.
Source: Modified from Fowler, K. R., Frank, L. J., and Williams, R. L., *IEEE Transactions on Instrumentation and Measurement*, Vol. 53, No. 4, August 2004, pp. 1065–1070, © 2004 IEEE. Used with permission from IEEE.
NOTE: Boxed column was the selected ADC.

centralized configuration with its point-to-point architecture would require more and heavier cables. A distributed configuration would place DC–DC converters near each experiment; this configuration was more fault tolerant—if any experiment failed, its converter would prevent it from dragging down the remaining experiments (Table 12.2) [1].

12.9.4   Cooling

The SET carrier had to provide a base plate for conductive cooling of its processor control unit (sometimes called the command and data handling unit or CMDH unit) and of the experiments. It was not to have any active cooling subsystem; power consumption and heat dissipation was to be limited.

12.9.5   Mechanical

The SET carrier had to have a platform for both the experiments and the processor control unit (or CMDH) unit. The general structure of the CMDH was a stacked "sandwich" chassis. This means that each circuit board had a machined aluminum band or ring that encased it, each "slice" of circuitry was stacked on top the next until all circuit boards were bolted together. This structure makes for efficient fabrication and a complete enclosure around all the circuit boards. The experiments, to study space environments such as micro-meteorite impacts, were to lie flat and side-by-side on the platform, exposed to the external space environment. In the distributed architecture, the DC–DC converters and ADCs would stack on a circuit board under each experiment's circuit board (Figure 12.3).

12.9.6   Software

We planned to write the source code in C, hold regular code reviews, and have defined, rigorous tests to verify functionality. The software design needed to be modular, and we planned to reuse code to support multiple different missions. Figures 12.4 and 12.5 give the planned structure of the software [1].

We believed a real-time operating system (RTOS) was necessary to aid development of the software and make the system more adaptable for different missions. A commercial RTOS seemed to make the most sense—shorter time to install and run, vendor technical support, and ease of software development. The choices quickly narrowed down because most commercial RTOSs had either too many features or architectures that required non-radiation-hard components in the interface; memory also constrained the size of the RTOS [1].

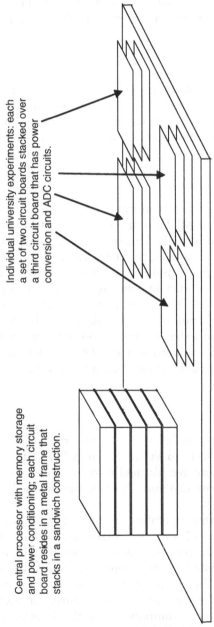

Individual university experiments: each a set of two circuit boards stacked over a third circuit board that has power conversion and ADC circuits.

Central processor with memory storage and power conditioning; each circuit board resides in a metal frame that stacks in a sandwich construction.

Deck provides mechanical support to the processor unit and experiments and strap down for the cables, and conducts heat to the host spacecraft.

FIGURE 12.3
A schematic of what the SET carrier might look like. (© 2007, Kim Fowler. Used with permission. All rights reserved.)

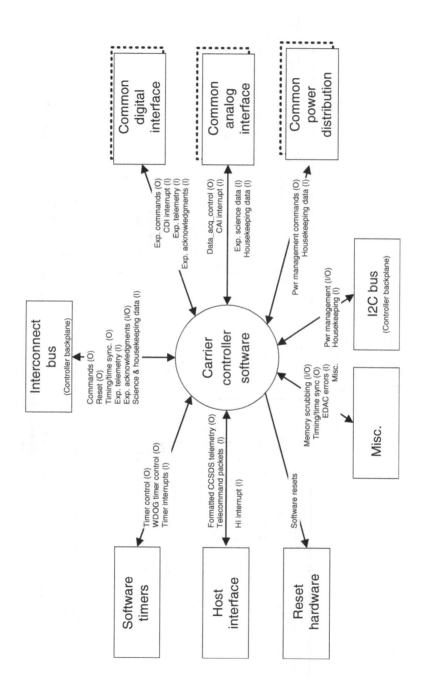

FIGURE 12.4

Context diagram for the software. (*Source:* Fowler, K. R., Frank, L. J., and Williams, R. L., *IEEE Transactions on Instrumentation and Measurement*, Vol. 53, No. 4, August 2004, pp. 1065–1070, © 2004 IEEE. Used with permission from IEEE.)

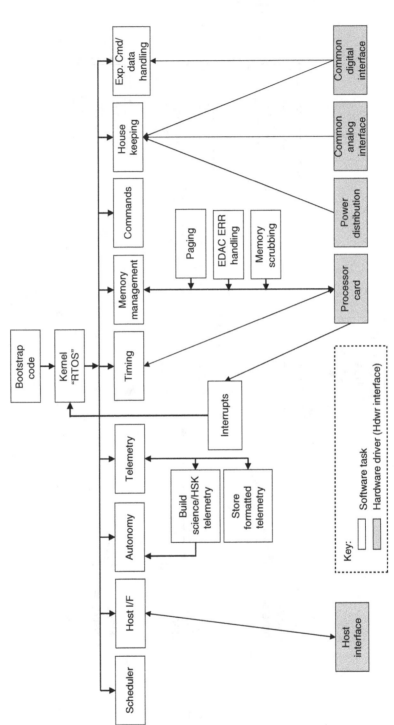

FIGURE 12.5

Software tasks required. (*Source:* Fowler, K. R., Frank, L. J., and Williams, R. L., *IEEE Transactions on Instrumentation and Measurement*, Vol. 53, No. 4, August 2004, pp. 1065–1070, © 2004 IEEE. Used with permission from IEEE.)

12.9.7 Ground Support Equipment

We planned for the ground support equipment or GSE to comprise nearly all COTS subsystems. We planned for a desktop computer or two to receive, display, store, and distribute the data. NASA would receive the telemetry from the spacecraft and feed the data to the GSE, which then distributed the data to the university experimenters. We seriously considered COTS software to display the data and store the values in a database, but we had not made a decision when the project left JHU/APL.

12.9.8 Buy vs. Build

As is true for most spacecraft, most of the SET carrier was going to be custom; the requirements for radiation hardness, low weight, and low power consumption all drive a custom design. We had planned for COTS RTOS software, GSE, and GSE software.

12.10 Tests

As already mentioned, JHU/APL did not complete the SET carrier, and so this case study has very little on the later stages of development other than a statement of what had been planned. Chapter 10 covers the tests that the SET carrier would have needed for its components and subsystems. The only difference is that the tests would also need to account for the mechanical, electrical, and data interfaces with the university experiments. The simulators would need to represent the following:

- Host spacecraft
- Experiments
- Unfinished portions of the SET carrier
- Telemetry to the host spacecraft

12.11 Integration

As already mentioned, JHU/APL did not complete the SET carrier, so this case study has very little on the later stages of development other than a statement of what had been planned. Chapter 10 covers the integration tests that the SET carrier would have needed; as with the component and subsystem tests, integration would have had to account for the mechanical, electrical, and data interfaces with the university experiments.

12.12   Manufacturing and Fabrication

As already mentioned, JHU/APL did not complete the SET carrier, so this case study has very little on the later stages of development other than a statement of what had been planned. Chapter 10 covers the detail that the SET carrier would have needed for its manufacturing, fabrication, tests, and integration.

Some differences between fabricating a single instrument and fabricating the SET carrier are as follows:

- The SET carrier platform would have been redesigned and built anew for each different host spacecraft.
- After the SET processor control unit was in place and verified, then each individual experiment card would be added to the SET carrier and tested. The design team would then verify the entire flight-quality SET carrier and the experiment cards with an end-to-end system test.
- All these steps would have taken place in the Fabrication and Integration phase before taking the SET carrier to the satellite for integration with the host.

12.13   Support

As already mentioned, JHU/APL did not complete the SET carrier, so this case study has very little on the later stages of development other than a statement of what had been planned. Chapter 10 covers the detail that the SET carrier would have needed for its support during spacecraft integration, launch, and orbital mission.

12.14   Disposal

There is very little of which to dispose. The spacecraft falls out of orbit and burns up in the atmosphere. The fabrication facility must handle all its materials according to NASA and OSHA (in the United States, i.e., the Occupational Safety and Health Administration) regulations.

12.15   Liability

The primary liability is the risk of a mission failing and therefore not getting any scientific data. This risk is borne equally by the university experimenters, the SET carrier developers (NASA and JHU/APL), the contractor for the host spacecraft, and the launch vehicle (NASA).

12.16   Summary

12.16.1   Emphases

The distributed architecture for the SET carrier produces the best balance between weight, power, fault tolerance, and integration ease. It also provides the same interface for all experiments, which makes communications within the carrier and development of the experiments a little easier.

All ICs and components in space flight must be radiation-hard. The system must also survive the shock and vibration of launch and wide temperature swings on orbit.

12.16.2   Gotcha's

Some things were just out of our control. We could not finish the project because NASA moved it in-house to finish it in their facility. As far as we know, our technical abilities or management practices were not in question.

Acknowledgments

My thanks to Larry Frank and Robert Williams at JHU/APL for their collaboration on the design of the SET carrier.

References

1. Fowler, K.R., Frank, L.J., and Williams, R.L., Space Environment Testbed (SET): Adaptable System for Piggybacked Satellite Experiments, *IEEE Transactions on Instrumentation and Measurement*, Vol. 53, No. 4, August 2004, pp. 1065–1070.
2. Fowler, K.R., ESC-207: Mission-Critical and Safety-Critical Development, *Embedded Systems Conference Silicon Valley 2006 Conference Proceedings*, April 4, 2006, pp. 13–14.
3. Personal communications with colleagues at JHU/APL, 2002–2005.

13

Case Study 10—Programmer for Implanted Stimulators

13.1 Concept and Market

13.1.1 Who, What, Why, How, Where, and When

Chronic pain in the lower back and legs may be treated in several different ways. Since the 1980s, one particular form of treatment has gained favor—electrical stimulation of the spinal cord, which blocks pain signals traveling up the spinal cord to the brain. A small but important market has grown for implanting these stimulators in patients with chronic pain. After release from the hospital, a patient may program the desired stimulation through a transmitter that encodes parameters and then couples them through radio frequency (RF) to the implant. Figure 13.1 illustrates the general configuration.

Adjusting the stimulator for each patient, however, is extremely tedious because there are many possible combinations of parameter values. Just eight electrodes represent 6050 combinations of polarity, while 16 electrodes represent over 62 million possible combinations. A sophisticated programmer/transmitter, for use by medical staffs and patients, can ease the burden of adjustment by selecting appropriate subsets of stimulation parameters from the wide variety available. The programmer that my company designed was a pentop computer that allows patients to tailor their own treatment by drawing simple lines and touching screen buttons. Figure 13.2 shows a prototype of such a programmer.

The programmer was to be used primarily by patients in physicians' offices. A physician or assistant taped the programmer's antenna over the site of the implant, set up the session on the programmer, and then handed the programmer to the patient. Patients responded to simple instructions and answered questions posed on the screen; they also drew outlines of where they felt both pain and stimulation effects.

The programmer was intuitive and very easy for patients to use; patients were intimately involved in their own treatment, an important factor. It reduced the time spent by medical personnel to adjust stimulators from hours to minutes. The data were more reliable.

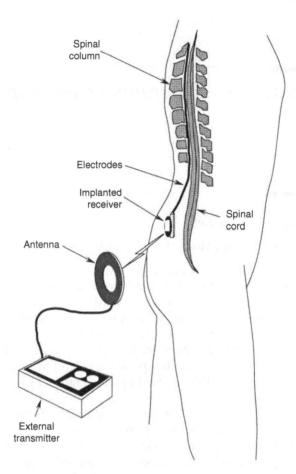

FIGURE 13.1
The general configuration for communicating with and programming an implanted stimulator.
(© 1996, Oxford University Press, used with permission.)

I helped develop the concept for this device over 13 years with partners in university medicine at The Johns Hopkins Hospital, Baltimore, MD. In 1998 I co-founded Stimsoft in Maryland, with two other partners to develop the programmer into a commercial device. The device and all its intellectual property assets sold in 2003.

13.1.2 Economics

Tens of thousands of stimulators are implanted each year around the world to treat chronic pain. (While I do not have exact numbers now, surgeries for implants may be approaching hundreds of thousands each year.) Physicians who implant stimulators could use a sophisticated programmer, which means that potentially thousands of units would be produced per year.

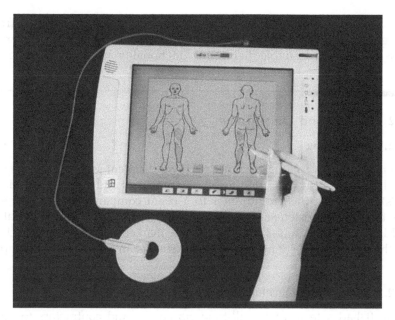

FIGURE 13.2
One example of a programmer used by patients for adjusting an implanted stimulator. (From the private collection of Kim Fowler, used with permission. All rights reserved.)

Worldwide, I estimate that the market might take about 5,000–10,000 programmers as a first run. I am purposely vague because I cannot confirm the exact numbers now. Marketing estimates have shown a low-volume market by any measure.

Although many of us hear about the high cost of medicine for patients, the ongoing effort to contain medical costs makes medical devices extremely cost-sensitive. This situation is further complicated by the desire or expectation that medical devices, such as this programmer, should last up to 10 years before replacement.

This is a contradictory and nearly impossible set of constraints for developing and selling a medical device:

- Low cost
- High margin
- Robust and rugged

13.1.3 Surveys and Focus Groups

To determine the utility and necessary features for the programmer, we surveyed targeted physicians at conferences as well as focus groups of physicians and medical staff. They considered the programmer a necessary complement to an implant. Their responses revealed that the programmer must be easy to use, require little training, and not cost them anything (they expected the manufacturer of the implants to supply a programmer free of charge to physicians). They clearly indicated that less functionality for ease of use was more desirable than more functionality and more training.

13.2   People and Disciplines

13.2.1   Marketing

The marketing of the programmer had two different stages. The first stage involved three principals of Stimsoft, the president and both vice presidents (VPs), marketing the programmer to a client company. The second stage was to potential customers—physicians who deal with chronic pain patients and implant stimulators. For this stage, the marketing group comprised both the three principals of Stimsoft and a marketing team from the client company.

13.2.2   Design and Development

Stimsoft had a team of 12 full-time employees and contractors to do the design and development of the programmer. That team comprised four software engineers, one hardware engineer who also wrote software, a software tester, a Food and Drug Administration (FDA) process and regulation specialist, a training specialist, a specialist for documentation and training, and an office administrator who doubled as receptionist. The two VPs also directed design of the architecture and reviewed technical progress.

13.2.3   Clinical Testing

We prepared for clinical testing by enlisting the participation of two separate medical centers outside of the Johns Hopkins Hospital system. At least one physician and two or three physician assistants in each center participated in the study of the programmer with patients.

Clinical testing is a standard and carefully monitored part of medical-device development. I will not elaborate further because it is out of the scope of this book. Careful communications with the FDA will clarify what is expected in setting up clinical tests.

In retrospect, we should have enlisted three or more medical centers and performed a more rigorous set of tests.

13.2.4   Management

The management was primarily the responsibility of the two VPs. One focused on software, while the other focused on hardware and systems development.

13.2.5   Manufacturing

Stimsoft outsourced the manufacturing and assembly of the first production run to Aubrey Group, a contract engineering and manufacturing firm. Aubrey Group had a team of about 25 people ranging from engineers to technicians and support staff. However, not everyone was involved in the production of Stimsoft's programmer and its antenna.

13.2.6   Sales, Distribution, Logistics

The client company eventually purchased the programmer and all its assets. They have a sales team numbering about 75 and a distribution channel to maintain the units.

13.3   Architecting and Architecture

13.3.1   Process

We used the V-model process to develop the programmer. During the early phases, we performed regular and in-depth code reviews. These reviews could force requirements changes if the team deemed it necessary. Everyone at Stimsoft was involved in maintaining a good process.

13.3.2   Parameters

The parameters fell into two basic categories: physical and subjective. The physical parameters included at least a 10-in. diagonal size screen, weight not in excess of 2 lb, resistance to disinfectant washes, sufficient battery reserves to last 8 hours, and the ability to communicate with a commercially-available stimulator implant. The subjective parameters included ease of use and comfort for the patient while resting the programmer in the lap.

13.3.3   Analysis

Most medical devices require thorough analysis to show both efficacy and safety. We certainly went through a lot of analysis to develop the programmer. First, it rested on a foundation of 13 years of university research, with data from nearly 1000 patients. Second, we attended focus groups in two different cities to refine our understanding of physician and patient desires. In one city, just the two VPs went to observe the proceedings of the focus groups; in the other city, most of the team went to observe. Third, the two VPs surveyed selected physicians at two different conferences to demonstrate prototype operations on pentop computers and to assess their acceptance of the new device. Fourth, the two software leads and the hardware engineer performed fault tree analysis (FTA) and failure modes effects analysis (FMEA) during the early phases of the project. Finally, Stimsoft undertook clinical studies to show efficacy and safety.

13.3.4   Architecture

The programmer was designed to be a stand-alone instrument to support a lone physician's office. It could store and print out data collected from a number of patients. It also could connect to the Internet through an analog telephone connection to download data. We had plans to eventually develop ways to analyze the large amounts of collected data, provide sophisticated analysis, and generate strategies for programming more effective stimulation patterns.

Figure 13.3 illustrates the architecture of the programmer. It comprised a personal computer (PC)-compatible pentop running Windows CE™. It communicated through a digital data stream with the transmitting antenna. It also supplied power to the antenna. The antenna was a commercially available subsystem that required no modification. The programmer also had a recharging cradle and an infrared (IR)-linked thermal printer. The final piece was a modem and telephone connection to download data to a central server.

The VPs, two software leads, and the hardware engineer defined the basic architecture of the programmer. The entire team was involved in refining the architecture.

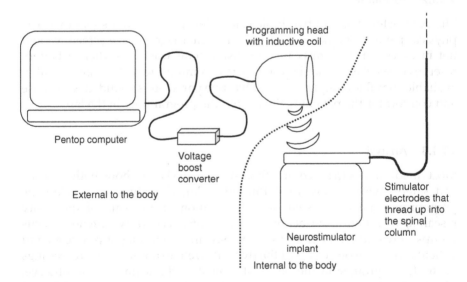

FIGURE 13.3
Block diagram of the architecture of the programmer. (© 2006 by Kim Fowler, used with permission. All rights reserved.)

13.3.5   Interfaces

The programmer had five types of interfaces: mechanical, electrical, IR, telephone, and human. The mechanical interface consisted of the pentop computer and its cradle. The electrical interface consisted of the pentop computer, its cradle for recharging, the power supply (a "wall wart"), and the antenna cable and power connections to the computer. The IR interface connected the pentop computer to its printer. The telephone interface connected the pentop computer to a telephone line through the cradle. Finally, the human interface was between the computer and the patient or medical staff with a stylus.

The human interface, primarily the graphical user interface (GUI) was critical to the design of the programmer. It was difficult to design because it was the focus of all the features and utility of the device.

13.4   Phases

13.4.1   Concept

The concept phase covered the business case, goals, objectives, and constraints. We determined the basic architecture of the programmer and its subsystems and components. We prepared and reviewed software prototypes of the GUI screens. We performed an initial risk analysis that included both FMEA and hazard analysis, which was essentially an FTA. We began all the standard documents (outlined in Section 13.6) particularly the plans: system, software, hardware, test, and configuration management.

We interviewed and selected our contract manufacturing firm (of the four that we researched and interviewed, three were very good). This contract firm would eventually build the production units. We also selected a commercial off-the-shelf (COTS) vendor for our pentop computer. We selected and installed Rational Rose® for the configuration management and design control system. We also set up our documentation system.

This phase was rather difficult for us to define. We were starting up the company, hiring people, and leasing equipment and office space. Suffice it to say the phase took about 2 years. This duration is too long for most projects.

13.4.2   Planning and Scheduling

The planning and scheduling phase lasted about a year and covered the business case, goals, objectives, constraints, and initial designs for the system, hardware, and software. We refined the basic architecture of the programmer and its subsystems and components. We reviewed storyboards of GUI operations and then refined and tested the software prototypes of the GUI screens. We attended and observed two different focus groups comprising medical personnel to better learn their desires. We had to find a new COTS vendor and pentop computer because the original vendor dropped their product line.

We continued the risk analysis that included both FMEA and hazard analysis. We continued revising and updating all the standard documents and began the product description document.

During this phase we began the design of the programmer through block diagrams, schematics showing logic diagrams, interface circuits, packaging plans, configuration and layout sketches, and preliminary analyses. We began estimates for cost, weight, power, and size, as well as the electronic designs.

We specified the software requirements, design, structure, state diagrams, design language (C++), and development systems. We also began developing Unified Modeling Language (UML) use cases and entering them into Rational Rose®, the configuration management and design control system. The software developers began writing code, performing code reviews, and testing each module.

13.4.3   Design and Development

The design and development phase lasted about 2 years but was not officially closed (we sold the programmer and company assets before completing the phase). During that time we completed the business case, goals, objectives, constraints, and the designs for the system, hardware, and software. We finalized the architecture of the programmer and its subsystems and components. We attended two different conferences for physicians who specialized in pain management and surveyed selected physicians. We demonstrated system and software prototypes of the GUI screens. We completed the risk analysis that included both FMEA and hazard analysis.

We continued revising and updating all the standard documents and completed a number of new ones. In particular, I wrote the design transfer plan that outlined the manufacturing of the programmer. Another principal in the company prepared the clinical plan. We also finished the design documents: software, hardware, and system. Finally, we recorded all the test results.

We completed the software requirements, the UML use cases, design, structure, and state diagrams. We did not complete all the software code reviews or testing each module before selling the company.

We worked with the contract manufacturer to design the DC–DC converter for the antenna and to build the units: computer, inline converter, and antenna. The contract manufacturer built 25 production prototype units. This might be considered a pilot run, which usually occurs in the next phase.

13.4.4   Controlled Release

The controlled release phase should last about a year. (We had sold the company's assets before entering this phase.) During that time the clinical trials and results are completed and presented. All documents are finished. Submission for FDA approval is made. The manufacturing begins.

13.4.5   Commercial Release

The commercial release phase swings into high gear after receiving FDA approval. Then sales and support begin, including manufacturing, distribution, and technical support. (Again, we had sold the company's assets before entering this phase.)

13.5   Scheduling

Scheduling, for a start-up company, is chaotic at best. We attempted to make top-down deadlines confirmed by bottom-up planning but events beyond our control often adjusted our timelines. Between requirements from the client, meeting with the FDA, contracting for manufacturing, and just dealing with human personalities, preparing and keeping to a schedule proved nearly impossible.

You should always be prepared to take more time than you expect. Do not expect less than 4–6 years from concept to FDA approval and first product.

13.6   Documentation

All phases had a standard set of documents that needed updating. Each phase also had some documents unique to it (see Table 13.1).

Everyone who generated any document had to make sure that all records were stored on the server, where they were backed up daily. Documents were also printed on paper and stored in a file cabinet. Even drafts of documents were stored in the filing cabinet as soon as they were generated. The office administrator made sure that documents were stored in the correct place under the appropriate phase.

Stimsoft had two documentation specialists and a specialist for FDA process and regulation who determined the basic types of documentation that we needed.

Some ways that we used to keep and track documents:

- Used colored folders to identify each phase. The colors we chose were: yellow, gray, red, green, and purple.

TABLE 13.1

Listing of Documents Developed for the Programmer
Project

Standard documents for all phases

- Project plan
- Development plans
  - System development plan
  - Software development plan
  - Hardware development plan
- Configuration management plan
- Requirements plan
- Specification plan
- V&V plan
- Document plan
- User manual
- Quality assurance phase review
- Marketing
- Technical reviews
- Technical communications (memos, letters, email notes)
- Document control forms

Phase 1: Concept
- Vision
- Risk plan
  - Risk assessment: FMEA and HA
  - Business risk

Phase 2: Planning and scheduling
- Product description
- Risk plan
  - Risk assessment: FMEA and hazard analysis
  - Business Risk

Phase 3: Design and development
- Product description
- Design transfer plan (includes manufacturing)
- Clinical plan
- Code standards
- Test results
  - Recorded errors
  - Test metrics
- Traceability

(Continued)

TABLE 13.1

Continued

- Design documents
  - Software design document
  - Source listings
  - Hardware design document
  - System design document

Phase 4: Controlled Release
- Product description
- Clinical results
- FDA submission

Phase 5: Commercial release
- Product description
- Design history file (DHF)
- Device master record
- Version description document (VDD)
- Publications
  - Brochures
  - Training materials
- FDA approval
- Training plan

- Divided up the filing cabinets into phases.
- Marked all documents "Obsolete" from the previous phase once a newer draft showed up in the current phase.
- Marked all current documents "Draft" until we signed off after review.
- Developed and adhered to a style guide.

13.7   Requirements and Standards

13.7.1   Market

The programmer was a Class II, 510K device that required premarket approval, or PMA. In "FDA-speak" this meant that we had to show through clinical trials that it was safe and effective.

The market comprised a number of people: users, influencers, and customers. The users were both medical staff in physicians' offices and patients with chronic pain. The medical staff, including the doctors, had to know how to set and run the programmer. Patients needed to perform rudimentary tasks, such as drawing circles on body outlines and pressing YES/NO buttons. Influencers were the medical staffs that needed or used the programmer; they ultimately were the most critical component in causing a purchase of a programmer. The hospitals or medical institutions, who were to purchase the programmers, were the actual customers, but they did not initiate the purchase process—the influencers did that.

13.7.2 Design and Development Standards

There were a number of standards to which the programmer had to adhere. Some of the more prominent ones are listed in Table 13.2 and include electromagnetic compatibility (EMC), electrostatic discharge (ESD), and product safety [1–3].

TABLE 13.2

Listing of Some of the Standards for the Programmer Project [1,2]

Category	Origin	Standard	Description
EMC	CENELEC, European Standards	EN 55011	Industrial, scientific, and medical radio frequency equipment—radio disturbance characteristics—limits and methods of measurement
		EN 61204-3	Low-voltage power supplies, DC output
		EN 61000	Electromagnetic compatibility, parts 2 and 4
		IEC 61000	Electromagnetic compatibility, parts 1 and 3
ESD	European Standards	IEC 61340-1	Electrostatics Part 1: Guide to the principles of electrostatic phenomena
		IEC 61340-5-1	Electrostatics Part 5-1: General requirements protection of electronic devices from electrostatic phenomena
		IEC 61340-5-2	Electrostatics Part 5-2: Protection of electronic devices from electrostatic phenomena, user guide
	USA	IEEE C62.47	IEEE guide on electrostatic discharge (ESD), characterization of the ESD environment
		IEEE C62.48	IEEE guide on interactions between power system disturbances and surge-protective devices
		IEEE C62.64	IEEE standard specifications for surge protectors used in low voltage data, communications, and signaling
		UL 1449	Transient voltage surge suppressors

(Continued)

TABLE 13.2

Continued

Category	Origin	Standard	Description
Safety	CENELEC	EN 61204	Low-voltage power supplies, DC output—safety requirements
	European Standards	IEC 60086	Primary batteries
		IEC 60601-1	Medical electrical equipment, part 1—general requirements for safety
		IEC 60601-1-2	Medical electrical equipment, part 1—general requirements for safety, Section 2—EMC
		IEC 60601-1-4	Medical electrical equipment, part 1—general requirements for safety, Section 4—programmable electrical medical systems
	USA	UL 60601-1	Medical electrical equipment, part 1—general requirements for safety
	ISO	ISO 11197	Medical electrical equipment—particular requirements for safety of medical supply units
		ISO/IEC Guide 63	Guide to the development and inclusion of safety aspects for medical devices
		ISO/TR 16142	Medical devices—guidance on the selection of standards in support of recognized essential principles of safety and performance
Biocom-patibility	USA—FDA	21 CFR 58	Prescribes practices for nonclinical laboratory studies to support applications to the FDA for medical devices
	ISO	ISO 10993-1	Use the Blue Book Memorandum from the FDA for testing for neurotoxicity and immunotoxicity of materials

CENELEC: European Committee for Electrotechnical Standardization; IEC: International Electrotechnical Commission; IEEE: Institute of Electrical and Electronics Engineers; ISO: International Organization for Standardization; UL: Underwriters Laboratories; SAE: Society for Automotive Engineers.

13.7.3 FDA Approval

All medical devices to be sold in the U.S. must receive FDA approval before commercial sales are permitted. Not only did we follow the FDA Design Control Guidance and perform clinical trials (albeit insufficiently for the first go around), but we also met with the FDA to talk with them and develop a professional relationship. We started early with the meetings to find out what the FDA expected and wanted. This was and still is a good practice to smooth the path to approval.

The company president, two VPs, FDA regulation specialist, software leads, and hardware design engineer were all involved in the meetings. The entire team, including business administrator, were regularly briefed on FDA approval and good processes.

13.7.4 Preparing Requirements

The requirements for the programmer had a long and varied path. University research over 13 years formed the basis for many requirements. We then refined the requirements through client marketing, focus groups, and surveys. These activities relied on the medical staffs (influencers in the sales domain) to provide insight as to need, utility, and cost.

The president (a neurosurgeon) and the VPs were the primary authors of the requirements. The engineering team, particularly the training specialist and two software leads, contributed to the effort by translating the requirements into specifications. The FDA process and regulation specialist guided our efforts.

13.8 Analysis

13.8.1 Feasibility

Thirteen years of university medical center research amply demonstrated the feasibility and utility of the programmer. The research covered nearly 1000 patients who tried and used various prototypes, resulting in many medical papers being published by the principals of Stimsoft and other medical personnel [4–10].

13.8.2 Focus Groups

During the planning and scheduling phase we attended and observed two different focus groups comprising medical personnel. The main result was the requirement that the programmer's operation be obvious and easy. Medical staff did not want to take much time in training—typically less than half an hour.

The client, who would eventually buy the programmer and its assets, set up the focus groups. The client contracted a company that specialized in running focus groups to hold the meetings. We traveled to two different cities to facilities owned and operated by the contracted focus group company.

13.8.3 Surveys

Three of us, the president and two VPs, attended two different conferences for physicians who specialized in pain management. The VPs demonstrated a prototype of the programmer and surveyed selected physicians. These physicians confirmed utility of the programmer and supplied suggestions for minor changes (Tables 13.3 and 13.4). The biggest concern that we uncovered in these surveys is that physicians expected the company supplying the implantable stimulators to supply programmers free of charge to them in exchange for their recommending and implanting the stimulators.

TABLE 13.3

Example of Physicians Surveyed for Programmer Functionality and Utility

Physician	Number of Patients Implanted/ Year	Average Time Spent Programming Each Patient (min)	Length Patients Remain in Care (Years)	Office Visits Per Year	Who Programs?	Program More if Less Complicated (1 = yes, 0 = no)	Program More if Less Time Intensive (1 = yes, 0 = no)	Acceptable Cost of Programmer ($)	Comments
Dr. A	120	60	ind.	3	Clinical nurse	1	1		Would rather nurses rather than reps program. Saving time would mean more money for . . .
Dr. B	40	10	0	2	Himself and TC	1	1		Tries to see as little of patients as possible. Implants them and returns to referring physician
Dr. C	100	8	ind.	1	Himself	0	0		Most patients only 1 visit/year. 10% visit 1/month
Dr. D	50	60	ind.	2	TC or rep	0	0	$0	Trials and percutaneous leads. Programming needs to be less intimidating and require less training

(Continued)

TABLE 13.3
Continued

Physician	Number of Patients Implanted/ Year	Average Time Spent Programming Each Patient (min)	Length Patients Remain in Care (Years)	Office Visits Per Year	Who Programs?	Program More if Less Complicated (1 = yes, 0 = no)	Program More if Less time Intensive (1 = yes, 0 =no)	Acceptable Cost of Programmer ($)	Comments
Dr. E	35	60	ind.		Nurse or rep	0	0	$2,500	Rather buy outright, not pay for use. "Do the right thing wizard - 'bingo' - will pay $2,500 . . ."
Dr. F	75	420	ind.	2	Nurse or rep	1	1		
Dr. G	25	45	ind.	4	Clinical staff	0	1		Has mandatory visits every 3 months to fine tune stimulator, it works well for his patients
Dr. H	12	15	4	2.5	Himself	0	0		Be careful about reimbursement issues. Physicians can only charge for "face-to-face" time
Average	57.1	84.8		2.4		0.4	0.5		
St. Dev.	37.8	137.4		0.9		0.5	0.5		

ind. = indefinite; TC = therapy consultant; rep = representative.

TABLE 13.4

Example Portion from a Survey for the Usability of the Programmer

User Test Survey—Summary Example		
Patient Tutorial Questions	*Yes*	*No*
Will these instructions be clear to your patient population?	7	1
Will pressing the buttons be easy for your patients?	8	0
Do you think the function of the OFF button will be obvious to your patients?	4	4
Will drawing on the body maps be easy for your patients?	7	1
Do you think the use of the front and back body maps will be obvious to your patients?	4	4
Do you think the Right and Left labels on the body maps will be effective with your patients?	6	2
Will marking the rating scales be easy for your patients?	8	0
Do you think your patients will be able to adjust the stimulation amplitude easily?	8	0
Do you think showing this tutorial to your patients will improve their comfort level with neurostimulation?	8	0
Average	6.7	1.3
Std. Dev.	1.7	1.7
% Agreement	83%	17%

Comments
- Have a way to stop and then resume the tutorial
- Voice was pleasant
- Several commented on how well worded it was
- One physician thought graphics were "too Mickey Mouse"
- One physician wants branded styluses—one with Alabama, the other with Auburn

Patient Tab Questions	*Yes*	*No*
Does the Patient Tab provide the basic information you need?	6	1
Do the buttons on the right side of the Patient Tab make sense to you?	8	0
Is the Leads information on the Patient Tab clear to you?	8	0
Does the automated Read Implant function make sense to you?	8	0
Are you likely to use the Remarks section?	8	0
Average	7.6	0.2
Std. Dev.	0.9	0.4
% Agreement	97%	3%

Comments
- Undo Changes button was not intuitive
- Training is a concern. If staff is to use programmer then they must be trained
- Consider have more information collection: male/female selection, age, diagnosis field (could use Remarks section) with boxes to check like radicularopathy, CRPS, spinal stenosis, and so forth, suggested CPT codes
- Read Implant button is not intuitive. Use "Read Serial Number" instead or consider "Read Generator"
- Use "Exit" instead of "LogOff" for button

(Continued)

TABLE 13.4

Continued

Protocol Tab Questions	Yes	No
Is it clear how you select the pulse parameters?	5	7
Is it clear how you select the electrode parameters?	3	4
Is it clear how you select thresholds?	4	3
Is the Trial Count box helpful?	5	1
Average	4.3	3.8
Std. Dev.	1.0	2.5
% Agreement	53%	47%

Comments
- The three columns for pulse parameters have a lot of information. The screen can be confusing, it needs training or clarification or simplification
- Explain "thresholds" or train personnel
- AOI threshold not clear at all
- Explain difference between Drawing and Rating under the Threshold tab
- Electrode parameters not clear, as was contiguous. Need training
- OFF polarity not clear–need training to explain
- Sets of 5, 6, 7, or 8 not should not show for a quad lead (1 × 4)
- Trial Count Box was not intuitive, it needs explanation or training
- Need a tutorial for physicians! Do not assume that they will do any better than patients; they are very busy, so unless programmer functions are explained they will not use it

Data Review Tab Questions	Yes	No
Is the Pain Drawing Sub Tab clear to you?	6	0
Is the Stim Data Sub Tab clear to you?	6	0
Was it easy to scroll through the pain drawings?	6	0
Is the Results Sub Tab clear to you.	6	0
Does the Results Sub Tab provide you with the information you need to choose the best settings?	6	0
Average	6.0	0.0
Std. Dev.	0.0	0.0
% Agreement	100%	0%

Comments
- Use the label "Pt." for patient rather than Pat. This is particularly true for Pat Rating
- Pain drawing was not clear because the color changed from when it was drawn to its review
- Would like an edit function for the pain drawing (Consider for programmer rev. 2)

Prescribe Tab Questions	Yes	No
Do you like the representation of the electrode polarities and pulse parameters that give the best settings?	4	0
Would you consider using the Retest Best Setting List Now during a patient appointment?	3	0
Is it clear to you how to select a prescription?	3	0
Average	3.3	0.0
Std. Dev.	0.6	0.0
% Agreement	100%	0%

(Continued)

TABLE 13.4

Continued

Comments		
• Would like to download all five prescriptions		

Stimulator Tab Questions	Yes	No
Press the Control Sub Tab within the screen. Is the Control Sub Tab clear to you?	2	0
Press either the Limits Sub Tab or the Next Button. Is the Limits Sub Tab clear to you?	2	0
Do you feel that you could easily edit a setting for the patient's prescription?	2	0
Would you feel comfortable programming a patient using the IPG Tab?	2	0
Would you feel more comfortable if the TC/CES did the programming?	0	1
Average	1.6	0.2
Std. Dev.	0.9	0.4
% Agreement	89%	11%

Comments
• Programmer needs to indicate when its done transmitting to the stimulator (if not instantaneous)
• Wants direct control, not ramp of stimulation
• Stim OFF needs to be bigger

Wrap Up Questions	Yes	No
To move between screens, do you prefer using the Tabs (Yes) or the Next and Back buttons (No)?	7	1
Which tab format would you prefer: this one where you see all the selections (Yes), or one where you see only one set of selections at a time (No)?	7	1
Is the wording on the various tabs easy to read?	6	2
Are the graphics on the various tabs easy to understand?	8	0
Would you be willing to use programmer to gather and review patient data?	8	0
Would you be willing to use programmer to program your patients?	7	1
After a stimulation session is set, the tablet is given to the patient. Would you be comfortable letting patients use programmer to go through the stim trials independently after they have completed the Patient Tutorial?	8	0
Do you think programmer will facilitate programming stimulators?	8	0
Average	7.4	0.6
Std. Dev.	0.7	0.7
% Agreement	90%	10%

Comments for patient screens
• Takes too long to start
• Give reminder of OFF button function
• Reduce different sizes of fonts
• Most fonts too small
• Pastel colored titles are too light! Use only one dark color for consistency

(Continued)

TABLE 13.4

Continued

- OFF button should be relabeled as Stim OFF (could have multiple meanings—tablet OFF or session OFF or stim OFF)
- Use VAS (for visual-analog scale) instead of "Rating"
- Button operation was not always reliable—get click sound but no activation
- Some inconsistency between operation of VAS and the stimulation amplitude slider. VAS is a slash, stimulation amplitude is a slide (We will consider it for the Rev. 2 programmer)
- Patients will confuse left vs. right and front vs. back
- Would like 3-D figure drawings and verbal instructions for patients (This is a tall order. We will consider it for the Rev. 2 or later programmer)
- No easy way to show pain on side of body. Many patients have lateral pain
- Ask three (3) questions about pain—least, most, and average pain on an average day
- Ask when worst pain occurs (that is when stimulator is most effective)
- Consider limiting session to run for 30 min; some patients fatigue after that
- Add a cross-hatched bar or spinning ball to indicate saving of data and remaining time

Surveys contain surprises in comments and anecdotal evidence; some of them can serve you well, others are nonsense. Here are just a few that we collected:

"Most, if not all, [medical or physician's] practices do not have the time to optimize the programming for neurostimulation."

"Consider a rubber grip for the stylus, so that older patients could grip it more easily."

"Programmer would save clinical staff time in high-volume practices."

"Major, huge, monster advance!"

"We are stuck in a company X mindset for doing things. Another company is coming out with a totally different way of thinking [for programming neurostimulators]."

"In the South [southern USA], if someone begins with, 'Bless their heart,' then you know that they are about to 'unload' on them."

13.8.4 Heuristics, Calculations, and Numerical Simulations

We used minor spreadsheet calculations and simple heuristics in choosing components. For instance, we gave more credence to a vendor with many customers and sales of a particular pentop; this, unfortunately, turned out to be wrong. We should have investigated the market design cycles with more diligence than just accepting assurances from the vendor's salesmen that products would not be made obsolete. An old Cold War heuristic comes to mind—"Trust, but verify."

We did not rely heavily on simulations. The software engineers used some UML software modeling to prepare the earliest version of the software. One engineer used a software tool to generate a finite-state machine in the embedded software for the antenna. Another used a software simulation to emulate communications with the printer.

13.8.5 Storyboarding

The GUI was critically important to the success of the programmer. A training specialist worked with the entire design team to storyboard the operation of programming a stimulator to understand how best to order the screen features and then sequence through them. Most of the design team talked to medical staff and discussed how we might train them to use the programmer and comprehend its capabilities; afterwards, storyboarding again helped us with the proper sequence of training instructions.

13.8.6 Testing

Testing also played into our feasibility analysis. We used years of clinical research with three different versions of prototype programmers to form the basis of the commercial design.

13.9 Design Trade-Offs

13.9.1 Requirements

The requirements comprised several different types of parameters, subjective and physical. The subjective parameters included ease of use, comfort for the patient while holding the programmer in the lap, and simplicity in training. The physical parameters included size, weight, materials, and function.

Some examples of physical requirements follow. The screen, for instance, had to be at least 10 in. (25 cm) diagonally. The weight of the programmer was not to exceed 2 lb (about 1 kg). It also had to resist disinfectant washes. Functionality required many different conditions; a very basic requirement was that it had to communicate with a commercially-available neurostimulator implant; it had to communicate with a commercial printer through an IR link; its battery reserves had to run as long as 8 hours.

We had to follow the FDA "Design Control Guidance for Medical Device Manufacturers" [11]. This drove how we developed requirements.

The entire design team was involved in developing and refining the requirements.

13.9.2 Hardware

We made a number of design trade-offs with the hardware:

- Chose a COTS pentop computer (more on this decision in the section Build vs. Buy).

- Chose a thermal printer that had an IR communications link. It was a cheap peripheral that printed out a receipt form of record—good enough for an immediate record of a programming session.

- Developed a custom communications link with the COTS antenna. It had a custom DC–DC converter to boost the 5 VDC power to 9 VDC required by the COTS antenna.

- Designed a tab for the recharging cradle to prevent attachment of the antenna while the pentop was in the cradle. Our client and the FDA wanted the system not to operate while recharging so as to avoid any leakage paths from the wall power outlet to a patient.

The two VPs, the hardware design engineer, and two software leads contributed to these decisions.

13.9.3 Power

The programmer had battery power for two reasons—electrical isolation and portability. The system would not be allowed to operate while recharging so as to avoid any leakage paths from the wall power outlet to a patient. Furthermore, the "wall wart" for powering the recharging cradle had to be medical grade; that is, its transformer had to be electrically isolated so that no leakage currents could flow from the power outlet, through the primary windings of the transformer, to the DC circuitry following the secondary windings.

We strove for 8 hours of power, but usually only got 4 hours of use out of the battery pack. We considered redesigning the case to increase the size of the battery pack, but this would have necessitated custom design in both the case and the recharging cradle; we were trying to minimize custom design. We also considered making the batteries easy to swap out, but this would have incurred extra training for medical staff and another potential area of problems should someone install them in the programmer incorrectly.

The two VPs and the hardware design engineer were primarily responsible for these decisions.

13.9.4 Cooling

This was an easy choice—no fans. We did not want the maintenance burden of cleaning dust filters. Moreover, a fan would have seriously reduced the splash resistance of the programmer. Most pentop computers dissipate low enough power to not need fans.

13.9.5 Software

We chose to code the software in C++; it provided a robust methodology for the user interface. The pentop computer ran Windows CE®, which made porting the software from an engineer's desktop to the pentop much easier.

The user interface took the vast majority of time to develop and test. The need for intuitive use while being fault-tolerant demanded care and thoroughness. The GUI went through years of development; we had one instance where a tester said, "You know, when you trace around the outside perimeter of the screen three times and then touch the center of the screen, the computer crashes." Who thinks of doing these sorts of things?! It just points to the need for good design, careful review, and good testing.

We used careful design and development processes, such as regular code reviews, tests, and field tests, to prepare the software. The two VPs, the five software engineers, and the one software tester were all involved in developing software.

13.9.6 Buy vs. Build

This project turned out to be a great example of the challenge to decide between buying and building.

The pentop computer, antenna, and real-time operating system (RTOS; Windows CE)® were all chosen to be COTS. For the antenna and RTOS, this was a good choice; the antenna from the client would be in their inventory for a long time, as it would be used in other products with the implanted stimulators; the version of Windows CE® that we used would freeze with FDA approval and we could not update it. The problem with COTS arose with the pentop computer—the original vendor dropped their product line, even though they promised that they would stay in the market, which forced us to find a new COTS vendor and pentop computer.

One critical concern, which is important for low-volume medical devices, is component obsolescence. If a device is going to be on the market for more than 3 years, possibly as long as 10 years, larger subsystems may have to be custom-designed rather than purchased COTS. This is true for devices that use a either a pentop computer or a personal digital assistant (PDA) for its platform; pentop computers and PDAs have very short life cycles, typically about 6 months. If a low-volume medical device uses a COTS platform, such as a pentop computer or PDA, then either you have to buy sufficient quantity of units up front and store them in inventory over the life cycle or get a contractual agreement from the vendor to stock them for you. Of course, either of these options will cost your company more than buying COTS parts that do not account for obsolete parts or inventory.

Part of the problem for low-volume medical devices is that they must get FDA approval before they can sell. FDA approval does not allow for changes in the product once approved. If a COTS vendor changes a major subsystem that your company purchases for its medical product, you could be prevented from continued sales with a new vendor-supplied subsystem in the product.

The added cost of a custom design may not be far-fetched after all, particularly for the platform. A custom design allows you to get exactly the form factor and user interface that you want for the platform; it also allows you to uniquely brand the enclosure for your product. Finally, custom design does not fall prey to vendor-enforced obsolescence quite as easily as does COTS.

A custom design for the pentop computer, or at least choosing a military vendor with the capacity for long-term inventory and control, would have been better for us. We could then get sufficient battery capacity to run 8 hours. We could have designed the cradle to recharge the computer while preventing any patient programming (to avoid leakage paths), all the while without removing the antenna.

The two VPs and all the design engineers, both hardware and software, were involved in making the buy versus build decisions.

13.9.7 Manufacturing

The only real choice for us was to outsource the manufacturing. We found and retained a medical design and manufacturing contract firm to produce the final product. A number of factors went into choosing a firm:

- A proven track record
- Demonstrated understanding of FDA Design Control Guidance
- Appropriate references from previous clients
- Examples of good documentation from previous projects
- A well-thought-out proposal

I searched a medical design magazine for companies who claimed to do medical design and manufacturing. After surveying the contenders, I called each company and asked questions from a checklist comprising material from the first four aforementioned bullet points. From this, I rated each company and chose the four top-rated ones. Finally, I traveled with a consultant, chosen as an impartial third party, to my original deliberations, to meet each of the four candidate companies. We had a checklist of questions that we followed while talking to each company. The good news was that three of the four companies we talked to were very capable for the task. We ultimately settled on Aubrey Group in Irvine, CA.

13.10   Tests

13.10.1   Informal

Like most companies, we had some informal tests to check prototypes in the lab and demonstrate their use in the field. Conversations following those demonstrations with some technical field representatives from the client could be viewed as informal tests. Most of the design team was involved at some point in informal testing. Most of these tests were completed before the preliminary design review.

13.10.2   Peer Review

We held regular and rigorous reviews of hardware and software designs. Code reviews, in particular, were systematic. One person, usually a VP, acted as moderator and collected the action items that arose. The code's author would present the software module—its intent and implementation. Two other software engineers would review the code. These reviews occurred through all phases of software development (i.e., up through the design and development phase; we never got through the controlled release phase before selling the company and the client took over).

13.10.3   Subsystem Tests—Hardware

We had no real need for these types of test because the hardware was chosen as COTS. We relied on the certifications and declarations of the vendors for functionality, survivability (shock and vibration), and EMC.

13.10.4   Subsystem Tests—Software

The software had a defined and rigorous development cycle. It was tied into Rational Rose,® the design control system, and Clearquest,® the software test suite. Any time a software engineer deemed a module complete, it went to verification by the test engineer. These tests occurred all the way through all phases of software development (i.e., up through the design and development phase; we never got through the controlled release phase before selling the company and the client took over).

13.10.5   Simulators

The software engineers used several software simulators during the planning and scheduling phase, sometimes called the preliminary design phase. These simulators allowed engineers to continue designing code before peripheral units, such as the antenna and printer, were available.

We did not rely on the simulators alone. Once the actual peripheral units were available, the software engineers verified their code with software tests as described earlier.

13.11   Integration

13.11.1   System

As the contract manufacturer prepared units consisting of pentop computers connected to antennas through their built-in DC–DC converters, the software engineers exercised all the modules in the system with well-established tests. We did not complete these system tests before the company sold.

13.11.2   Environmental

We had an outside test house subject the units to shock, to a drop test of 3 ft (1 m) to a concrete floor, and to some minimal vibration. We found that the pentop computer really needed a better shock-resistant case. Again, because the company sold before all the tests were completed, we did not complete these. A custom design for the pentop computer or a military computer would have survived shock better. The hardware design engineer was primarily responsible for these tests.

We did some preliminary testing for EMC at a company that had an anechoic chamber for "precompliance" testing. The indication was that the pentop computer with the attached low-frequency antenna would probably pass EMC certification.

13.12   Manufacturing

13.12.1   Fabrication and Assembly

Aubrey Group, the contract manufacturing firm, fabricated the DC–DC converter and then attached it to the antenna. They also assembled the antenna and pentop computers into a completed programmer unit. They build 25 programmers in the first run before Stimsoft sold.

13.12.2   Tests

A suite of system tests were in design but never finished. Had the programmer gone on to commercial release, the tests would have been a part of manufacturing and would have been performed by the contract manufacturer.

13.13   Support

13.13.1   Logistics

Stimsoft was designing this programmer to be distributed by the client firm, who eventually bought us out. Each unit would have to be recorded and tracked throughout its life cycle. Stimsoft did not have the sales or support channels for such a massive undertaking.

13.13.2   Maintenance

About the only maintenance foreseen for the programmer was regular cleaning. This would be a disinfectant wipe-down by the medical staff that owned the unit.

13.13.3   Technical Support

When problems arose, Stimsoft was to provide troubleshooting for the client. We planned to use the design engineers in the beginning, but if enough support was demanded, we would then plan to hire some support staff— medical and engineering.

13.14   Disposal

The client firm would retrieve the units once a programmer failed or had reached end-of-life. Most likely the Waste from Electrical and Electronic Equipment (WEEE) directive would have come into play for recycling components of the programmer.

13.15   Liability

All medical devices have some degree of safety concerns. This programmer primarily had several safety concerns, none of which had to do with its intended operation for adjusting implanted stimulators. The concerns had to do with leakage currents, a warm computer potentially burning the lap of a patient, and failed batteries either exploding or leaking. Stimulators do not cause damage, only discomfort, if programmed inappropriately.

Regardless, a recall of the programmer to fix anything could cost thousands (or millions if large enough in numbers and late enough in market penetration) of dollars and potentially take years of effort. Then there would be answering to the FDA and undergoing audits to prove that the fixes were sufficient.

Even if the programmer operates correctly but a patient sues, everyone could come under scrutiny. Even the corporate veil is not always sufficient to protect the officers.

. Still when all is considered, a safe device, low probability of failures, and its utility, the programmer's liability is fairly low. It is not as low, however, as some of the other case studies, such as a hobby device or possibly the military equipment or satellite subsystems.

13.16   Summary

13.16.1   Emphases

Any medical device must follow the FDA's Design Control Guidance, which is freely available on the Web along with some other useful documents [11]. We at Stimsoft focused on setting up good processes and complete documentation. We had both documentation specialists and an FDA specialist onboard to help smooth the process.

You need a good team to build a good product. Do not overlook the importance of documentation and training specialists or of a good business administrator. You will also need, from time to time, a good FDA consultant to help you through the process.

We found that feasibility and thorough analyses, such as surveys, prototype demonstrations, and university research, all helped in refining the requirements and architecture of the programmer.

Finally, we focused on the GUI because it was so important to acceptable operation of the programmer. It had to be fault-tolerant and intuitive. The training specialist helped us immensely with her storyboarding efforts. Surveys and demonstrations of the prototype all helped refine the GUI.

13.16.2   Gotcha's

The constraints for developing a medical device are often contradictory. They usually need to be low-cost, robust, and rugged. They also tend to be low-volume sales so require high margin in the ratio of sales price to cost. This type of market is very difficult to understand.

Even though we focused on the GUI, it still proved difficult to build and code. It required the most effort of any part of the project. I am still not sure that we got it right. One of the problems is that it is human nature to add more features. Sometimes simpler is better.

The whole situation around whether to buy or to build is still out for consideration. We decided to buy the pentop computer to save time and effort in development. It saddled us with limitations—an unreliable COTS vendor, insufficient battery reserves, and insufficient shock resistance. On the other hand, a custom design or a military version would have been far more expensive, at least in the beginning of the market cycle. I believe that in the final analysis, we should have bitten the bullet and designed a custom platform.

References

1. EMC Standards, *Compliance Engineering*, Vol. 21, No. 1, 2004 Annual Reference Guide, pp. 75–82.
2. ESD Standards, *Compliance Engineering*, Vol. 21, No. 1, 2004 Annual Reference Guide, pp. 103–105.
3. Product Safety Standards, *Compliance Engineering*, Vol. 21, No. 1, 2004 Annual Reference Guide, pp. 165–172.
4. Fowler, K.R., Neurological Stimulation System, *Proceedings of the AAMI 21st Annual Meeting*, April 1986, p. 27.
5. North, R.B. and Fowler, K.R., Computer-Controlled, Patient-Interactive, Multichannel, Implanted Neurological Stimulators, *Applied Neurophysiology*, Vol. 50, 1987, pp. 39–41.
6. North, R.B., Nigrin, D.J., Szymanski, R., and Fowler, K.R., Computer-Controlled, Patient-Interactive, Multichannel, Implanted Neurological Stimulation System: Clinical Assessment, *Pain*, Suppl. 5, 1990, p. S83.
7. Fowler, K.R. and North, R.B., Computer-Optimized Neurological Stimulation, *Proceedings of the Annual International Conference of the IEEE Engineering in Medicine and Biology Society*, Vol. 13, No. 4, 1991, pp. 1692–1693.
8. North, R.B., Fowler, K.R., Nigrin, D.J., and Szymanski, R., Patient-Interactive, Computer-Controlled Neurological Stimulation System: Clinical Efficacy in Spinal Cord Stimulator Adjustment, *Journal of Neurosurgery*, Vol. 76, 1992, pp. 967–972.
9. North, R.B., Nigrin, D.J., Fowler, K.R., Szymanski, R., and Piantadosi, S., Automated "Pain Drawing" Analysis by Computer-Controlled, Patient-Neurological Stimulation System, *Pain*, Vol. 50, 1992, pp. 51–57.
10. North, R.B., Sieracki, J.M., Fowler, K.R., Alvarez, B., and Cutchis, P. N., Patient-Interactive, Microprocessor-Controlled Neurological Stimulation System, *Neuromodulation* Vol. 1, No. 4, 1998, pp. 185–193.
11. U.S. FDA, *Design Control Guidance for Medical Device Manufacturers*, March 11, 1997, relates to FDA 21 CFR 820.30 and sub-clause 4.4 of ISO9001. You can find it at http://www.fda.gov/cdrh/comp/designgd.pdf.

14

Case Study 11—Implanted Medical Devices

14.1 Concept and Market

14.1.1 Who, What, Why, How, Where, and When

This chapter is an amalgamation of several different medical devices that I have encountered. It will focus on devices that surgeons implant in patients to provide years of therapy (e.g. pacemakers, stimulators, and drug pumps). They are Class III devices that require premarket approval, or PMA. They are either life-sustaining or safety-critical devices and must be proved to be so through clinical trials to be both safe and effective. Figure 14.1 illustrates just one of these types of implantable medical devices.

Implanted devices that stimulate tissue electrically often are preferable to drug therapies or surgery because they have far fewer side effects and are reversible—they can be removed or turned off if they cause problems. Implantable drug pumps provide better, closed-loop control of drug delivery.

Surgeries to implant these devices occur every day in hospitals around the world. Some might be simple outpatient surgeries, such as stimulator implants. Others might require a week-long hospital stay.

14.1.2 Economics

Tens of thousands (or hundreds of thousands for pacemakers) are implanted each year to treat a variety of serious health issues. Each device and its implantation surgery can cost anywhere from US$40,000 to US$100,000. Each device can cost between US$5,000 and US$80,000.

Specialized devices generally exist in the realm of low-volume manufacturing. If a company builds 20,000 devices in a year and sells them for US$20,000 per unit, this amounts to gross sales of US$400MM per year—a nice business! More common devices, such as pacemakers, where a company may build 200,000–1,000,000 devices each year, edge into high-volume manufacturing with a decided emphasis on high quality. These might be lower in cost, say closer to US$5,000 per unit, but at these quantities the gross sales might approach US$5B per year.

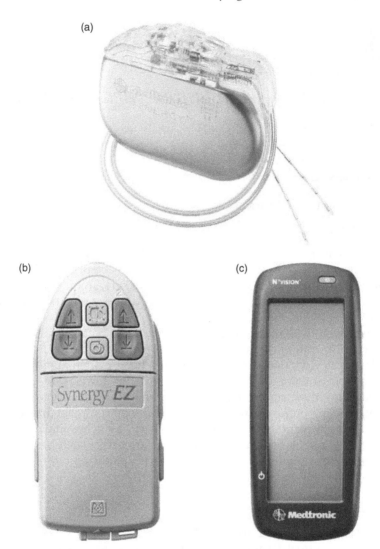

FIGURE 14.1
One example of an implantable medical device. (a) The implanted medical device, this is a Synergy® Neurostimulation System. (b) The patient's programmer, a Synergy® EZ Patient Programmer, that communicates through an inductively coupled signal with the implant. (c) The programmer, the N'Vision®, used by a physician or assistant to adjust the implant operation. (Photographs provided courtesy of Medtronic, Inc.)

14.2   People and Disciplines

14.2.1   Marketing

Companies that develop and sell these implanted devices have large, sophisticated marketing groups. People in these groups may not be physicians but they are knowledgeable and highly specialized. They often spend significant portions of their time educating and training both physicians and medical staffs in the use of the devices that they sell. Some of these folks accompany surgeons into operating rooms to observe and even advise on implantation of the devices.

Each of these marketing representatives (or sales representatives or clinical consultants or technical consultants—companies give them a number of different names) provide the "face" of the medical device company to the medical staffs and physicians to which they sell the devices. They can be quite helpful to physicians who have little time on their hands to learn new techniques or the latest features of particular medical devices.

Marketing "reps" usually handle between 100 and 400 implants each year. Obviously surgeons who handle more implants on average need less hand-holding. Consequently, a surgeon who only implants 5 or 10 devices each year will need more attention from the "rep" than a surgeon who routinely implants 100 or 200 or more devices per year. In well-developed therapies, such as heart pacemakers, surgeons or hospitals that implant thousands of devices each year may only need occasional visits from the "rep."

The numbers of marketing representatives needed depends on the frequency of implantation and the expertise of the physicians. For newer devices, such as brain stimulators, a "rep" may oversee every surgery performed in a hospital for the first few years until the company believes that the medical staff is proficient with implanting and using their devices. This means that if the company sells 20,000 devices in a year, it may need a marketing staff approaching 200 people and "reps." For a well-established market, such as cardiac pacemakers, where possibly as many as 400,000 devices are implanted each year worldwide, a company may only need 400 or 600 "reps" to peddle their devices.

These surgeons are a valuable source of feedback into the utility and efficacy of the medical devices that they implant. This is a prime example of where the customer can really provide insight into necessary requirements. Furthermore, many companies' marketing groups will sponsor conferences to help educate physicians into the use of their devices. They will also fund medical studies into efficacy of their devices to help with Food and Drug Administration (FDA) approval.

Medical marketing is a large, diverse, and fascinating field. You will want to draw on marketing's insight and experience when developing implanted medical devices; besides, you cannot avoid them.

14.2.2 Design

The design group for an implanted medical device will comprise a number of people [1]:

- Team lead or management
- Biomedical engineers

- Systems engineer or architect
- Electronic hardware engineers
- Software and firmware engineers
- Information technology (IT) specialists (which may be a corporate function)
- Industrial designers
- Mechanical engineers
- Manufacturing engineers
- Technicians
- Documentation specialists
- Regulatory specialists and regulatory affairs representative
- Quality assurance (QA)

Additional important groups:

- Clinical
- Packaging
- Purchasing

Each of these people will probably overlap in function and expertise. Some people, particularly in smaller organizations, may wear several "hats" and do several different jobs simultaneously.

The team lead manages the business of the team and assures that the project moves forward. They primarily care for the business case and for getting the appropriate people and expertise on staff for the project.

The biomedical engineers will most likely have the greatest interaction with the clinicians, physicians, medical staffs, and marketing team. They usually are responsible for the project requirements and assure that testing addresses the metrics. They tend to have more interaction with the clinical staff than anyone else.

The systems architect or engineer may be the team leader in small projects. In larger projects, the systems architect can play more of an advisory or mentoring role to the design team. A systems architect helps everyone see the "big picture" during development and can ease the functional integration of the different specialties.

The electronic hardware engineers and technicians shepherd the development of the circuitry of the medical device. The team may need a range of expertise—digital, microprocessor, analog, and radio frequency (RF). They handle component selection, circuit board design and fabrication, and event tree analysis (ETA), fault tree analysis (FTA), and failure modes and effects analysis (FMEA); they prepare testing and record the results; they write instructions for assembly and prepare the bill of materials. In some cases, a team might even design the integrated circuit chip for a processor or application-specific integrated circuit (ASIC) for the medical device.

The software engineers design, prepare, code, and test the software embedded in the device. They may also program the software of ancillary devices to control the device. They cooperate with the hardware engineers in ETA, FTA, and FMEA and the testing of the device. IT specialists are becoming more important all the time; many medical devices transmit data that must be stored in databases on servers and manipulated over great distances via networks and wireless connections.

Mechanical engineers and technicians develop the materials and enclosure of the medical device. The team may need a range of expertise—packaging, mechanisms, and disposables. They select materials, advise the machining and fabrication of the enclosure, cooperate with the ETA, FTA and FMEA analyses, and help write both the instructions for assembly and the bill of materials. If any mechanisms are to be incorporated in the device, such as a pump, the mechanical engineers will have significant design duties beyond the enclosure; in such a case, they will have input to the testing, as well.

Documentation specialists, regulatory specialists, and QA personnel are critical to the development of a medical device, as may be evident from Chapter 2. FDA approval requires diverse if not mounds of documentation; hence, a documentation specialist can ease the process considerably for the team. The same is true for a regulatory specialist. QA may take on these roles in small projects or organizations, but their focus tends to be more toward appropriate tests that adhere to standards.

14.2.3   Clinical Testing

Implanted devices must undergo clinical testing to demonstrate efficacy and safety. A separate team of physicians, nurse consultants, and clinical specialists handle clinical testing.

Clinical testing is a standard and carefully monitored part of medical device development. I will not elaborate because it is out of the scope of this book. Careful communications with the FDA will clarify what is expected in setting up clinical tests.

14.2.4   Management

The management team ranges from the president of the company, through vice presidents, directors, and team leaders. Each company is different and handles the management of projects differently. I will not elaborate because this subject is also out of the scope of this book.

14.2.5   Manufacturing

Many companies building implanted medical devices have in-house manufacturing facilities. They do this to control and monitor quality and production. Very small companies might outsource the manufacturing and assembly of a medical device to a contract engineering and manufacturing firm.

Manufacturing includes machining and welding the enclosure, fabricating and assembling the circuit boards, sealing up the penetrations and cable feed-throughs, and sterilizing the final product and its packaging. Much of the assembly is manual and labor-intensive; only for high-volume products would more automated lines be used.

14.2.6 Sales, Distribution, and Logistics

Administrative personnel and technicians provide the support to get the product to the customer and keep it working properly. Sales, distribution, and logistics are often tightly integrated with the marketing group.

14.3 Architecting and Architecture

14.3.1 Process

A medical device is safety-critical; its development must reflect that reality. The V-model process is most typical in developing an implanted medical device. Most medical devices cannot be reprogrammed once released; such would be contrary to FDA approval, so spiral development is ruled out beyond the design phase.

Software is a particular concern. It needs to be right, and it helps to be fault-tolerant too. This forces you to perform regular, in-depth code reviews in the early phases before clinical testing. All requirements must be tested and the results approved as satisfactory before the product goes into clinical testing.

Chapter 2 goes into more depth on processes.

14.3.2 Parameters

The parameters divide into three or four basic categories: clinical (or therapeutic), physical, user, and subjective. Some example categories of these parameters follow:

- *Clinical*—dosage, current limit, rates, frequency, protocols, and patterns of therapy
- *Physical*—volume, weight, power consumption, battery life, data format, telemetry encoding, radiated power, electromagnetic compatibility (EMC) limits, biocompatibility of materials
- *User*—input needed, button size on controller, programming options
- *Subjective*—comfort, limitation of mobility, ease of use

These parameters all feed into the requirements. Producing good requirements and specifications usually turns out to be one of the most difficult activities to complete, often requiring many iterations times before the end of the project.

14.3.3 Analysis

All implanted medical devices require thorough analysis to show both efficacy and safety. Analyses include

- University research—proof of concept, possibly via in-vitro and in-vivo animal studies
- Medical studies—often university clinical studies under the authority of a review board
- Surveys of physicians and medical staff—need, utility, and desires
- Focus groups—particularly if the device needs significant human input to operate
- Hazard analyses—FTA, FMEA, ETA
- Risk management, including margin and risk analyses (risk management is a more comprehensive term now that ISO 14971 has been published and is a key standard)
- Trade-off studies for feasibility analysis
- Clinical studies

14.3.4 Architecture

The architecture of implanted medical devices needs considerable attention. It must satisfy a number of goals, requirements, and objectives.

One way to view the architecture of an implanted medical device is that of an instrumentation system: input (sensing), processing, and output (actuation, delivery; Figure 14.2). The input may be any number of different things, such as a glucose sensor for an insulin pump or an electrical probe for neural impulses for a pacemaker. The input may also be a simple command from an external programmer, and then the implanted device runs autonomously until the next program change. Output can be several different responses: pumped drugs or electrical stimulation or mechanical extension or compression or even a simple telemetry signal.

One goal for implanted medical devices is for the output (chemical, pharmaceutical, or electrical) to have the desired effect with minimal side-effects. Another goal is to minimize the sensor's effect on the biological quantity being sensed. You do not want a glucose sensor producing chemicals that contaminate the sensed levels in the blood or corrosion distorting the pickup of minute neural impulses.

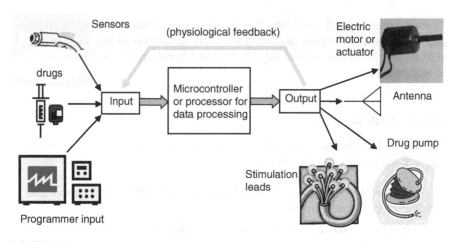

FIGURE 14.2
An instrumentation view of implanted medical devices. (© 2006 by Kim Fowler, used with permission. All rights reserved.)

Other important goals for implanted medical devices include dependability, manufacturability, and safety. The devices must be dependable, which means they are reliable, fault-tolerant, testable, and safe. The architecture should lend itself to manufacturing. Minimizing mechanical penetrations will reduce the chance of fluid invasion that can destroy the device.

As described in Chapter 2, a number of techniques can help toward reliable and fault-tolerant design: stress margins, redundancy and error checking, interlocks, and fail-safe (or trapdoor) operation.

Making a design testable can help check for functionality as specified; "testable" means that all portions or subsystems of the device can be reached and observed. Testable design assures conformance to the specifications but does not guarantee against design errors. Sometimes reuse of previously successful circuits and subsystem modules can provide a measure of confidence, but full system tests are still needed for each new application.

14.3.5 Interfaces

Implanted medical devices have three (or possibly four) types of interfaces: biological, mechanical, and electrical (and human, if the device is programmable). None of these is necessarily more important than the others; they all must receive considerable attention during design and development.

The biological interface includes therapeutic delivery, side effects, and sterilization. Clearly these concerns carry over into the mechanical and electrical interfaces, too. The best way to view the biological interface is what the device does to the tissues and fluids of the patient. Some basic questions to settle during design:

- Is the therapy (e.g., drugs, electrical stimulation, compression) appropriate? Will it work for years or for the time required? Will the body accommodate to the therapy and eventually render it ineffective?
- What are the side effects of the therapy? Can they be reversed or eliminated?
- What are the effects of the leaching of materials and chemicals out of the implant?
- What level of sterilization is needed to render the implant safe or acceptable?
- How is the device attached? How much movement is allowed?

The mechanical interface includes therapeutic delivery, sterilization, fluid invasion, and corrosion. The size, weight, and attachment points of the device also represent part of the mechanical interface. Another way to view the mechanical interface is in terms of what the tissues and fluids of the patient do to the device. Some basic questions to settle during design:

- Is the delivery of therapy (e.g., drugs, electrical stimulation, compression) reliable? Will it work for years?
- Could the delivery be easily obstructed or blocked by tissue reaction (e.g., fluid buildup plugging a pump outlet or scar tissue isolating a stimulation lead)?
- How long will the seals last before fluid invades?
- Are the materials sufficiently biocompatible?
- What are the corrosion products? What are the rates of corrosion?
- How heavy is the implant?
- How is it held or attached in place?

The electrical interface includes therapeutic delivery, battery capacity, signals between subsystems, and telemetry. Some basic questions to settle during design:

- Is the delivery of therapy (e.g., electrical stimulation, motor actuation for pumping drugs or compression) reliable? Will it work for years?

- Are the signals and data formats between subsystems robust and fault-tolerant?
- How long will the battery last for different rates of discharge and operation?
- Is the telemetry to an external receiver reliable for various depths of implantation?
- What are the potential leakage paths? How can they be isolated?

The human interface exists on either end of the instrumentation architecture, at the input and at the output. The patient or medical staff affects input by programming the device. The external programmer has some sort of graphical user interface (GUI), which is important to the utility of the entire system including the implanted device. Remember, *people prefer understanding and ease-of-use over performance*—this principle is a direct determinant of success in a patient's adopting the device. The output is the therapeutic result that patients sense or feel. If a patient perceives results, then perception also is important to the market success of adopting the device.

14.4 Phases

14.4.1 Concept

The concept phase covers the business case, goals, objectives, and constraints. All basic university and medical research should be complete to a degree that allows you to make basic assumptions about feasibility. You should determine the basic architecture of the device and review simulation prototypes if available. You should set up and perform initial analyses for risk and hazard (ETA, FTA, and FMEA). You should have notified manufacturing (or selected a contract manufacturing firm) and support logistics, who will eventually build and support the production devices. You should establish the standard documents (see Table 14.1, which is very similar to that outlined in Section 13.6 and in Table 13.1), particularly the plans: system, software, hardware, test, and configuration management.

During the concept phase, you should demonstrate that the requirements of the project are understood and that the proposed approach will meet these requirements. Example items to be addressed in the concept design review (CoDR) are as follows:

- Business case, goals, objectives, and constraints
- Research—university, medical, literature, patent searches
- Review initial drafts of standard documents and phase specific documents—Table 14.1

TABLE 14.1

Sample Listing of Documents Needed for Developing Implantable Medical Devices

---

Standard documents for all phases
  Project plan
  Development plans
    System development plan
    Software development plan
    Hardware development plan
  Configuration management plan
  Requirements plan
  Specification plan
  V&V plan
  Document plan
  User manual
  Quality assurance phase review
  Marketing
  Technical reviews
  Technical communications (memos, letters, email notes)
  Document control forms

Phase 1: Concept
  Vision
  Feasibility and tradeoff analyses
  Risk plan
    Risk assessment: ETA, FTA, FMEA
    Business risk

Phase 2: planning and Scheduling
  Product description
  Feasibility and tradeoff analyses
  Risk plan
    Risk assessment: ETA, FTA, FMEA
    Business risk

Phase 3: Design and development
  Product description
  Feasibility and tradeoff analyses
  Design transfer plan
  Clinical plan
  Code standards
  Test results
    Recorded errors
    Test metrics
  Traceability
  Design documents
  Software design document
  Source listings
  Hardware design document
  System design document
  Final parts list

Phase 4: Controlled Release
  Product description
  Clinical results
  FDA submission
  Submissions for UL, CE, IEC certifications
  Training plan

(Continued)

TABLE 14.1

Continued

Phase 5: Commercial release
   Product description
   Design history file
   Device master record
   Version description document
   Publications
      Brochures
      Training materials
   FDA approval
   UL, CE, IEC certifications
   Training plan

- Requirements
  - Clinical
  - Function
  - Performance
  - Safety
  - FDA approval
  - Certifications
- System architecture
  - Clinical concept
  - Major subsystem components and interfaces: hardware, software, mechanical, material, biological, and pharmaceutical
  - Features and feature management
  - Operations concept
  - Training, support, and logistics
- Analyses
  - Feasibility: design constraints and major trade-off studies performed
  - Risk assessment
  - Margin
- Staffing
- Quality assurance
- Schedule

You will be able to constrain the baseline design following the closure of the action items resulting from the CoDR. You can then purchase long-lead items, development support equipment, breadboard parts, and materials.

14.4.2 Planning and Scheduling

The team prepares the detailed design during the planning and scheduling or preliminary design phase. This means that all the design descriptions are prepared in an initial state:

- Software—operational storyboards, software use cases, software size estimates, software requirements, design, structure, logic flow diagrams, computational loading, design language, and development systems
- Hardware—block diagrams, signal flow diagrams, schematics showing logic diagrams, and first interface circuits
- Mechanical—packaging plans, mechanism schematics, subsystem layout, material selection

The design descriptions include estimates of weight, power, volume, reliability, and longevity. The analyses, modeling, and any early test results should be completed, too.

The planning and scheduling or preliminary design review (PDR) is held when the design advances sufficiently to begin some breadboard testing or the fabrication of engineering models. The PDR presents the design descriptions. The PDR should also present the analyses, the estimates, and the bases for both. Example items to be addressed in the PDR are as follows:

- Business case, goals, objectives, and constraints
- Closure of action items from the CoDR
- Completion of research, trade-offs, and feasibility
- Review of initial drafts of standard documents and phase specific documents—Table 14.1
- Requirements
  - Clinical
  - Function
  - Performance
  - Safety
  - FDA approval
  - Certifications
- System architecture
  - Clinical concept
  - Major subsystem components and interfaces: hardware, software, mechanical, material, biological, and pharmaceutical
  - Features and feature management
  - Operations concept

- – Training, support, and logistics
- Analyses
  - – Risk assessment
  - – Margin in safety, weight, power
  - – Mechanical/structural design, weight, thermal, and life tests
  - – Electrical, power, electromagnetic interference (EMI)/EMC
- Software requirements and design—operations, bug rates, code reviews
- Design verification, test flow and test plans
- Staffing
- Quality assurance
- Schedule

The completion of the PDR and the closure of any action items generated by the review become the basis for the start of the detailed design effort and the purchase of parts, materials, and equipment.

14.4.3 Design and Development

During the design and development, or critical design phase, the team prepares the detailed design for the design freeze, which will lead to fabrication and manufacture of the product in the next phase. This means that all the design descriptions are prepared in a complete and comprehensive form:

- Software—operational storyboards, software use cases, software size estimates, software requirements, design, structure, logic flow diagrams, computational loading, design language, and development systems
- Hardware—block diagrams, signal flow diagrams, schematics showing logic diagrams, and first interface circuits
- Mechanical—packaging plans, mechanism schematics, subsystem layout, material selection

All estimates of weight, power, volume, reliability, and longevity will be in final form, too.

The design and development or critical design review (CDR) should present all the same basic subjects as the PDR but in final form. The CDR should include all of the items specified for a PDR but updated to the final form, plus the some additional items: Example items to be addressed in the CDR are as follows:

- Business case, goals, objectives, and constraints
- Closure of action items from the PDR

- Summary of research, trade-offs, and feasibility
- Review initial drafts of standard documents and phase specific documents—Table 14.1
- Requirements
  - Clinical
  - Function
  - Performance
  - Safety
  - FDA approval
  - Certifications
- System architecture
  - Clinical concept
  - Major subsystem components and interfaces: hardware, software, mechanical, material, biological, and pharmaceutical
  - Features and feature management
  - Operations concept
  - Training, support, and logistics
- Analyses—updated and final
  - Risk Assessment: ETA, FTA, and FMEA
  - Margin in safety, weight, power
  - Mechanical/structural design, weight, thermal, and life tests
- Electrical, power, EMI/EMC
- Software requirements and design—operations, bug rates, code reviews
- Design verification, test flow and test plans
  - Test results
  - Test history of the hardware
- Previous anomalies, deviations, waivers, and their resolution
- Staffing
- Schedule
- Quality assurance

Completion of the CDR and resolution of all the action items generated by it constitutes the baseline design.

14.4.4 Controlled Release

The controlled release, or production, phase occurs prior to manufacturing. Its purpose is to assure that the design of the device has been validated through the test, verification, validation, and acceptance program, and that all deviations, waivers and open items have been satisfactorily closed and that the project, along with all the required operating procedures, documentation, logistics, and support equipment is ready for production. This is also the stage where the company makes all of the submissions for certifications and FDA approval.

The Controlled Release Review (CRR) is an important review to confirm that the design is ready for production. Here are some example items:

- Business case, goals, objectives, and constraints
- Closure of action items from the CDR
- Completion confirmation of the Design Transfer Plan
- Review of initial drafts of standard documents and phase specific documents—Table 14.1
- Design verification, test flow and test plans
  - Test results
  - Test history of the hardware
  - Previous anomalies, deviations, waivers and their resolution
  - Rework/replacement of hardware, regression testing, or test plan changes
  - Compliance with the test verification matrix
  - Measured test margins vs. design estimates
  - Trend data
  - Could-not-duplicate failures should be presented along with assessment of the problem and the residual risk that may be inherent in the item
- Training plan
- Staffing
- Schedule
- Quality assurance
- Clinical and field studies

14.4.5 Commercial Release

The commercial release phase is the stage-gate from manufacturing. It means that the pilot production is complete and has been analyzed. All documentation is in final form. All certifications and FDA approval have passed. The final review should cover

- Business case, goals, objectives, and constraints
- Product launch by sales and marketing

- Closure of action items from the CRR
- Review initial drafts of standard documents and phase specific documents—Table 14.1
- Training Plan
- Staffing
- Schedule
- Quality assurance

14.4.6 Logistics, Maintenance, Disposal

This phase remains for the life cycle of the product; it can be as long as 10–20 years. It covers inventory and distribution, training medical staff to use and implant the device, providing technical troubleshooting and support, and eventual disposal of retired or defective devices. Any further discussion is beyond the scope of this book.

14.5 Scheduling

Scheduling is, or should be, a combination of top-down and bottom-up planning. Top-down plans are set by marketing and upper management to meet market conditions. It should be supported by a detailed bottom-up plan. As mentioned in the last chapter, medical product development always takes more time than you expect. Some projects can take upwards of 10 years from concept to FDA approval.

Bottom-up planning can begin from the activities called out in each phase in the previous section. Carefully estimate how much effort and time will be needed for each one. Then use a program like Microsoft's Project® to total the activities together; it will give you a good bottom-up estimate of time and effort. Table 14.2 has some estimates of efforts that a project like this might encounter—a combined effort totaling more than 45 years is not unrealistic—it is probably low!

14.6 Documentation

14.6.1 Purposes

Documentation serves a number of purposes in medical devices. It is considered part of the labeling by the FDA. It serves in training medical staff and patients and in supporting use of the device after sale and distribution.

TABLE 14.2

Some Estimates of Effort for an Implantable Medical Device

Activities	Phase 1: Concept	Phase 2: Planning and Scheduling (Preliminary)	Phase 3: Design (Critical)	Phase 4: Controlled Release	Phase 5: Commer. Release
Business case, goals, objectives, constraints	800	800	400	400	200
Research (preparing results)					
University	400	200			
Medical	800	400			
Literature	400	200			
Patent searches	200	200			
Requirements					
Clinical	400	400	200		
Function	800	800	400		
Performance	1000	1000	600		
Safety	1000	1000	600		
FDA approval	400	600	1000		
Certifications	200	200	200		
System architecture					
Clinical concept	400	400	200		
Subsystem and interfaces: hardware	1000	1000	400		
Subsystem and interfaces: software	2000	2000	1000		
Subsystem and interfaces: mechanical	400	400	200		
Subsystem and interfaces: material	200	200	100		
Subsystem and interfaces: biological	400	400	200		
Subsystem and interfaces: pharmaceutical	1000	1000	600		
Features and feature management	100	200	1000		
Operations concept	400	400	200		
Training, support, and logistics	100	100	400		

Analyses					
Feasibility: constraints and trade studies	2000				
Focus groups	500	1000			
Surveys	600	800			
Margin	600	600	600		
Weight, thermal, and life tests	400	400	400		
Electrical, power, EMI/EMC	400	600	600		
Staffing	200	200	200	200	200
Quality assurance	200	400	600	400	400
Schedule	200	200	200	100	40
Software requirements and design	1000	1000	2000	1000	200
Hardware requirements and design	600	600	1000	600	100
Mechanical requirements and design	400	400	1000	400	100
Design verification, test flow and test plans		400	1000	2000	
Standard documents					
Project plan	120	40	20	20	20
Development plans					
System development plan	200	100	40	40	20
Software development plan	200	100	40	40	20
Hardware development plan	100	80	40	20	20
Configuration management plan	40	20	5	5	5
Requirements plan	40	40	20	10	5
Specification plan	40	40	20	10	5
V&V plan	200	400	100	40	5
Document plan	20	20	5	5	5
User manual	100	100	400	400	400
Quality assurance phase review	80	80	40	20	20
Marketing	400	200	200	40	40
Technical reviews	600	600	800	600	400

(Continued)

TABLE 14.2

Continued

Activities	Phase 1: Concept	Phase 2: Planning and Scheduling (Preliminary)	Phase 3: Design (Critical)	Phase 4: Controlled Release	Phase 5: Commer. Release
Technical communications	2000	2000	2000	2000	2000
Document control forms	80	40	20	5	5
Phase documents					
Vision	40				
Risk plan					
Risk assessment: ETA, FTA, FMEA	800	600			
Business risk	200	200			
Product description	1000	200	400	400	200
Feasibility and tradeoff analyses		600	200		
Design transfer plan			200		
Clinical plan			400		
Code standards			40		
Test results					
Recorded errors			2000		
Test metrics			600		
Traceability			200		
Design documents					
Software design document			1000		
Source listings			80		
Hardware design document			600		
System design document			200		
Final parts list			80		
Clinical results				400	
FDA submission				2000	
Submissions for UL, CE, IEC certifications				1000	

						Total
Training plan	40	40	40		400	200
Design history file					80	200
Device master record						200
Version description document						40
Publications						
Brochures						400
Training materials						600
FDA approval						200
UL, CE, IEC certifications						200
Subtotals, efforts (h)	25,760	24,000	25,090	12,635	6,450	93,935
Subtotals, efforts (months)	153.3	142.9	149.3	75.2	38.4	559
Subtotals, efforts (years)	12.4	11.5	12.1	6.1	3.1	45

CENELEC: European Committee for Electrotechnical Standardization; IEC: International Electrotechnical Commission; IEEE: Institute of Electrical and Electronics Engineers; ISO: International Organization for Standardization.

Note: numbers are in hours of effort (© 2006 by Kim Fowler, used with permission. All rights reserved).

14.6.2 Types

There are informal and formal types of documentation; all of which are archived for potential audits by the FDA in the future. The informal documentation includes notebooks, notes, and e-mail messages. The formal documentation includes letters, memos, project documents, manuals, brochures, and presentations.

All phases have a standard set of documents that need updating. Each phase also has some documents unique to it (Table 14.1).

You will need specialists in documentation, FDA process, and regulation/certification to determine the basic types of documentation that you will need.

14.6.3 General Formats for Documents

The major documents, project documents and manuals, have a template from which to begin preparation. They follow the general format:

- Title page
- Author page and revision number
- Table of contents
- Introduction with purpose, scope, and overview
- Body of document
- Appendices
- Glossary of terms

In test plans, the body of the document outlines what is to be done and how, by whom, when, and where. Design documents state what has been done and how, by whom, when, and where. Templates of documents may be found at Reference 2.

A user's manual is a challenging piece to prepare. One of the most important sections is the introduction titled "Getting Started," which often facilitates a user's understanding of the operation. Many times it is the only section that users read. Prepare it carefully.

The best guidance for writing manuals for medical devices is the FDA's "Write It Right" document [3]. You can find it online.

Any documents or records generated should be stored whether as paper files or as microfiche or as electronic files on a server or all three. Certainly, files should be backed up daily on a server; having a secure storage facility that can resist fire, floods, and disasters is important, too.

14.7 Requirements and Standards

14.7.1 Market

These are Class III devices that mean that they need PMA from the FDA. Some of these devices, such as pacemakers and nerve stimulators, may be eligible for 510(k) designation, which will speed their FDA approval cycle somewhat.

The users generally are patients, although, if the implanted device has an external programmer, then medical staff in physicians' offices can be included in the user pool. Influencers are the medical staffs that implant or operate the devices; they initiate the purchase process. Hospitals or medical institutions, which purchase the devices, are the actual customers. Most marketing people recognize that the influencers—the medical staffs—are the important group to reach for market penetration.

14.7.2 Design and Development Standards

There are a number of standards to which an implantable medical device must adhere. Some of the more prominent ones are listed in Table 14.3 and include EMC and product safety [4,5]. The UL 60601-1 (USA) standard is slightly more stringent than IEC 60601-1; if you develop a device according to UL 60601-1 it generally satisfies IEC 60601-1.

As a point of interest, Table 14.4 lists standards specific to implantable neurostimulators. This is an example of standards that apply to a specific device type.

14.7.3 FDA Approval

All medical devices must receive FDA approval before commercial sale. You must follow the FDA Design Control Guidance and perform clinical trials to develop an implantable medical device [6–8]. It is a good practice to meet with the FDA and to develop a professional relationship with them; they will help you better understand what the FDA wants and expects to see when you submit for approval. Meeting with the FDA and understanding their concerns will smooth the path to approval.

Your entire team needs to understand the import of FDA approval and good processes. Regular briefings on progress and quality processes will help keep awareness of FDA requirements.

14.7.4 Preparing Requirements

The requirements for an implantable medical device can take a long time and much effort—and probably should. You can still use the suggested techniques in Chapter 1 to speed development. Many people, studies, and parameters contribute to the preparation of the requirements.

Sections 14.3.2 and 14.3.3 indicate some of the concerns that need to be covered in the Requirements and Specifications Document. The Reference 2 has a template that might help, as well. Requirements must address concerns that include clinical (e.g. dosage and protocols for therapy), function—which includes user issues (e.g. programming options) and

TABLE 14.3

Listing of Some of the Standards for an Implantable Medical Device and its Support Equipment (Programmers, etc.)

Category	Origin	Standard	Description
EMC	European Standards	EN 55011	Industrial, scientific, and medical radio frequency equipment—radio disturbance characteristics—limits and methods of measurement
		EN 61204-3	Low-voltage power supplies, DC output
		EN 61000	Electromagnetic compatibility, parts 2 and 4
		IEC 61000	Electromagnetic compatibility, parts 1 and 3
	USA	IEEE 139	IEEE recommended practice for the measurement of radio frequency emission from industrial, scientific, and medical (ISM) equipment installed on user's premises
		IEEE 299	IEEE standard for measuring the effectiveness of the electromagnetic shielding of an enclosure
		IEEE C63.011	Limits and methods of measurement of radio disturbance characteristics of industrial, scientific, and medical (ISM) radio frequency equipment
		IEEE C63.18	Recommended practice for an on-site ad hoc test method for estimating radiated electromagnetic immunity of medical devices to specific radio frequency transmitters
Safety	European Standards	EN 61204	Low-voltage power supplies, DC output—safety requirements
		EN 45502-1	Active implantable medical devices—general requirements for safety
		EN 45502-2-1	Particular requirements for pacemakers
		EN 45502-2-2	Particular requirements for implantable defibrillators
	IEC	IEC 60086	Primary batteries
		IEC 60601-1	Medical electrical equipment, part 1—general requirements for safety
		IEC 60601-1-2	Medical electrical equipment, part 1—general requirements for safety, section 2—EMC (applicable to non-implantable parts of implantable devices)
		IEC 60601-1-4	Medical electrical equipment, part 1—general requirements for safety, section 4—programmable electrical medical systems
	USA	UL 60601-1	Medical electrical equipment, part 1—general requirements for safety
		IEEE C95.1	IEEE standard for safety levels with respect to human exposure to radio frequency electromagnetic fields, 3 kHz–300 GHz

		IEEE C95.3	Recommended practice for measurements and computation of radio frequency electromagnetic fields for human exposure, 100 kHz–300 GHz
		IEEE C95.3	IEEE recommended practice for measuring potentially hazardous electromagnetic fields of RF and microwave
	ISO	ISO 11197	Medical electrical equipment—particular requirements for safety of medical supply units
		ISO 14708	Implants for surgery—active implantable medical devices—Part 1: General requirements for safety, marking, and for information to be provided by the manufacturer
		ISO/IEC Guide 63	Guide to the development and inclusion of safety aspects for medical devices
		ISO/TR 16142	Medical devices—guidance on the selection of standards in support of recognized essential principles of safety and performance
		ISO 14971	Application of risk management to medical devices (note: this is becoming a very important document)
Biocompatibility	USA–FDA	21 CFR 58	Prescribes practices for nonclinical laboratory studies to support applications to the FDA for medical devices
	ISO	ISO 10993-1	Use the Blue Book Memorandum from the FDA for testing for neurotoxicity and immunotoxicity of materials

CENELEC: European Committee for Electrotechnical Standardization; IEC: International Electrotechnical Commission; IEEE: Institute of Electrical and Electronics Engineers; ISO: International Organization for Standardization.

TABLE 14.4

Referenced Standards from ISO 14708-3 (Implantable Neurostimulators)

Standard	Description
ANSI/AAMI PC69:2000	Active implantable medical devices—Electromagnetic compatibility—EMC test protocols for implantable cardiac pacemakers and implantable cardioverter defibrillators
ISO 780:1997	Packaging—Pictorial marking for handling of goods
ISO 8601:2000	Data elements and interchange formats—Information interchange—Representation of dates and times
ISO 10993-1:2003	Biological evaluation of medical devices—Part 1: Evaluation and testing
ISO 11607:2003	Packaging for terminally sterilized medical devices
ISO 14155-1:2003	Clinical investigation of medical devices
ISO 14971:2000	Medical devices—Application of risk management to medical devices. Amendment 1:2003
ISO 15223:2000	Medical devices—Symbols to be used with medical device labels, labeling and information to be supplied. Amendment 1:2002
IEC 60068-2-14:1986	Environmental testing—Part 2: Test. Test N: Change of temperature
IEC 60068-2-47:2005	Environmental testing—Part 2-47: Tests—Mounting of specimens for vibration, impact and similar dynamic tests
IEC 60068-2-64:1993	Environmental testing—Part 2: Test methods—Test Fh: Vibration, broad-band random (digital control) and guidance. Corrigendum 1:1993
IEC 60601-1:2005	Medical electrical equipment—Part 1: General requirements for basic safety and essential performance
IEC 60601-1-2:2001	Medical electrical equipment—Part 1: General requirements for safety—2 Collateral standard: Electromagnetic compatibility—Requirements and tests. Amendment 1:2004
IEC 60601-2-27:2005	Medical electrical equipment—Part 2–27: Particular requirements for the safety, including essential performance, of electrocardiographic monitoring equipment
IEC 61000-4-3:2002	Electromagnetic compatibility—Part 4–3: Testing and measurement techniques—Radiated, radio frequency, electromagnetic field immunity test

This Standard is at the Draft International Stage and is not Yet Published as of November 2006.

subjective issues (e.g., comfort and utility), performance (e.g., volume, weight, power consumption, battery life), safety, FDA approval, and certification.

While marketing might initiate the requirements, the systems architect or the team leader is the primary author of the requirements. Many people contribute significantly to preparing the requirements. Table 14.2 estimates as much as 10,800 hours (over 64 months or 5 years of effort!) to prepare requirements over three phases, which indicates just how important requirements are.

You should implement a dedicated process for requirements that takes input from patients, physicians, medical staffs, customers, marketing, standards, and FDA regulations. The process should have a clearly defined flow from intent, through the codification of intent, through review, and end with a clear connection to validation testing. Your requirements process will be company specific and can be policy that is used in project after project.

14.8 Analysis

14.8.1 Feasibility

Feasibility derives from a number of sources. Marketing often initiates the concept. That concept should then be subjected to a variety of different studies and analyses: university research, focus groups, surveys, heuristics, calculations, numerical simulations, and testing.

Research in a university medical center is very important, if not mandatory; prototypes may be tested for safety, efficacy, feasibility, and utility. The research will result in medical publications that benefit from peer review.

14.8.2 Focus Groups

Focus groups discuss issues and potential product concepts to clarify need, operations, and utility. The selected participants discuss concerns and specific issues related to the treatment field of the device. A facilitator directs the discussion and asks questions to find out how and why the device might be useful or how it may not be useful.

While participants discuss and answer questions, your team observes from behind a one-way mirrored wall. Your team does not participate at all, except to pass the occasional note to the facilitator if a point needs to be clarified.

In most cases, your company will contract a company that specializes in running focus groups to hold the meetings. The contracted company provides the facilities, the facilitator, and the discussion outline; they are expert in running focus groups and know how to avoid bias in obtaining the answers. Often the participants are compensated in some way—stipend, food, or entertainment tickets—to take several hours of their time to attend a session.

Focus groups tend to be most useful during the first two phases of a project. They are one way to help clarify issues during the "fuzzy front end."

Marketing often sets up the focus groups. I strongly advocate that designers and engineers get to sit in on one or more sessions; seeing potential customers and patients respond to questions and your concepts is always eye-opening!

14.8.3 Surveys

Surveys serve much the same purpose as focus groups but can cover a wider audience, more quickly. Again, the science of surveys is an expertise that most of us do not have. Contracting a company to run your surveys can make a lot of sense; they know how to prepare a survey that avoids bias and still gets useful answers for you.

Surveys tend to be most useful during the first two phases of a project. They are one way to help clarify issues during the "fuzzy front end."

Marketing often sets up the surveys. Designers and engineers should participate to ensure that no important issues are overlooked.

14.8.4   Heuristics, Calculations, and Numerical Simulations

These types of analyses tend to be very company specific. I would urge you to collect heuristics and codify a number of basic, but useful, tools to perform your analyses. Basic spreadsheet calculations, such as Table 14.2, can help you quantify your effort.

You can purchase commercial programs for modeling thermal heat paths or EMI. These can be useful simulations if your company designs a number of different products; often generating your own models is eventually more useful (such as done by Keithley Instruments in Chapter 6).

14.8.5   Storyboarding

Storyboarding is a form of prototyping that can be very useful for a system or device that has a user interface. A storyboard or a series of storyboards helps designers plan sequences of operations; the storyboard can quickly clarify functions and show where potential problems and unforeseen interactions might exist.

User interfaces may refer to either patients or physicians (and their medical staffs); each has their set of concerns in learning how to use a new interface. A training specialist who understands storyboarding can really assist the design team to understand the necessary operations and the appropriate order sequence of those operations.

It is also good for the design team to talk to medical staff and patients to learn how they perceive the device and its interface. Training on the use of the device and its interface is one of the sequences of actions that you need to take seriously.

14.8.6   Testing

Testing is important for feasibility analysis, too. You have a variety of tests that you might conduct to assure feasibility:

- University research into basic physiology
- Prototypes for engineering studies of function, durability, and reliability
- Materials research and environmental tolerance (the human body is extraordinarily corrosive)

- Animal trials, where cats, dogs, or pigs receive the implants and are studied for effects
- Clinical research with possibly different versions of prototypes

Remember, prototypes are just that—they are not close to the finished product in development. Feasibility shown by prototype is only about 5% of the effort to develop the final marketable device (see Table 14.2)!

14.9   Design Trade-Offs

14.9.1   Requirements

Implantable medical devices have many different types of requirements. They span the gamut from clinical concerns, physiological issues, and sterilization to materials, physical parameters, software operations, and subjective perceptions.

While the team lead or systems architect may own the requirements, everyone on the team contributes to preparation of the document. Requirements require a significant effort during the first three phases of development.

Clinical concerns: The clinical requirements all have to do with how the device is handled and used by medical staff. The requirements include implantation procedures, storage and shelf life, indications of when to use the device, and indications of when not to use the device.

Physiology: The physiological requirements have to do with the expected results when implanting and using a device:

- Physiological reactions that can be expected
  - Types of response
  - Ranges of response
- Durations of response
- Side effects
- Long term effects
  - Accommodation of stimuli
  - Change in response
  - Hypersensitivity
  - Chemical imbalances or changes
- Ionic deposits.

Sterilization: Sterilization of the medical device is a first step and an important requirement. There are several different types of sterilization—autoclaving (steam heat and high pressure), chemical washes, and chemical gases. Sterile packaging and handling is also critical to safe and proper implantation of devices. The manufacturer usually performs the sterilization and packaging and must know and follow these requirements.

The medical personnel that handle the purchased devices must know how the packaged devices are stored before surgery and how they are delivered and opened during surgery. Sterile procedures must be articulated in the requirements.

Materials: The materials within a medical device must protect the patient from any chemical interactions that could proceed from the device. The materials also must be inert and highly resistant to corrosion. All device enclosures must be welded closed to form hermetic seals to prevent the leakage of body fluids into the device and the leaching of toxic substances out of the device and into the body.

Physical parameters: The physical parameters include mechanical size and weight, electrical operations and power reserves, the user interface, fluid hydraulics and reservoirs, and communications links and protocols.

Some mechanical requirements might include:

- Weight not to exceed
- Size not to exceed
- Volume not to exceed
- Attachment points to body

Some electrical requirements might include:

- Operational modes
- Power consumption for different operational modes
- Battery power, energy content, and reserves
- Power down or sleep modes
- Reliability
- Fail safe operation and fault tolerance

Some user interface requirements might include (these all have to do with the external programming or communications device):

- Sequences of operations—intuitive and clear
- Size of keypad
- Size of display
- Haptic feel
- Resist disinfectant washes

Some fluid hydraulic requirements might include:

- Reservoir capacity
- Power consumption for different operational modes
- Pump size and power delivery
- Reliability
- Fail safe operation and fault tolerance

Some communication requirements might include:

- Data formats
- Transmission protocols and error correction
- Memory requirements
- Power levels

Software operations: The requirements on the software operations are usually the most extensive of all. Some software requirements might include:

- Operational modes
- Therapy delivery and control
- Power management
- Communications
- Fail safe operation and fault tolerance

Subjective parameters: The subjective parameters usually have to do with comfort levels for the patient, ease of use, and ease of training.

14.9.2   Hardware

Probably the biggest concern for implanted electronics is power consumption. I know of at least one manufacturer who designs and fabricates their own custom ICs, particularly small microcontrollers for implanted medical devices. They want sufficiently low power consumption so that the implanted device will function for years before draining the battery.

Another major concern is fault-tolerant operation. Should anomalous behavior occur, whether externally generated or due to internal failure, the device should operate or shut down in a safe way. One example of this need is the communications link—it should be robust. I know of another situation where a particular model of device behaved erratically when patients walked between the security monitors in a particular department store.

EMC and EMI need attention while designing circuits, too. Proper layout of components, power and return planes, and signal traces will cover the majority of the shielding concerns. See References 9 and 10. Printed circuit boards (PCBs) need to resist moisture absorption and be durable to allow high-density traces and circuits; ceramic substrates with gold-plated traces, rather than FR4 and copper, can achieve these goals.

Primarily electronic design engineers do this work, although they obviously take input from software engineers and mechanical engineers.

14.9.3 Power

Battery capacity and power consumption, as just stated in the previous section, are critical issues to the operational duration of an implanted device. Most research has been into batteries with higher energy density. One concern with these high-energy batteries is the risk of generating extreme heat or even explosions with a short circuit fault. Designers need to put in place appropriate safeguards against excessive current demand and heat build-up.

Interestingly, recharging batteries within implants has not been popular even though it is feasible; only a very few companies have incorporated rechargeable batteries in their implantable devices. Apparently, the additional burden of equipment and strapping on a belt with a recharging coil and waiting (or sleeping) has been considered too time-consuming or too much effort for physicians and patients to do. It may also carry liability too great for medical manufacturers to spend the time and money to gain FDA approval—they do not seem willing to produce a device that might put patients at risk should they forget to recharge their implanted devices. It all comes down to cost versus benefit—does the benefit of a longer-lasting device that reduces the number of surgeries and lowers the total cost per patient over many years sufficiently balance the increased risk of mistakes in recharging?

14.9.4 Software

Software in an implantable medical device is safety-critical development if there is any! Development processes need to include thorough studies that support detailed and complete requirements, careful design, properly-conducted code reviews, tests, and field tests.

Some companies build their own custom, time-slice real-time operating systems (RTOSs) that use a cyclic executive; these are straightforward to analyze [11]. A time-slice RTOS can guarantee that no tasks are starved or left undone; the trade-off for that guarantee is efficiency.

A certified compiler is a worthwhile investment for medical-software development. If your company fabricates its own processors, then your team or someone in your company will have to write and test a compiler and linker. If you decide to purchase the RTOS, there are several companies that provide certified software products for safety-critical markets.

Software engineers who develop the software must work closely with the biomedical engineers to assure correctness in operations. They will have to work with the electrical engineers who develop the hardware to assure proper and efficient handling of peripheral devices and input/output (I/O). Finally, the software team must work closely with the test team to do unit and system testing.

14.9.5 Hardware vs. Software

As complexity of medical devices increases more functions will implement in software. Hardware accelerates specialty functions. The balance is somewhat dictated by how thoroughly the software modules and hardware subsystems can be verified.

One aspect of fault tolerance is use of a hardware or mechanical function as a safety check or limit for the software operation. Conversely, software can check and verify hardware function. An example might be a restriction in the delivery tube that will not allow fast delivery of drug. Another might be a hardware counter that independently prevents an overdose of operations when it times out. The watchdog timer is a simple example of a hardware check on software operation. Another example might be a thermal sense switch on the battery that opens should temperature exceed a maximum limit and shuts down the device to prevent a failure from overheating.

The systems architect needs to work closely with the team to specify the appropriate trade-offs between hardware and software. Most of this work is performed in Phase 1, concept development.

Until recently, upgrades were not allowed in implantable medical devices; once in the patient's body, the device was not going to be changed. The software and hardware are FDA-approved and-certified as a system—it can not be changed after approval. If you do want to change something, you must take the device through a whole new development cycle and set of approvals. The implantable medical device market does not have "service release" policies to fix problems, like automobiles in Chapter 7.

That situation has changed in the past few years. Some programmers and implantable devices now have the ability to receive software downloads for upgrades and fixes. Currently, these changes are limited to patches.

14.9.6 Buy vs. Build

Most subsystems within an implantable medical device are custom designed and built. The basic components may be purchased, although in some cases, even the processors can be custom application-specific integrated circuits (ASICs). The concerns for specificity, reliability, longevity, obsolesce, and inventory in commercial off-the-shelf (COTS) products all disappear with custom designs. The trade-off is that cost, particularly due to non-recurring engineering (NRE), increases greatly.

The COTS items in an implantable medical device might include:

- Electronic components and ICs
- Electric motors
- Batteries—but these often are closely specified by outside vendors

The custom-design subsystems in an implantable medical device usually include:

- ASICs
- PCBs
- Wires, cables, and connectors
- Electrodes
- Pumps, reservoirs, and "plumbing"
- Sensors
- Enclosure

If you do purchase COTS components, beware of obsolescence and inventory. Most likely, the vendor would not stock the selected components for the lifetime of your product. Your company will need to do large one-time buys and then place the components in inventory. This situation is a risk for long-term medical devices with long development cycles.

14.9.7   Mechanical

The mechanical aspects of an implantable medical device are very important and affect the electrical and software design. Generally, the enclosures are hermetically-sealed, welded titanium cases; this is necessary for long-term implantation. Most companies that build implantable devices have captive facilities and resources for the mechanical fabrication and assembly. Most people view these captive facilities as necessary to maintain quality and to assure that the expertise remains available to them.

14.9.8   Manufacturing

For the same reason that mechanical machining and fabrication usually remains in-house, most companies perform the manufacturing of implantable medical devices in-house. Many people see that in-house manufacturing better assures quality.

14.9.9   Test

Implantable medical devices have many types of tests in development. Manufacturing will also have tests. But implantable medical devices do not generally have tests or diagnostics for failure; these certainly are not needed for maintenance or repair.

14.9.10 Maintenance

Implantable medical devices are not maintained (and certainly not repaired!); they are replaced if nonfunctional—for example, depleted battery or failed component. This is generally true for external devices used for communication or programming, although some companies repair these external devices in their service departments.

14.10 Tests

14.10.1 Formal and Informal

One of the main activities in designing a medical device is testing. There are formal tests, which are planned, executed, and rigorous; the following sections contain more about formal testing.

Otherwise, informal tests, such as laboratory bench tests on prototype concepts, help elucidate operations, interactions, and feasibility. The tests comprise mock-ups and prototypes of circuits and modules. Most often the goal is to clarify or demonstrate a single, important aspect of operation. Most informal tests take place early in development, usually in Phase 1, concept development. Biomedical, electrical, and software engineers perform these informal tests.

14.10.2 Laboratory Tests

Laboratory tests are the first in a long series of tests for developing a medical product. They might be animal tests, clinical tests, or formal versions of bench tests. The tests invariably are under the authority of a clinical or research review board. The tests can use early prototypes of the proposed device to demonstrate efficacy or safety. They can also reveal physiological operations, interactions, and feasibility.

One example of a laboratory test might be implanting a new model of drug pump in an animal and then studying the long-term results. This type of research can indicate whether the implanted device functions as intended over a long period of time. It can also show if problems might occur, such as clogging of the drug-dispensing port.

Physicians, university researchers, and biomedical, electrical, and software engineers can together or separately participate in laboratory tests and research. Often, the tests are conducted in university laboratories, hospitals, or clinics early in Phase 1, concept development.

14.10.3   Inspections, Code Walk-Throughs, and Peer Reviews

Some of you may be familiar with inspection as a part of quality control in manufacturing. Another form of inspection includes reviews and code walk-throughs. These are formal activities that are planned, executed, and rigorous for all aspects of design: software, hardware, and mechanical modules and subsystems. The primary goal of inspections is to find problems through examination of the operations, interactions, and potential consequences.

Inspections are meetings where subsets of the design team review schematics, software listings, and subsystem assemblies. The reviewers are electrical, software, and manufacturing engineers and technicians; they discuss the intent, the execution of the specifications, the operations, interactions, and potential consequences. They are similar to design reviews but have a narrower scope. Inspections also occur much more frequently than design reviews.

Inspections and code walkthroughs generally occur in the first three phases of development. Chapter 2 and Reference 12 discuss templates for recording the concerns and actions items that arise from inspections and code walkthroughs.

14.10.4   Design Reviews

Design reviews, like inspections, are formal, planned, and rigorous. Their primary purpose is to confirm that the project is adhering to the specifications. Often, specific or major operations, interactions, and consequences are presented and discussed.

Design reviews are larger meetings and conclude each phase of development. Sometimes, intermediate design reviews occur during a phase. Major design reviews include all members of the design team and cover the entire system: hardware, software, mechanics, and clinical issues. Smaller design reviews can include designated members of the design team and can focus on specific issues or subsystems.

Design reviews occur in all phases of development. Team members prepare presentations to address the major concerns. The entire team and possibly reviewers from outside the design team review the presentations, which summarize progress and test results, schematics, and software listings for the project.

14.10.5   Subsystem Tests—Hardware

Subsystem tests for the project's hardware are a formal portion of the test plan. The test plan prescribes the tests of the hardware subsystems; these include tests for functionality, reliability, EMC, and EMI. These tests can tie into a design control system, such as those described in Chapter 2, to record the results that help confirm adherence to the specifications.

Electrical and mechanical engineers and technicians perform subsystem tests on engineering models and initial units from a pilot production run.

They run these tests in both laboratories and special test facilities. EMC and EMI tests, for instance, usually require special chambers and equipment. Subsystem tests take place in Phases 2 and 3.

14.10.6 Subsystem Tests—Software

Subsystem tests for the project's software are a formal portion of the test plan. The test plan prescribes the tests of the software subsystems; these include tests for functionality and fault tolerance and recovery. These tests can tie into a design control system, such as those described in Chapter 2, to record the results that help confirm adherence to the specifications.

Software engineers and technicians perform subsystem tests on engineering models and initial units from a pilot production run. They run these tests in laboratories and any special facilities that might be required. Subsystem tests take place in Phases 2 and 3.

14.10.7 Environmental

Environmental tests for the project's hardware and mechanics are a formal portion of the test plan. The test plan prescribes the tests of the hardware, mechanical, and enclosure subsystems; these include tests for shock, vibration, fluid immersion, thermal cycling, and a Highly Accelerated Life Test (HALT). These tests can tie into a design control system, such as those described in Chapter 2, to record the results that help confirm adherence to the specifications.

Electrical and mechanical engineers and technicians perform subsystem tests on engineering models and on initial units from pilot production run. They run these tests in both laboratories and special test facilities. Shock and vibration tests, for instance, require a shake table, while thermal cycling and HALT tests require thermal environmental chambers. Subsystem tests take place in Phases 2 and 3.

14.10.8 Manufacturing

Tests for manufacturability usually require a pilot run to produce the first units of the medical device. These tests demonstrate that the device can be fabricated and produced in volume. They are usually conducted on the actual manufacturing and assembly line during Phase 3 by manufacturing, electrical, and mechanical engineers and technicians.

14.10.9 Simulators

Simulators fill several roles. They can be body simulators that allow surgeons to practice implant surgery. They can be computer models that indicate physiological responses to therapy; these might be pharmokinetics of drugs from drug pumps or neural mapping for stimulators or heart operation with pacemakers. The goal of using simulators is to understand operations and interactions early and then iron out the problems quickly. Physicians and biomedical engineers usually lead the effort. They conduct these simulations in the laboratory or in their offices during the first three phases.

14.11   Integration

14.11.1   Hardware

Integration tests for the project's hardware are a formal portion of the test plan, which prescribes integration of the hardware subsystems. Integration follows the completion of the subsystem tests and strives for correct functionality in the hardware.

Electrical and mechanical engineers and technicians perform integration tests on engineering models and initial units from pilot production. They run these tests in both laboratories and special test facilities. EMC and EMI tests, for instance, usually require special chambers and equipment. Integration takes place in Phase 3.

14.11.2   Software

Integration tests for the project's software are a formal portion of the test plan, which prescribes integration of the software modules. Integration follows the completion of the subsystem tests and strives for correct functionality in the software.

Software engineers and technicians perform integration tests on engineering models and initial units from pilot production run. They run these tests in both laboratories and any special facilities that might be required. Integration takes place in Phase 3.

14.11.3   System

Integration of the system is a formal portion of the test plan, which prescribes integration of the mechanical, hardware, and software subsystems. Integration follows the completion of the subsystem tests and strives for correct functionality in the system. Please note that most projects do not have a clean division between integrating the software and the hardware. Often a project will have to integrate a group of hardware modules and software packages as a subsystem. The system grows by adding subsystems together and testing each subset of combined subsystems.

Integration is performed on engineering models and initial units from pilot production. Integration tests run in both laboratories and special test facilities to confirm functionality, reliability, and fault tolerance and to confirm adherence to specifications. These tests can tie into a design control system, such as those described in Chapter 2, to record the results that help confirm adherence to the specifications.

Integration takes place in Phase 3. Selected members from across the entire design team often help with integration.

14.11.4 Environmental

Beyond the environmental tests for the hardware subsystems, you must perform environmental tests on the entire system. The test plan prescribes the tests of the system; these include tests for shock, vibration, fluid immersion, thermal cycling, and HALT. These tests can tie into a design control system, such as those described in Chapter 2, to record the results that help confirm that the system adheres to the specifications.

Electrical and mechanical engineers and technicians perform integration system tests on engineering models and on initial units from pilot production. They run these tests in both laboratories and special test facilities. Shock and vibration tests, for instance, require a shake table, while thermal cycling and HALT tests require thermal environmental chambers. The final environmental tests of the system take place in Phase 3.

14.11.5 EMI and EMC

Beyond the EMI and EMC tests for the hardware subsystems, you must perform EMI and EMC tests on the entire system. The test plan prescribes the EMI and EMC tests of the system; these include tests for conducted and radiated emissions and susceptibility. These tests can tie into a design control system, such as those described in Chapter 2, to record the results that help confirm that the system adheres to the specifications.

Electrical engineers and technicians perform EMI and EMC tests on final system as represented by the engineering models and by the initial units from pilot production run. They run these tests in both laboratories and special test facilities. EMC and EMI tests usually require special chambers and equipment. The final EMI and EMC tests of the system occur in Phase 3.

14.12 Manufacturing

14.12.1 Electrical and Electronic

Manufacturing for the electronics of an implantable medical device includes fabricating the PCBs, fabricating the cables and wiring, and assembling the components onto boards. Highly skilled technicians and trained personnel manufacture the production medical device. Most manufacturing of medical devices is done by hand-assembly. Sometimes, a product and its volume of production will indicate pick-and-place automated assembly and solder reflow ovens. Generally, cables and wiring are almost always done by hand. Manufacturing occurs in Phases 4 and 5.

14.12.2 Mechanical

Manufacturing for the mechanical aspects of an implantable medical device includes fabricating the mechanical enclosure and mechanisms and assembling the completed circuit boards, motors, actuators, cables, and wiring into the enclosure. Again, most manufacturing of medical devices is hand assembly by highly-skilled technicians and trained personnel. Machining of the mechanisms and enclosures may be accomplished with computer-controlled machines, such as mills and lathes, but the mechanical components are not stamped in a mass-produced fashion.

14.12.3 Assembly

Most manufacturing of medical devices is done by hand-assembly. This mode of manufacturing is dictated by the small production runs for medical devices where robots are too inflexible. Manual labor by skilled and trained personnel is more expensive, but humans are more adaptable and flexible than robots.

Most medical companies have their own captive manufacturing plant. They do this to maintain quality in the product and assure retention of skilled resources.

14.12.4 Tests

Many different tests can be run on the assembly line. The three most prominent are visual inspections, manual checks of test points, and automatic test with automatic test equipment (ATE). Technicians and trained personnel perform these tests. These tests usually intersperse between various steps in the fabrication and assembly of a medical device.

Visual inspections check for correct assembly. Are components installed in the right orientation? Are solder joints of the right shape and reflectance? Are labels correctly applied and legible? Is there any evidence of cracks or flaws?

Electrical tests can be either manual or ATE operations. Test personnel can probe test points to verify the correct range of voltage levels or for the correct signals. ATE can run a sequence of complex tests quickly to confirm a major portion of functionality.

14.13 Support

14.13.1 Launch

After FDA approval, the product launch occurs in Phase 5, commercial release. Sales, marketing, and clinical representatives train physicians both in the use of the medical device and in surgical implantation techniques; they also train medical staff to program the device. The company participates in medical trade shows, sponsors medical conferences, and sponsors ongoing university medical research on the devices.

Product launch usually begins with distribution of marketing literature and with training of sales representatives. In some cases, the sales representatives will introduce or "talk-up" a new product for weeks and months before it is available so that customers are ready to start using it at the time of launch.

14.13.2 Logistics

The company must set up a carefully controlled distribution of product. The distribution includes devices, literature, advisory boards, Web sites, and 24-hour telephone help lines. Sales teams and clinical representatives continue training of new customers, physicians, and medical teams. This effort continues for the life of the product.

Some implantable devices, such as drug pumps, have chambers that need refilling or replenishment. The company and its distribution channels need to supply or support the replenishment.

14.13.3 Technical Support

Technical support is a critical part of the distribution of the device. Physicians and their medical staff need instruction for handling, implanting, and dealing with anomalies. Trained personnel handled routine concerns; service engineers step in if a significant problem or recall occurs. Sometimes, members of the original design team get called in to help with a problem.

Implantable medical devices generally have no means for maintenance other than replacement of a failed device. The only support that might be considered maintenance is if some substance needs replenishment.

Technical support continues for the life of the product. Most companies maintain Web sites and toll-free numbers to serve physicians, medical staff, and patients.

14.14 Disposal

Medical waste disposal is a major concern and tightly controlled. In the United States, the FDA requires companies to recover their devices at the end of life and maintain life-cycle records. Trained support personnel perform these tasks, which, like technical support and logistics, continue for the life of the product.

14.15 Liability

14.15.1 Economics

Recalls of defective medical devices because of a manufacturing problem or design flaw can sink a company. The entire team needs constant awareness of the liabilities and the potential downsides of design choices. Even with better processes and more regulation, medical device liability remains high with society because the complex systems that compose medical devices have undergone dramatic increases in unforeseen circumstances and use.

A major flaw can surface at any time; the most problematic seem to arise several years after product launch. These kinds of problems also seem to be the most difficult to diagnose and fix.

14.15.2 Safety

Safety liability refers to any physical or psychological harm to a patient. Harm can even be perceived and not necessarily be organic. While you cannot prevent perceived harms, you can reduce their possibility of occurrence through good design, good practices, and good procedures. Carefully educating everyone in the treatment regime, from physician to patient, can help reduce incorrect use.

14.15.3 Legalities

Legal liability includes any or all of the above. Most companies in the United States are corporations. The corporate structure provides a measure of protection (but not complete protection) from lawsuits for individuals within a company. Even with the corporate shield, you can still suffer costly lawsuits and defamation in the public media and press, which ultimately might lose the company.

14.16 Summary

14.16.1 Emphases

Any medical device must follow the FDA's Design Control Guidance, which is freely available on the web along with some other useful documents [3–8]. This is particularly true for implantable medical devices.

Set up good practices and procedures and complete documentation; specialists in documentation and in dealing with the FDA can smooth the process. Implantable devices almost always require a large team of people with many different skills.

As always, good people can make any process work. No process, regardless how good it is, can make bad people work well.

14.16.2 Gotcha's

Developing an implantable medical device has many pitfalls. The market tends to have low-volume sales but high prices. The tradeoffs are endless—therapeutic effectiveness, power, weight, longevity, and cost. Attaining fault tolerance is a never-ending battle to make devices safer. Development usually takes 4–8 years or longer!

Acknowledgment

My thanks to Curt Sponberg at Medtronic, Inc. for reviewing this chapter and providing insight into some of the standards for medical devices.

References

1. Pazemenas, V., Rapid Development for Medical Products, *IEEE Instrumentation & Measurement Magazine*, Vol. 3, No. 2, June 2000, p. 35.
2. Document templates may be found at www.cool-stream.com
3. *Write It Right, Recommendations for Developing User Instruction Manuals for Medical Devices Used in Home Health Care*, FDA Center for Devices and Radiological Health, August 1993.
4. EMC Standards, *Compliance Engineering*, Vol. 21, No. 1, 2004 Annual Reference Guide, pp. 75–82.
5. Product Safety Standards, *Compliance Engineering*, Vol. 21, No. 1, 2004 Annual Reference Guide, pp. 165–172.
6. U.S. FDA, *Design Control Guidance for Medical Device Manufacturers*, March 11, 1997, relates to FDA 21 CFR 820.30 and sub-clause 4.4 of ISO9001. You can find it at http://www.fda.gov/cdrh/comp/designgd.pdf
7. Sawyer, D., *Do it by Design, An Introduction to Human Factors in Medical Devices*, FDA Center for Devices and Radiological Health, December 1996.
8. *Medical Device Use-Safety: Incorporating Human Factors Engineering into Risk Management*, FDA Center for Devices and Radiological Health, July 18, 2000.
9. Brooks, D., *Signal Integrity Issues and Printed Circuit Board Design*, PTR Prentice Hall, Upper Saddle River, NJ, 2003.
10. Bogatin, E., *Signal Integrity: Simplified*, PTR Prentice Hall, Upper Saddle River, NJ, 2004.
11. Laplante, P., *Real-Time Systems Design and Analysis, 3rd Edition*, IEEE Press, Wiley-Interscience, Piscataway, NJ 2004, p. 91.
12. Design review templates can be found at www.cool-stream.com

15

Summary Comparisons Across the 11 Case Studies

15.1 Comparing the Case Studies

This chapter wraps up the 11 case studies with tabulated comparisons. Each of the following sections in the chapter, 15.2 through 15.13, aligns with one area across the case studies. These areas are as follows:

- Market
- People and disciplines
- Architecting and architecture
- Scheduling
- Documentation and processes
- Requirements and standards
- Analyses
- Design trade-offs
- Test and integration
- Manufacturing
- Support and service
- Liability

Each section is self-contained. This chapter does not have a summary or conclusion; each section acts as a summary for that particular concern across the case studies.

Not all the comparisons are exact or strictly objective. As an example, some products may have little integration while others have a medium amount of integration. This is a subjective evaluation that I have made based on a few products that have very extensive programs of integration.

15.2   Market

Each product in these 11 case studies has a worldwide market. Every market has specific concerns ranging from the volume of sales to the expected longevity of the product to its profit margin with a myriad of parameters in between. Table 15.1 lists some of these concerns.

The sales volume depends on the particular model of product. Simple, cheap implantable devices will sell in far greater volume than complex, expensive devices. In some cases the sales volume is considered proprietary and is unknown, though you can begin to guess at the volume by looking at the size of the staff for design and support. Longevity is either a vague sort of range or a median value; clearly some products will last twice or three times the duration shown.

TABLE 15.1

Comparison of Market Concerns for the 11 Case Studies

	Sales (units/year)	Longevity (years)	Comments and Concerns
Large appliances	3–5 million	20–40	Slim profit margins; objectives are low cost and high reliability
Small office telecoms	180,000–250,000	3–5	"High-mix, low-volume" production requiring flexible facilities
Lab instruments	45,000–60,000	5–20	"High-mix, low-volume"; 500 different products
Automobile ECM	200,000–1,000,000	5–10	Development takes 2 years, updates may occur over 5 years; must archive source code for 10 years after final update
Oil field flowmeters	[a]	15–30	Sales follow oil industry ups-and-downs
Military equipment	50–100	20	Requirements depend on operational arena: O-Level, I-Level, D-Level
Space instrument	1	5–15	Requirements depend on mission profile
Commercial space system	10–40	0.001–1	Growing market, more products being sold every year
Satellite subsystem	1	1–3	Planned to build up to 8 units to amortize cost
Neurostimulator programmer	2,000–10,000	10	Low cost (many doctors expected it to be free), must be rugged and robust
Implantable devices	40,000–5 million	2–8	High reliability demanded; individual product sales range between US$5K and US$40K

[a] Proprietary number, unknown how many units are sold each year.

15.3 People and Disciplines

Many different people can aid the development of any one product. The case studies focused on the developers and designers and those directly involved in the definition of the product. The case studies did not include all staff throughout the life cycle of a product. For larger volume production projects, I did not include manufacturing, assembly, test, delivery, and administrative personnel. For the smaller projects, such as satellite and space instruments, I included everyone—it is difficult to separate developers from technicians on these kinds of projects. Table 15.2 attempts to list the people most prominently involved in development.

TABLE 15.2

Comparison of Responsibilities Within Development Teams for the 11 Case Studies

	People Per Team	People Involved in Definition, Design, and Development
Large appliances	8–30	Four software developers, 3 hardware engineers, 1 CAD operator, 1 technician, several control engineers (part time)
Small office telecoms	3	Very small engineering team—2 or 3 engineers, usually 1 hardware engineers and 1 or 2 software developers
Lab instruments	3–20	Software and hardware engineers, marketing, R&D scientists, manufacturing engineers, application support
Automobile ECM	8	Originally mechanical engineers but software, firmware, and hardware engineers are changing the composition of the team
Oil field flowmeters	10	PhD physicists, fluid dynamicists, firmware and hardware engineers, manufacturing engineers
Military equipment	6–30	Many different types of engineers, technicians, training and documentation specialists
Space instrument	8–30	Program manager, lead, systems, and many different types of engineers, support technicians
Commercial space system	6	One lead or systems engineer, 1 software, 2 hardware, 2 technicians
Satellite subsystem	35	Program manager, Lead, Systems, and many different types of engineers, support technicians
Neurostimulator programmer	12	Two vice presidents, 4 software engineers, 1 hardware engineer, 2 software testers, 1 technical writer, 1 trainer, 1 support staff
Implantable devices	50–70	Huge, varied team—many specialties

Interestingly, the job descriptions seemed very specific for high-volume products, implanted medical devices, or space instruments. People have very clear delineations of responsibilities in those types of projects. Smaller companies and projects required people to "wear multiple hats" in developing their products.

15.4 Architecting and Architecture

Architecture plays the key role in defining a product. Chapter 1 briefly explained some of the higher-level concepts such as distributed, centralized, modular, monolithic, loose coupling, and tight coupling. Another concern, often overlooked by designers, is the human interface; it should be considered in the basic concept and definition of the product. One lower-level (or bottom-up) concern is the type of processors that will be incorporated in the product.

Table 15.3 provides an overview of some architectural features in the 11 case studies. As many products become more complex, their architecture tends to become more distributed. Most products still rely on some sort of central processing, however, good modular design often helps with

TABLE 15.3

Comparison of Architectural Concerns for the 11 Case Studies

	Distributed vs. Centralized	Modular vs. Monolithic	Loose vs. Tight Coupling	Types of Processors	Human Interface
Large appliances	Cent	Mono	t	μC	s, e
Small office tele-coms	Cent	Mono	t	μC, μP	s, m
Lab instruments	Cent	Mod	t	μC, μP, DSP	s, e
Automobile ECM	Cent	Mono	t	μC	n
Oil field flowmeters	Dist	Mod	L	μC, μP	n
Military equipment	Cent, dist	Mod, mono	t, L	μC, μP, DSP	c, *
Space instrument	Cent	Mod, mono	t	μP	c, *
Commercial space system	Cent	Mod, mono	t	μC, DSP, FPGA	c, *
Satellite subsystem	Dist	Mod	L	μC, μP	s, *
Neurostimulator programmer	Cent	Mono	t	μP	s, c, e
Implantable devices	Cent	Mono	t	μP	n

s = simple to use; c = complex operations. Development effort: e = extensive; m = some; n = none; t = tight; L = loose. * = specialized for trained personnel or varies.

future upgrades and with integration; these are two disparate concerns aided by modular design.

The seeming paradox in the human interface for space systems is that all spacecraft have some sort of ground support system, which usually has a complicated interface to display the variety of desired parameters but relatively little effort is spent developing it. Military equipment in depot-level facilities can be similar to spacecraft support equipment; consequently, they, too, may have complex configurations for the human interface. Generally, both situations involved highly trained and skilled operators, so their interfaces do not require all the niceties of a commercial consumer product.

15.5 Scheduling

Many products have tight constraints on time-to-market because of mandated deadlines, hence they use top-down scheduling. Some complex products have such a long development cycle, with certifications and approvals along the way, that they use bottom-up types of scheduling. Table 15.4 shows the comparisons of schedules across the 11 case studies.

15.6 Documentation and Processes

Many companies still use proprietary processes. Nevertheless all companies document their products to some degree. Table 15.5 provides the summary comparisons of documentation and processes for the 11 case studies.

TABLE 15.4

Comparison of Scheduling for the 11 Case Studies

	Top-Down	Bottom-Up
Large appliances	✓	
Small office telecoms	✓	✓
Lab instruments	✓	✓
Automobile ECM	✓	
Oil field flowmeters	✓	✓
Military equipment	✓	
Space instrument	✓	✓
Commercial space system	✓	✓
Satellite subsystem		✓
Neurostimulator programmer		✓
Implantable devices		✓

TABLE 15.5

Comparison of Documentation and Processes for the 11 Case Studies

	Document Type	Process Model
Large appliances	ISO	V + S
Small office telecoms	Proprietary	W + S
Lab instruments	Proprietary	S
Automobile ECM	Proprietary	V
Oil field flow meters	ISO	W
Military equipment	ISO	W
Space instrument	Proprietary	W
Commercial space system	CMMI*	V + S
Satellite subsystem	Proprietary	W
Neurostimulator programmer	FDA	V
Implantable devices	FDA	V

ISO = International Organization for Standardization; CMMI = Capability Maturity Model Integration; * = currently implementing; FDA = Food and Drug; W = waterfall; V = V-model; S = spiral.

TABLE 15.6

Comparison of the Standards Used in the 11 Case Studies

	Standards
Large appliances	UL, CE
Small office telecoms	UL, CE
Lab instruments	UL, CE
Automobile ECM	SAE, OSEK, CAF, EPA
Oil field flowmeters	UL, ATEX, CSA, GOST
Military equipment	Military-Standards, SOW
Space instrument	SOW, NASA
Commercial space system	SOW, NASA
Satellite subsystem	SOW, NASA
Neurostimulator programmer	FDA
Implantable devices	FDA

UL = Underwriters Laboratory; CE = Conformite Europeene marking; SOW = Statement of Work; FDA = Food and Drug Administration.

15.7 Requirements and Standards

Each market has different requirements and different standards. Most consumer-oriented products have to be certified by either Underwriters Laboratory (UL) or have Conformite Europeene (CE) marking. Interesting, military equipment and spacecraft designs are moving away from strict standards and toward reliance on statements of work (SOW) to define the necessary requirements. Table 15.6 lists some of those standards to which these products conform.

15.8 Analyses

Designers and development teams determine the feasibility of product concepts through analyses. A variety of different analyses might apply to any given product. Table 15.7 lists some of the more typical analyses used by companies. Not all companies use all six different types of analysis, and only two companies in this book use all six in driving their carefully controlled processes for laboratory instruments and medical devices. I have since found that a large telecoms company does all these analyses as well for designing products.

While Agar Corporation does not use heuristics explicitly in designing flowmeters, they essentially have a handbook of rules for designing circuits and systems to avoid sparks and undesired ignitions. Several companies rely on detailed prototypes, which approximate the final product, to prove feasibility and utility. Aerospace development will often use either an engineering development unit or an engineering model to confirm earlier analyses.

15.9 Design Trade-Offs

Design trade-offs require many dimensions. This section has nine subsections to compare how the 11 case studies make their individual assessments for design.

TABLE 15.7

Comparison of the Different Analyses Used in the 11 Case Studies

	b	h	c	r	s	p
Large appliances	✓	✓	✓	✓		✓
Small office telecoms	✓	✓	✓			
Lab instruments	✓	✓	✓	✓	✓	✓
Automobile ECM		✓	✓	✓		✓
Oil field flowmeters						✓
Military equipment				✓		✓
Space instrument		✓	✓	✓	✓	*
Commercial space system		✓	✓	✓	✓	*
Satellite subsystem		✓	✓	✓	✓	*
Neurostimulator programmer				✓		✓
Implantable devices	✓	✓	✓	✓	✓	✓

b = business case; h = heuristics; c = calculations; r = risk analysis (includes FMEA, FTA, or ETA); s = simulations; p = prototype testing; * = engineering development unit or engineering model.

15.9.1 Various Goals

This subsection has seven high-level considerations for defining the design of a product. These extend the architectural considerations in Section 15.4. Table 15.8 lists these considerations and the comparisons across the case studies.

15.9.2 Processor Elements

Processing varies widely between different products. Table 15.9 compares the processors and processing elements used in the case studies. It is more specific than the architectural considerations in Section 15.4.

15.9.3 Other Circuit Concerns

There are other circuit issues, besides choice of processor, that drive design. Table 15.10 lists some of these concerns.

15.9.4 Cooling

One of the specific issues in system design is cooling. Table 15.11 lists how each of the case studies approaches cooling.

15.9.5 Power

Another of the specific concerns in system design is power conversion and distribution. Table 15.12 lists how each of the case studies approaches power conversion and distribution.

TABLE 15.8

Comparison of the Different Analyses Used in the 11 Case Studies

	C	LT	Sa	s/w	D	H	P
Large appliances	✓	✓	✓	✓	✓		
Small office telecoms	✓						
Lab instruments				✓	✓		
Automobile ECM	✓	✓		*	✓	✓	
Oil field flowmeters		✓	✓	✓	✓	✓	✓
Military equipment		✓		✓	✓	✓	
Space instrument		✓		*	✓	✓	✓
Commercial space system				*			✓
Satellite subsystem				*	✓	✓	✓
Neurostimulator programmer	✓	✓	✓	✓	✓		
Implantable devices		✓		✓	✓	✓	✓

C = cost sensitive; LT = long term; Sa = safety; s/w = careful software processes; * = wide variation or implementing better processes; D = dependable; H = harsh environments; P = low power consumption.

TABLE 15.9

Comparison of the Different Processors Used in the 11 Case Studies

	Bits = 8	16	32	Processors
Large appliances	✓	✓	✓	Highly integrated microcontrollers with many peripherals
Small office telecoms			✓	Control by microprocessors. Extensive use of FPGAs
Lab instruments	✓	✓	✓	Microcontrollers and microprocessors
Automobile ECM			✓	Full featured microcontrollers for algorithm sophistication with many integrated peripherals
Oil field flowmeters	✓	✓	✓	Microcontrollers: 8051, Microchip PIC, Atmel AVR
Military equipment	✓	✓	✓	Some COTS (depot level equipment), others have industrial-range environmental specifications
Space instrument	✓	✓	✓	Radiation-hard microcontrollers and processors; generally older architectures; low level of integration
Commercial space system			✓	Analog Devices Blackfin DSP acts as controller and FPGAs
Satellite subsystem		✓		Radiation-hard 80196 microcontroller
Neurostimulator programmer			✓	General purpose microprocessor running Windows CE
Implantable devices	✓	✓	✓	Microcontrollers—some are custom built

TABLE 15.10

Comparison of the Different Circuit Issues Used in the 11 Case Studies

	Comments and Concerns
Large appliances	20–40 year lifetimes. Problems with vendors: (1) changing fabrication parameters—smaller design rules and lower voltages—causing EMI problems, and (2) cannot specify flash memory retention greater than 20 years. Prefer single-sided PCBs to reduce cost
Small office telecoms	Most hardware focuses on data flow. Program loads are between 1 and 5 MB
Lab instruments	Forward-looking platform management saves inventory and manufacturing costs and speed development. Circuit boards are the focus of serious design efforts to reduce EMI of low-level signals
Automobile ECM	Need greater than 10 year lifetimes. Concern for ignition cycles (key on to key off); plan for 100,000–200,000 cycles. Problem is writing EEPROM every time key turns off. Multilayer PCBs to control EMC

(Continued)

TABLE 15.10

Continued

	Comments and Concerns
Oil field flowmeters	Avoid sparks, reduce power dissipation and charge storage. Longevity problems with vendors (see problems of large appliances in first line above). Sheath and tie down cables—no abrasion from vibration
Military equipment	Commercial components and subsystems for Depot Level, industrial grade components for Operational Level
Space instrument	Radiation hardness, resistance to outgassing. Memory is not dense, large, or fast. Electromechanics are challenge for launch vibration. Cables and connectors are a perennial source of problems
Commercial space system	Commercial components for very-short-term missions or low-radiation environments
Satellite subsystem	Distributed architecture for fault isolation and easing integration and for reducing cable complexity and weight
Neurostimulator programmer	Designed to use COTS. Outsourced up-converter design and manufacture. Avoid excessive heat build-up in laps of patients
Implantable devices	Power consumption is biggest concern. Dependability, fault tolerance, EMC, and moisture resistance are also important

TABLE 15.11

Comparison of Cooling Used in the 11 Case Studies

	Comments and Concerns
Large appliances	No real concerns—low-density power dissipation
Small office telecoms	Aim for low-power dissipation. Cooling airflow from chassis fans already in chassis racks
Lab instruments	Passive, convective cooling. Avoid fans because of EMI and added complexity and cost
Automobile ECM	Conduction only. Plan for extremes in temperature
Oil field flowmeters	No fans, strive for low-power dissipation. Enclosures must be sealed in volatile gaseous environments
Military equipment	O-Level sealed, no fans. Depot Level uses chassis fans
Space instrument	Only conduction to external radiative surfaces
Commercial space system	Only conduction to external radiative surfaces
Satellite subsystem	Only conduction to external radiative surfaces
Neurostimulator programmer	No fan—seal to allow disinfectant wash downs
Implantable devices	None, strive for low-power dissipation

15.9.6 Software

Software development is the biggest concern for most design and development efforts. Table 15.13 lists how each of the case studies approaches software development and hosting.

TABLE 15.12

Comparison of Power Conversion and Distribution Used in the 11 Case Studies

	Comments and Concerns
Large appliances	Universal input, 220–240 VAC, 50–60 Hz. Test for 52 different AC line conditions
Small office telecoms	Universal input, 100–240 VAC, 50–60 Hz. COTS power converters; use either supplies that slide into a chassis or "wall warts"
Lab instruments	Universal input, 100–240 VAC, 50–60 Hz. Low-level measurements and EMC force custom design of converters
Automobile ECM	12, 24, or 42 VDC power; spikes, sags, dropouts, ripple, and noise. Most ECMs have internal regulators
Oil field flowmeters	Universal input, 220–240 VAC, 50–60 Hz. Keep AC–DC converter in remote, safe area; it produces 12 VDC
Military equipment	Variety—aircraft supplied or universal AC input
Space instrument	Major source of problems: DC–DC converters, load conditions, and distribution. Switching power transistors susceptible to radiation concerns (total dose and SEE)
Commercial space system	Raw +28 VDC power (ranges from 22 to 34 V), use DC–DC converters
Satellite subsystem	Distribute raw power to point-of-load (POL) converters for fault isolation and reduction of cable weight
Neurostimulator programmer	Battery capacity needs to be greater than 8 hour of office time. Could only get 4–6 hour at best from COTS pentop. Custom design could increase battery capacity
Implantable devices	Battery capacity and power consumption are critical issues. Avoid excessive current demand and heat build-up. Rechargeable batteries—cost benefit, only recently becoming more prevalent

TABLE 15.13

Comparison of Software Development Used in the 11 Case Studies

	Comments and Concerns
Large appliances	Software in C. Custom time-slice RTOS to guarantee timeliness and amenable to analysis for 8-bit microcontrollers. Custom pre-emptive priority, interrupt-based RTOS for 32-bit systems. Re-use proven modules. Must be correct at production, no upgrades. Use very rigorous development processes
Small office telecoms	Software in C. Custom RTOS. No code review
Lab instruments	Software in C, moving to C++. Custom RTOS in older products. New products—OSE, WinCE. Web site for software upgrades
Automobile ECM	Huge variation in rigor between different companies. Most software in C with some assembly language. Both custom and commercial RTOSs

(Continued)

TABLE 15.13

Continued

	Comments and Concerns
Oil field flowmeters	Rigorous development. Software in C. Custom RTOSs for 8-bit microcontrollers. QNX and Linux RTOSs for 16- or 32-bit systems
Military equipment	Code reviews. O-Level use HP Basic or assembly. D-Level use Atlas or C languages. Both use custom RTOSs
Space instrument	Careful design; regular, mandated design reviews. Usually custom RTOS because commercial RTOS often have too many features and require too much memory; often time-slice architecture that allows analysis
Commercial space system	Moving to Green Hills Velosity® RTOS
Satellite subsystem	C code with regular, mandated design reviews
Neurostimulator programmer	C++ in Windows CE environment. Rigorous development with regular, mandated design reviews
Implantable devices	Safety-critical, rigorous development processes. Certified compilers. Custom and commercial RTOSs—must be simple and easily analyzed

TABLE 15.14

Comparison of Buy vs. Build Used in the 11 Case Studies

	Comments and Concerns
Large appliances	High-volume, low-margin markets. Every subsystem is custom designed and built to achieve cost, functionality, and reliability
Small office telecoms	All custom design, most power converters are COTS
Lab instruments	All custom design except some LCD display systems are COTS
Automobile ECM	Cycle between automobile company custom design and vendor design. Future leaning towards outside vendors and COTS
Oil field flowmeters	Older designs used COTS single-board computers. New designs are more distributed and custom-designed and built
Military equipment	O-Level custom design. Depot-Level use VME and VXI boards.
Space instrument	Custom design. GSE is COTS
Commercial space system	Commercial components used: ICs, video cameras, DC–DC converters
Satellite subsystem	Custom design. GSE is COTS. To be replicated 6–10 times
Neurostimulator programmer	COTS pentop computer but supplier problems—6-month market cycle; need 3–5 year inventory. Custom design with COTS subsystems would have been better in the long run
Implantable devices	Custom design: PCB (ceramic) and titanium hermetically sealed

15.9.7 Buy vs. Build

Another consideration for development is whether to buy or to design a custom subsystem or component. Table 15.14 lists how each of the case studies approaches the question.

15.9.8 Plan for Manufacturing and Assembly

Planning for manufacturing and assembly is another consideration for designing a product. Table 15.15 lists how each of the case studies approaches plans for manufacturing and assembly.

TABLE 15.15

Comparison of Plans for Manufacturing and Assembly Used in the 11 Case Studies

	DFx, x =	Comments and Concerns
Large appliances	M, A, T	Automatic manufacture and assembly in-house; test all buttons, inspect all displays, test all loads; software distinct between products; shave pennies everywhere
Small office telecoms	M, T	JTAG, POST, BIT, link tests for T1, DSL, ISDN; manual labor and automatic manufacture and assembly in-house
Lab instruments	M, A, T, f, t, i	f = flexibility, t = transfer, i = improvements; mfg in house, some outsource; test for quality—BIT and ICT (bed of nails, flying probe)
Automobile ECM	M, A, T	High-temperature plastic enclosures; "calibrations"—software loads done in manufacturing
Oil field flowmeters		Outsource fabrication but design and manufacture test jigs in-house. Assemble in-house
Military equipment	M	Manual fabrication, ATE for cables
Space instrument		Manual manufacturing and assembly in-house; problems with "churn" in revision of PCB fabrication
Commercial space system		Outsource manual manufacture and assembly
Satellite subsystem		Manual manufacture and assembly in-house, some PCBs in automated assembly
Neurostimulator programmer		Outsource manual manufacture and assembly
Implantable devices	M, A, T	Both manual and automatic assembly in house; welded titanium enclosures

M = design for manufacturing; A = design for assembly; T = design for test; f = design for flexibility; t = design for transfer (to other outsource manufacturers); i = design for improvement.

15.9.9 Plan for Diagnostics, Repair, Maintenance

Like planning for manufacturing and assembly in designing a product, another consideration is planning for service and repair. Table 15.16 lists how each of the case studies approaches plans for service and repair.

15.10 Test and Integration

Test and integration is an important point in the development of a device. It determines if the product meets its specifications. Table 15.17 lists the amount of effort with which each case studies approaches test and integration. Please note that this evaluation is a subjective rating. It is more relative than absolute in its comparisons.

15.11 Manufacturing

Manufacturing is another important point in the development of a device. Table 15.18 lists the types of manufacturing and assembly each case studies uses. The variation in commercial off-the-shelf (COTS) incorporation depends on the model of product and whether suitable components or subsystems are available.

TABLE 15.16

Comparison of Plans for Service and Repair Used in the 11 Case Studies

	Comments and Concerns
Large appliances	Local shops—replace modules only, no component repair
Small office telecoms	No field repair, replace entire unit; technical support fields first call
Lab instruments	Repair centers around world. Trains large customers to repair units themselves
Automobile ECM	"Service releases" to upgrade software; recalls if problem really bad. Background Debug Mode (BDM) to get additional, nonstandard diagnostics from the ECM for laboratory analysis
Oil field flowmeters	Large field sales staff and affiliates around the world. Considering wireless designs to do BIT without opening enclosures
Military equipment	BIT, calibration; users call engineers directly (small market)
Space instrument	Limited or no diagnostics and repair
Commercial space system	Limited diagnostics and repair
Satellite subsystem	Limited or no diagnostics and repair
Neurostimulator programmer	No field repair, replace entire unit; engineers diagnose problems
Implantable devices	No field repair, replace entire unit; extensive support staff

TABLE 15.17

Comparison of the Amount of Effort for Test and Integration in the 11 Case Studies

	l	m	e
Large appliances		✓	
Small office telecoms	✓		
Lab instruments		✓	
Automobile ECM		✓	
Oil field flowmeters		✓	
Military equipment		✓	
Space instrument			✓
Commercial space system			✓
Satellite subsystem			✓
Neurostimulator programmer		✓	
Implantable devices			✓

l = little; m = medium; e = extensive.

TABLE 15.18

Comparison of the Manufacturing in the 11 Case Studies

	u	a	i	o	DFx	COTS
Large appliances		✓	✓		✓	
Small office telecoms	✓	✓	✓		✓	
Lab instruments		✓	✓		✓	
Automobile ECM		✓	✓		✓	
Oil field flowmeters	✓	✓	✓	✓		*
Military equipment	✓	✓	✓	✓		*
Space instrument	✓		✓			
Commercial space system	✓		✓	✓		✓
Satellite subsystem	✓		✓			
Neurostimulator programmer				✓		✓
Implantable devices	✓	✓	✓	✓		

u = manual; a = automated; i = inhouse; o = outsource; DFx = design for x;
COTS = commercial-off-the-shelf subsystems; * = varies.

15.12 Support and Service

Support and service can be an important consideration in the development of a device. Table 15.19 lists the amount of effort with which each case study approaches test and integration. Please note that this evaluation is a subjective rating. It is more relative than absolute in its comparisons. Obviously third-party repair shops can fix a failed appliance—hence, the two check marks in "none" to "little" support and service. Some products are marked by their large, available support staff, in particular the oil field flowmeters and implantable medical devices.

TABLE 15.19

Comparison of the Available Amount of Support and Service
in the 11 Case Studies

	n	l	m	e
Large appliances	✓	✓		
Small office telecoms		✓		
Lab instruments			✓	
Automobile ECM		✓		
Oil field flowmeters				✓
Military equipment			✓	
Space instrument	✓			
Commercial space system	✓			
Satellite subsystem	✓			
Neurostimulator programmer		✓		
Implantable devices				✓

n = none; l = little; m = medium; e = extensive.

TABLE 15.20

Comparison of the Liability Across the 11 Case Studies

	E	S	L
Large appliances	✓	✓	✓
Small office telecoms	✓		✓
Lab instruments	✓		
Automobile ECM	✓	✓	✓
Oil field flowmeters	✓	✓	✓
Military equipment	✓		✓
Space instrument	✓		
Commercial space system	✓		
Satellite subsystem	✓		
Neurostimulator programmer	✓	✓	✓
Implantable devices	✓	✓	✓

E = economic; S = safety; L = legally protect patents and intellectual property.

15.13   Liability

Every product has liability. Every one of them has some sort of economic
liability (E column)—contractual or exposure to government recall. Some
have legal concerns with protecting their patents and intellectual property
(L—last column in the table). Table 15.20 lists the areas of appreciable liability each case studies.

16

Some Observations on Architectural Trade-Offs in Selected Real-Time Systems

16.1 Some Thoughts

This chapter sketches some of the attributes of three very different real-time systems and some of the considerations for their designs. One system records and indicates the available spaces in a car parking garage. Another is a data-acquisition system that collects biological data. The third application is a gun-fuzing system for the military.

Each of these systems requires attention to architecture, component selection, power supply and distribution, software development, packaging, manufacturing, installation, and maintenance. Each system addresses the design process differently from the other two.

16.2 Indicating System for a Parking Garage

16.2.1 Purpose and Description

Some parking garages at airports have arrays of sensors, light-emitting diodes (LEDs), and message boards to help people find available parking spaces. Each space has a sensor above it on the ceiling that detects the presence of an automobile (Figure 16.1). The sensor has a red LED that lights when an automobile occupies the space; it has a green LED that lights when no vehicle occupies the space. The green LED indicates that the space is available for another automobile.

At the end of each isle is a message board that displays the number of open spaces and an arrow pointing down the isle with the open slots. If a row is full with cars, the message board will display a red X, which indicates a full row and no available slots (Figure 16.2).

16.2.2 Issues or Concerns

The liability for this parking garage system is quite low; it facilitates people in finding parking spaces, but it does not prevent them from finding a space

FIGURE 16.1
A sensor above a parking space in an airport parking garage. (© 2006 by Kim Fowler, used with permission. All rights reserved.)

FIGURE 16.2
A message board in a parking system in an airport parking garage. (© 2006 by Kim Fowler, used with permission. All rights reserved.)

if it fails or malfunctions. What this means is that false indications of filled spaces can be acceptable; however, false indications of open spaces would be less than acceptable—it will upset customers by wasting their time driving down an aisle that indicates spaces but they find none. On the other hand, driving down an isle that indicates no spaces and finding one leaves a driver feeling lucky or fortunate.

16.2.3 Real-Time Calculations

This is a soft, real-time system because the deadlines are not as critical to system operation. However, if delays are noticeable, they may affect the success of the system. The system has a range of acceptable durations to detect automobiles, indicate their presence, and update the message boards.

What is acceptable? That really depends on operational studies that assess human reactions to the system and that determine traffic flow and volumes. To specify a parking garage system, you will need to specify the maximum delay to maintain customer satisfaction and minimize driver frustration. Here is one possibility:

- Consider a maximum potential holiday traffic volume of 2000 cars entering and leaving the parking garage every hour, then the average time between cars is $(3600 \text{ sec/h})/(2000 \text{ cars/h}) = 1.8$ sec. This might suggest a minimum time for detection.

- Is the minimum detection time, between 1.5 and 2 sec, worth the extra cost to design a system that handles holiday traffic once or twice each year? This trade-off of cost vs. user satisfaction (or frustration) will drive the design specifications. A maximum acceptable delay might be somewhere between 3 and 5 sec.

16.2.4 Architecture

The cost of different system configurations can significantly affect the choice of architectures. For the sake of illustration, I will assume an average system that accommodates 5000 parking spaces and has an operational life of 10 years.

There are several interesting points to note:

- The highest cost within the system may be installation. The labor for installing the system can be greater than the cable, conduit, mechanical suspension, and the total production cost of the sensors.

- Next in line for cost is the system of cable, conduit, and mechanical suspension. The physical configuration of the system directly dictates cost.

First, there will be one sensor above each parking spot. If the average parking garage has an average of 5000 spaces, then each system will need

5000 sensors. If marketing estimates sales of systems to 200 parking garages that hold an average of 5000 cars each, then your company will eventually need to manufacture one million (10^6) sensors plus spares for replacement of failed units.

What kind of network should connect the sensors—a proprietary design or a more typical configuration based on a current standard? Here is where the cost of cable and installation play a major role. If a proprietary network calls for a two-wire cable and conduit, containing both power and signal, costs about US$27 per car slot, and a parking space is 4 m wide, then a 5,000 car garage will cost about US$133,000 for the cabling and conduit alone. If you chose a more typical configuration with a four-wire cable plus power cable, which costs about US$47 per car slot for the cable and conduit, then the costs are closer to US$233,000 (Table 16.1).

A proprietary network design could lend itself to a smaller cable and conduit, which might be cheaper and might reduce the installation time. That cost comes right off the bottom line and improves the profit margin.

The trade-off is whether the extra cost of the standard four-wire configuration will ease maintenance enough and provide more satisfied users over the 10 years of operation over the proprietary network.

16.2.5 Hardware

The primary trade-off is determining the type of processing element that would be the most appropriate: microcontroller vs. application-specific

TABLE 16.1

Estimated Costs for Mechanical Support Hardware (Not Including Sensors) for Two Different Types of Parking Garage Systems (© 2006 by Kim Fowler, Used with Permission. All rights reserved)

| Item | | System #1: 2-wire | | | | System #2: 4-wire | | |
	Qty	Metric	Cost	Extension	Qty	Metric	Cost	Extension
Cable	1	Meter	$3.00	$12.00	1	Meter	$6.00	$24.00
Conduit	1	Meter	$3.00	$12.00	1	Meter	$5.00	$20.00
Suspension hardware	1	Set/slot	$2.50	$2.50	1	Set/slot	$2.50	$2.50
Subtotal/ parking space				$26.50				$46.50
Total installation hardware costs				$132,500				$232,500
Width per car slot	4	Meters						
Spaces in garage	5000							

integrated circuit (ASIC). Minimizing peripheral support components and chips is important too.

You might decide to purchase an 8-bit microcontroller to be the processor within each sensor. These can cost between US$0.50 and US$1 per chip in large quantities. If you chose this path, then your company will have to employ engineers to design, code, and test the software. The nonrecurring engineering (NRE) is the cost of writing the software and designing the circuitry.

Another possibility would be to build a fully-custom ASIC controller for each sensor—the NRE to design a chip can easily be as high as US $1,000,000. Fabrication can approach US$3 per chip or more. This totals to US$4 per chip in final delivered form. One problem is that fully custom ASICs are not easily revised, upgraded, or fixed, unless you incorporate a flash memory and upload critical parameters. (This would be similar to the "calibrations" process used in automobile ECMs in Chapter 7.)

Table 16.2 is a fairly detailed scenario that compares these two approaches. This is only one scenario, out of many different possible ones, for how you might design such a system.

16.2.6 Power

Power distribution and conversion is not a simple task. Thin, narrow-gauge wire is cheaper than thicker wire, but it has higher resistance per unit length; this causes a greater voltage drop down the cable. When cable runs are long, 80 m (about 260 ft) in this case study, the loss in a cable can be significant. These concerns motivate two important efforts; reduce the power consumption in each sensor and increase the voltage of the power delivery.

This case study assumes DC power distribution rather than AC to remove the need for a step-down transformer and its cost from each sensor. Regardless of voltage source, each sensor must have local regulation. Higher voltage distribution is more efficient than lower voltage. The trade-off is safety; lower voltage has a lower shock hazard.

Table 16.3 outlines the total voltage drop over 20 car park slots for several different supply voltages. Clearly higher voltages experience lower loss. You might settle on the 18 or 24 VDC power distribution because it allows longer runs in bigger parking garages and still is considered quite safe.

The calculations assume 4-m wide slots. These calculations were simplistic; they assumed that each sensor required 100 mW. Digital logic operates at 3.3 V and the total current is 30 mA: always-on two LEDs, each needing 10 mA, the processor and its support circuitry, which need 9 mA, and the ultrasonic transducer pulsed once per second at 40 mA for 25 μsec. The calculations also assumed that the power converter within each sensor is 80% efficient; this is a rough approximation because most power converters have a nonlinear characteristic or transfer function between efficiency and input supply voltage. Calculating power consumption and voltage drop in distributed systems with nonlinear loads is quite difficult!

TABLE 16.2

Estimated Sensor Costs for Two Different Types of Processing: Microcontroller vs. ASICs (© 2006 by Kim Fowler, used with permission. All rights reserved)

	Microcontroller-Based				ASIC-Based			
	Qty	Metric	Unit Cost	Extension	Qty	Metric	Unit Cost	Extension
NRE								
Project management	2000	Hours	$80.00	$160,000	2000	Hours	$80.00	$160,000
Preparing specifications	800	Hours	$60.00	$48,000	800	Hours	$60.00	$48,000
Simulations and feasibility	800	Hours	$60.00	$48,000	800	Hours	$60.00	$48,000
Chip design								$1,000,000
Software code (2 LOC/h)	5000	LOC	$30.00	$150,000	200	LOC	$30.00	$6,000
Hardware design	2000	Hours	$60.00	$120,000	400	Hours	$60.00	$24,000
Development and test	2000	Hours	$60.00	$120,000	2000	Hours	$60.00	$120,000
Documentation	800	Hours	$60.00	$48,000	400	Hours	$60.00	$24,000
Field tests	2000	Hours	$60.00	$120,000	2000	Hours	$60.00	$120,000
Manufacturing plans	800	Hours	$60.00	$48,000	800	Hours	$60.00	$48,000
Manufacturing test equipment	400	Hours	$60.00	$24,000	400	Hours	$60.00	$24,000
Customer meetings, product launch	1000	Hours	$60.00	$60,000	1000	Hours	$60.00	$60,000
Overhead—travel, meetings, and so forth				$75,000				$75,000
NRE subtotal				**$1,021,000**				**$1,757,000**
Cost of NRE per sensor								
Quantity produced	2E+05			$5.11				$8.79
Quantity produced	5E+05			$2.04				$3.51
Quantity produced	1E+06			$1.02				$1.76
Cost to produce 1 million sensors								
Cost (NRE + COGS) ($40/h burdened labor rate)								
Processor NRE cost	1	NRE unit	$1.02	$1.02	1	NRE unit	$1.76	$1.76
Processor unit cost	1	Part	$0.50	$0.50	1	Part	$3.00	$3.00
Peripheral component costs	1	Set	$8.40	$8.40	1	Set	$8.40	$8.40

Line item	Qty	Unit	Cost	Qty	Unit	Cost
Circuit board fabrication	1	Board	$5.00	1	Board	$5.00
Circuit board assembly	1		$4.00	1		$4.00
Enclosure cost	1		$4.00	1		$4.00
Final assembly and test	1		$3.33	1		$3.33
Cost per unit subtotal			**$26.25**			**$29.49**
Final cost of sensors for 5000 slots (total)			**$131,272**			**$147,452**

Cost (NRE + COGS) ($10/h burdened labor rate)

Line item	Qty	Unit	Cost	Qty	Unit	Cost
Processor NRE cost	1	NRE unit	$1.02	1	NRE unit	$1.76
Processor unit cost	1	Component	$0.50	1	Component	$3.00
Peripheral component costs	1	Set	$8.40	1	Set	$8.40
Circuit board fabrication	1	Board	$5.00	1	Board	$5.00
Circuit board assembly	1		$1.00	1		$1.00
Enclosure cost	1		$4.00	1		$4.00
Final assembly and test	1		$0.83	1		$0.83
Cost per unit subtotal			**$20.75**			**$23.99**
Final cost of sensors for 5000 slots (total)			**$103,272**			**$119,952**

Cost (NRE + COGS) (automated assembly)

Line item	Qty	Unit	Cost	Qty	Unit	Cost
Processor NRE cost	1	NRE unit	$1.02	1	NRE unit	$1.76
Processor unit cost	1	Component	$0.50	1	Component	$3.00
Peripheral component costs	1	Set	$8.40	1	Set	$8.40
Circuit board fabrication	1	Board	$5.00	1	Board	$5.00
Circuit board assembly	1		$0.20	1		$0.20
Enclosure cost	1		$4.00	1		$4.00
Final assembly and test	1		$0.20	1		$0.20
Cost per unit subtotal			**$19.32**			**$22.56**
Final cost of sensors for 5000 slots (total)			**$96,605**			**$112,785**

TABLE 16.3

Estimated Voltage Drop for Different Supply Voltages (© 2006 by Kim Fowler, Used with Permission. All rights reserved.)

Supply Voltage (Vs)	% Total Voltage Drop	Voltage Drop (ΔV)	Voltage at Last Sensor (Vd)
12	11.4	1.37	10.6
18	5	0.9	17.1
24	3	0.72	23.3
30	1.8	0.54	29.5

16.2.7 Software

Software should probably be in a high-level language, such as C, to speed development and improve archival maintenance. A simple real-time operating system (RTOS) will ease both development and maintenance.

This sort of system probably would not allow upgrades to the software in the sensors once in the field. The labor cost to open each sensor and upload updates would be fairly expensive. The opposite would be true for the message boards or central control units; they may be Web-based to allow software upgrades and changes.

Software development is a fairly simple process and not mission critical; consequently, while good practices should be encouraged, rigorous development is not as important. (I hate saying that, but the reality is good processes do cost a little extra.) The one caveat is that contractual liability still remains—therefore, I still encourage good software development.

16.2.8 Packaging

Packaging has to endure some environmental extremes, such as temperature (–40 °C to +50 °C), humidity, and dust. Furthermore, the packaging must last for many years. One line of defense is durable conformal coating for the components and circuit board. Encapsulating the components will also hinder repair, but for a replaceable unit—as opposed to repairable— this is not a concern.

16.2.9 Buy vs. Build

Sensors will be custom built considering the large number to be produced and their specific requirements. Greater quantity reduces the per-unit costs of each sensor; Table 16.2 illustrates just one example of some of the parameters that factor into producing sensors.

The same would probably hold for the message boards that hang from the ceiling at the end of each row of parking spaces. These are custom applications and are not widely available. The power consumption, size, and expense of the message boards would not be as critical as for sensors—though still important—because they are far fewer in number. In any case, you will have to perform a careful set of trade-off studies before settling on a particular approach, which must account for NRE, per-unit costs, environmental constraints, reliability, and maintenance.

If the processing is not distributed between the message boards, then the system will need a central control unit to update the parking status for the entire system. You should seriously consider buying a single-board computer (SBC) for the central control unit. Each system only needs one (remember the total market is expected to be 200). A standard form factor, such as PC104, might fit this application well and will have a wide variety of choices. Many vendors offer commercial or ruggedized SBCs; some of them will inventory these for you, as well.

16.2.10 Manufacturing

Table 16.2 shows that automated fabrication and assembly can be significantly cheaper than manual labor. Pick-and-place machines and automated solder ovens are one example of the automated manufacturing. If this is your company's only product, then employing a contract manufacturer would be a given. If your company has these facilities and has available capacity, then you could manufacture the systems in-house.

Design-for-manufacture (DFM), design-for-assembly (DFA), and design-for-test (DFT) all make sense when designing the sensors. These initiatives require some upfront costs but result in lower per-unit costs and higher quality that should lead to higher reliability.

16.2.11 Installation and Maintenance

If installation over each parking space takes 20 minutes to drill holes, anchor the suspension rods, and attach the conduit and cable, plus another 10 minutes to attach a sensor to the suspended mechanical structure and test it, then a 5000-car garage will require 2500 labor-hours to install a system (Table 16.4).

At a burdened labor rate (that is where overhead and management costs are included, along with employee benefits, to the salary) of about US$45/hour, the installation cost for the sensors alone is about US$112,500. (This estimate does not explicitly include the cost to install the message boards above each isle or to install the central processing unit or to commission the system. These costs, however, are already factored into the sensor installation.)

A team of eight installers would take slightly less than 2 months to install a system that accommodates 5000 spaces. There are two ways to employ a team of installers; one way is to maintain a traveling team of trained personnel; the other way is to contract help at each location and train the installers on location. A trained team would probably be faster and be more

TABLE 16.4

Estimated Labor Costs for Two Different Installation Crews (© 2006 by Kim Fowler, used with permission.)

Items	Case 1: Traveling Installers				Case 2: Locally Contracted Installers			
	Qty	Metric	Time	Cost	Qty	Metric	Time	Cost
Burdened labor rate	$45.00	Per hour			$45.00	Per hour		
Time per parking space	0.5	Hour	2500	$112,500	0.65	Hour	3250	$146,250

	Case 1: Travel Costs				Case 2: Travel Costs			
	Qty	Metric	Cost	Extension	Qty	Metric	Cost	Extension
Lodging	36	Days	$60	$2,160	47	Days	$60	$2,820
Per diem	36	Days	$40	$1,440	47	Days	$50	$2,350
Daily transportation or rental car	36	Days	$20	$720	47	Days	$50	$2,350
Travel allowance	4	Trips	$400	$1,600	5	Trips	$400	$2,000
Subtotal per person				$5,920				$9,520
People having to expense travel	8				1			
Total travel				$47,360				$9,520
Total installation expense (labor + travel)				$159,860				$155,770

Number of sensors = 5000.

consistent in quality. A local team would potentially be cheaper, but would require training, could be slower, and may not be as consistent in their quality of installation.

Case 1: The company has a team of trained installers who travel to each site, the estimate of travel cost would include the following components (all in 2006 U.S. dollars):

- Local hotel (about US$60 per night per person)
- Food and incidentals (US$40 per diem)
- Travel allowance every 2 weeks for a long weekend (about US$400 per roundtrip per person)
- Transportation costs (about US$20 per day per person)

The travel cost for a permanent team for 2 months of installation would be about US$50,000. Add this to the labor rate of the team and the total costs for labor and travel for installation of one system is about US$160,000 (Table 16.4).

Case 2: The company hires local contract installers; the following assumptions are made:

- One person travels to the site to train and then oversee the installation
- The same travel costs exist as in Case 1 for that one person (plus slightly higher per diem and a rental car)
- Local training and local inefficiencies extend installation time by 30%

The travel cost for one person (remember the 30% longer duration) would be about US$9500. Add this to the labor rate of a local contract team that takes 30% longer (about US$146,000) and the total costs for labor and travel for installation of one system is about US$156,000 (Table 16.4).

For the two scenarios developed in Table 16.1 through 16.4, the total system cost for a two-wire system could be about US$390,000, while the cost could be about US$490,000 for a four-wire system (Table 16.5). Is the $100,000 difference in cost worth the potentially more difficult and therefore more expensive maintenance? The sensors are quite cheap. You probably will just replace a failed sensor and not repair it.

These sensors may have to operate for up to 10 years, continuously, in all sorts of harsh environments. How do you diagnose a failure and replace the sensor? Will built-in-test (BIT) reduce maintenance? Probably not. Rudimentary diagnostics and unit replacement will probably be the most appropriate plan for maintaining this type of system.

TABLE 16.5

Total Estimated Costs for Two Different Types of Parking Garage Systems (© 2006 by Kim Fowler, used with permission. All rights reserved)

Cost Item	Two-Wire System	Four-Wire System
Cable, conduit, and suspension	$133,000	$233,000
Sensors	$97,000	$97,000
Installation	$160,000	$160,000
Total cost	$390,000	$490,000

16.3 Data Acquisition System for Biological Monitoring

16.3.1 Purpose and Description

As electronics shrink in size and more transistors and functionality cram into each chip, their potential usage expands. Some biological studies, for instance, need to follow animals to study migration or feeding habits; collecting various data is becoming cheaper and more available. Some science teams have even built small electronic modules that easily attach to animals to collect data.

These specialized data collection systems with their modules are generally inexpensive compared to other systems presented in this book. They also tend to have very small production runs.

16.3.2 Issues or Concerns

While these systems are low in cost, they have a potentially high return in scientific value. The liability is very low, essentially nonexistent. Should a system fail, only data are lost; it can be reacquired in a follow-on study. No creature is physically harmed; besides, the nature of scientific investigation is fraught with uncertainty and failure.

The main concern is to assure the accuracy of the data. Calibration and metrology become an important consideration in data assurance. In particular, the analog-to-digital converters (ADCs) and sensors are critical to correct acquisition and should be calibrated appropriately.

16.3.3 Real-Time Calculations

Acquisition times for measuring biological systems range from fractions of a millisecond for fast events, like nerve impulses, down to 20 or 50 or 100 times per second for other physiological processes, and slower still to once per second for other events. Many situations in biology need to be sampled less frequently at once per minute or once per hour.

16.3.4 Architecture

While the deadlines for calculations and measurement times in biological data acquisition may be thousands or millions of times longer than for many embedded systems, they still have some unique challenges for the designer of embedded systems.

First, power is always a concern—batteries cannot be too big for animals to carry. Often systems remain unattended for weeks or months in the wild. Solar power might work if cells can be situated to receive sunlight regularly.

Electronic components suffer environmental extremes and have their limitations. There are always problems with moisture leaking into the circuitry. Large temperature variations are inevitable; though if attached to a warm-blooded creature, there is some moderation of those extremes because the body provides some thermal control. Memory within the data acquisition system can only be so large before it runs out of space to hold more data.

Location and marking by global positioning system (GPS) is becoming prevalent. Obscuring of the antenna and degrading of the signal reception always remains a problem. Transmitting the data or location can relieve the limitations on memory capacity, but then there are power consumption concerns with radio transmissions.

Finally, anyone who has dealt with electronics in the outdoors knows that wires are always susceptible to breakage. Either they flex too much or they get caught on something. Also critters can nest in your enclosures, chew up things, or clog them with debris if the enclosures are not completely sealed.

16.3.5 Hardware

Typically, a low-power microcontroller is the center piece of each data acquisition module. The modules also need sensors and conversion components, both ADCs and digital-to-analog converters (DACs). Most biological quantities, such as temperature, pressure, speed, and physiological parameters, do not need either high conversion resolution (less than 14-bits) or high conversion speed (less than 4000-Hz sampling). A large number of different microcontrollers have 10-bit or 12-bit converters built-in that are sufficient in speed for these biological applications.

Data acquisition needs memory for data storage and sometimes a transmitter to send the data to a collection receiver. Memory has to be low-power and non-volatile; flash-memory components have the density to hold megabytes to gigabytes of data. Flash memories tend to withstand shock and vibration much better than hard disk drives. Hard drives might be useful in some situations, but they need to be protected from the elements or be ruggedized.

16.3.6 Power

Supplying reliable power for a long time is probably the most difficult part of designing a data acquisition system for biological monitoring. As mentioned previously, systems cannot be too big or too heavy for animals to carry. Moreover, the systems are often unattended for long periods of time.

All this means that systems must be designed to consume very low power. You may have to design them so that they regularly cycle from active data collection to sleep mode, which draws miniscule amounts of power. Furthermore, you will have to consider the operating voltage levels. Either the power conversion must be very efficient to provide a rock steady supply voltage or the components must be able to operate over a wide range of supply voltages as the battery voltage drops.

If the system does not consume too much power and if the situation allows, solar power might be able to recharge the batteries. Obviously, solar cells must be situated to receive sunlight to recharge the batteries.

These are just a few of the concerns with powering a data acquisition system. Clearly, power supply and distribution are major considerations in developing the system architecture.

16.3.7 Software

Data acquisition is one area where software development processes may not be critical. (I will probably regret saying that!) If you really want data assurance, then developing software carefully with code reviews, tests, field tests, and configuration management is still advisable. You never know when someone may want to review the system and process that collected the data.

16.3.8 Packaging

Packaging the modules is probably the biggest challenge in data acquisition for biological monitoring. You will have to seal against moisture, tolerate wide temperature swings, and protect against flexure in wire and cable. Encapsulating the components in an impervious compound such as epoxy will help resist water invasion, protect against shock, and reduce mechanical flexing.

The best design for wires and cables is not to have any. Recognizing that is impractical, you should provide strain relief, particularly where wires both exit a case and enter a sensor.

16.3.9 Buy vs. Build

Almost all data acquisition systems are custom-designed. The very specific requirements of these types of systems strongly indicate that only custom design will accomplish the task.

16.3.10 Manufacturing

Again, the very specialized nature of these systems and very small number of finished units would strongly indicate custom, manual fabrication. A contract manufacturer might be useful here; still their costs could easily add US$100s per unit cost. As most of these data acquisition systems are used for university research, consider student labor—here is a cheap pool of motivated personnel that can learn complex procedures quickly.

Automated test is not typically appropriate for this type of system. You could test each module after assembly and before encapsulation with standard laboratory gear. BIT might serve you here but probably is not worth the extra effort.

16.3.11 Installation and Maintenance

Installation and attachment are often performed by graduate student labor. These modules probably are not repairable because the encapsulation covers the components and cements them in place. Hopefully, each module is cheap enough to be a replaceable unit.

16.4 Gun Fuzing System

16.4.1 Purpose and Description

The military wants to strike targets accurately; this also applies to firing a large gun. One of the concerns is setting the timing on the fuzes of explosive projectiles as they leave the barrel of the gun during the firing so that the projectile explodes near the target. The problem is that minute differences in grain burning between successive rounds create very slight differences in velocity of the projectile, which leads to dispersion in detonation near the target. An embedded system can set the fuze based on a real-time measurement of the muzzle velocity of each shell. Here is one way to do it.

One solution is to measure the speed of the shell between two points just outside the muzzle and then use an antenna to set the fuze timing as the shell flies by. Assume that

- Nominal shell velocity is 1000 m/sec
- Shell velocities vary between 995 and 1025 m/sec
- Distance between the shell sensors, which measure the velocity of the projectile, is 10 cm
- Distance between the last shell sensor and the antenna is 10 cm
- Dispersion (allowable miss distance) in the explosion should only vary about 1 m around the target

16.4.2 Issues or Concerns

Liability for such a project is not high, but it is not as low as the parking garage example. While the military accepts risk in using weapon systems, the government will require a contractual liability with penalty clauses for nonperformance.

One technical issue is to assure that the adjusted fuze timing is transmitted to the shell. This is a matter of completing the calculations during the time the shell exits the barrel and then making sure that radio signal transmission and reception are robust.

Another concern is that the calculations achieve the appropriate time resolution.

16.4.3 Real-Time Calculations

The real-time calculations must complete in short order. Calculation time is constrained by the time to measure the speed of the shells, to calculate the adjustment to fuze timing, and to transmit the appropriate value to the shell exiting the gun barrel.

The minimum time, $T_{\text{min vel calculation}}$, to measure the shell velocity is constrained by the fastest shell velocity and by the spacing between the sensors. $T_{\text{min vel calculation}}$ for the fastest shells is 97 μsec.

$$T_{\text{min vel calculation}} \leq \frac{d_{\text{between sensors}}}{v_{\text{max}}} = \frac{0.1 \text{ m}}{1025 \text{ m/sec}} = 97 \text{ μsec}$$

The time, $T_{\text{min dispersion calculation}}$, to calculate the fuze correction is the fastest shell's flight time between the last sensor and the antenna. $T_{\text{min dispersion calculation}}$ for the fastest shells is 97 μsec.

$$T_{\text{min dispersion calculation}} \leq \frac{d_{\text{between last sensor and antenna}}}{v_{\text{max}}} = \frac{0.1 \text{ m}}{1025 \text{ m/sec}} = 97 \text{ μsec}$$

Calculating the necessary resolution in the timing to maintain accuracy requires an understanding of the dispersion. Assume that dispersion is evenly spaced over or through the target. Therefore, dispersion should go from 0.5 m in front of the target to 0.5 m behind the target. The minimum time resolution for dispersion occurs for the fastest shells.

$$\Delta T \leq \frac{d_{\text{dispersion}}/2}{v_{\text{max}}} = \frac{1 \text{ m}/2}{1025 \text{ m/sec}} = 490 \text{ μsec}$$

A final concern is the resolution of the measured velocity of each shell. If each shell nominally travels 1 m in 1 msec, then the accuracy of calculation should be down to a centimeter or better. Assuming the need for millimeter resolution, the timing values have to resolve to one microsecond or less.

16.4.4 Architecture

This gun fuse system is a fairly simple control and instrumentation system. Figure 16.3 illustrates a block diagram of the system. It has several simple sensors, a central processing unit, an output transmitter, and a network connection with the weapons system.

Good system engineering would develop the embedded system to calculate the nominal distance to the target and to fuse the shell in the barrel of the gun before firing. The system should transmit the dispersion correction as an increment to this nominal value when the shell leaves the muzzle. Should the fusing correction fail, then the shell's fuse will still be set to ignite near the target. This will allow the gunner to zero in on the target even if the correction system fails.

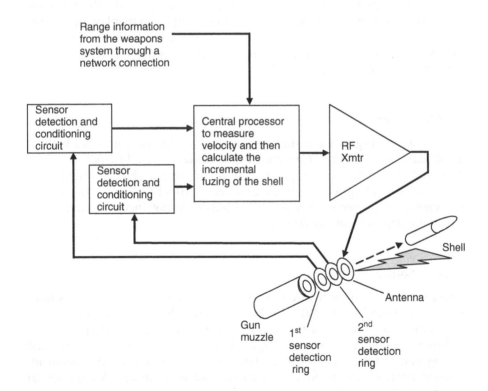

FIGURE 16.3
Block diagram of the control system for measuring a shell's muzzle velocity and correcting its fuze to eliminate dispersion. (© 2006 by Kim Fowler, used with permission. All rights reserved.)

Fortunately, you have flexibility in transmitting the value of the fuze correction into the shell. Remember that for the fastest shells the estimated transmit time is 97 μsec. Realistically you probably have more than twice this time because the free space propagation of the electromagnetic energy will transmit to the shell even after it has passed beyond the antenna. This gives you good margin in the time to set the fuze correction. On the other hand, you do not want an overly powerful transmitter, which could cause radio frequency (RF) interference in other equipment.

16.4.5   Hardware

The central processor will probably be a large microcontroller, like those found in automobile ECMs, or a single-board computer with a large microprocessor. The real-time constraints and deadlines are similar. The processor will need hardware to capture and compare sensor values that measure the shell's velocity and timers with 1 μsec resolution to time the shell velocity.

The processor will also need a network connection with the weapons system of the platform. At the very least this would be a display unit for the weapons officer. Most likely, it will have a connection to the range-finding system that will supply range, azimuth, wind velocity, and temperatures, all parameters necessary to set the fuze.

16.4.6   Power

If this is a mobile gun, the fusing system will have DC power from a battery, most likely at 24 or 42 V. Consequently, it needs to be a low-power device to minimize power consumption. If it is onboard a ship, it might have 120 VAC, 400 Hz power or it might receive 270 VDC power. Regardless, the choice of power supply and distribution is not inconsequential and needs appropriate consideration.

16.4.7   Software

You might write the software in C or in Ada 2005. Many developers have experience with C. Ada 2005, on the other hand, has nice features that can speed development, but it tends to need a bigger processor [1,2].

This is a military project; consequently the development processes must be rigorous: code reviews, tests, field tests, configuration management, version control, and archives to support potential audits. A commercial RTOS would be most appropriate for this type of system.

16.4.8 Packaging

A system for a gun must follow military standards for temperature, shock, vibration, condensation, and salt-spray. This generally means that it must be quite rugged and sealed.

Another consideration is the gun recoil, which will make cabling to the sensors and transmitter difficult to maintain. The mechanical movement of the barrel requires stress-relief coils to reduce breakage of the cable due to stress factures. The sensors must withstand the shock of the hot gasses expelling shells at the muzzle.

16.4.9 Buy vs. Build

The sensors and transmitter for this system will have to be custom designed and manufactured. The embedded system for controlling the fuze setting should probably be a COTS single-board computer. The mechanical enclosure could be purchased if the number of systems is small, on the order of several thousand or less for the total market life.

16.4.10 Manufacturing

Manufacturing can be performed in several different ways. If a company has shop and assembly resources and manufacturing personnel and the capacity to add this system, then it could be built in-house. Contract manufacturing would be a good option if your company does not have the manufacturing capability or the capacity.

This sort of project does not lend itself to a great deal of automation in manufacturing. Some pick-and-place machines for component assembly and circuit board fabrication might be appropriate. Automated testing would probably not be cost competitive. Having a skilled technician test specific points on finished products is more likely to fit this type of work.

16.4.11 Installation and Maintenance

A government contractor would probably employ a skilled technician to install this system. Diagnostics could include BIT to locate a failure quickly. Repair would be replacement of modules and circuit boards; components would be repaired at a military depot.

16.5 Summary

Three real-time, embedded systems were discussed in this chapter: a parking garage collection of sensors and message boards to indicate available parking slots, a data acquisition system for biological monitoring, and a gun fusing system. Each one has real-time constraints but in very different time frames. The parking system needs to respond within a few seconds. The timing of the data acquisition system ranges from low milliseconds to many hours between samples and responses. The gun-fuzing system must respond in microseconds to meet its deadlines.

Each system has similar areas to consider in design: architecture, hardware, power, software, packaging, manufacturing, installation, and maintenance. Still, each system has different emphases and final solutions for these design concerns:

- The parking garage and data acquisition systems need low power microcontrollers to reduce power consumption and ease power distribution and supply. The gun-fuzing system needs a powerful and fast processor to complete the calculations in time.
- The parking garage system is a custom design. It would benefit from automated manufacturing and test. The gun-fuzing system would be well suited for contract manufacturing. The data acquisition system would be custom built by students or by a contract manufacturer.
- The parking garage and data acquisition systems tend toward using less skilled labor for installation, while the gun-fuzing system needs skilled contractors to install its equipment.
- Repair for the parking garage and data acquisition systems consists of replacing failed units. The gun-fuzing system would also be repaired by module replacement, but it also has access to depot repair to replace specific components, if needed.

References

1. Rogers, P., Programming Real-Time with Ada 2005, *Embedded Systems Design,* September 2006, Vol. 19, No. 9, pp. 23–45.
2. McCormick, J.W., Instrumentation Education Through Model Railroading, *IEEE Instrumentation & Measurement Magazine,* October 2006, Vol. 9, No. 5, pp. 40–45.

17

Some Observations about Consumer Appliances

17.1   Concept and Market

17.1.1   Economics

We use appliances everyday to speed up routine chores. They include electric razors, coffeemakers, toasters, blenders, food mixers, microwave ovens, washers, and dryers. Even cell phones, personal digital assistants (PDAs), and laptop computers might fit into the appliance category. I will focus, however, on the more traditionally understood appliances in my observations for this chapter.

Appliances occupy a unique place in design and development. First, thousands or millions of each model are sold every year. They compete in an extraordinarily cost-sensitive environment, and yet consumers expect them to last 5–10 years. Larger appliances are expected to last even longer.

Profit margins are very thin. Manufacturers look at every aspect of design and production to cut costs—even to shaving pennies from the shipping container. This chapter will only introduce you to some of the ways consumer appliances are produced.

17.1.2   Features Wanted

Consumers purchase appliances emotionally. Perception of quality and purpose and style play a huge role in marketing and selling appliances. Features and their perceived utility are a key component in marketing an appliance.

Intuitive operation counts as one of the most important aspects of features. Now here is a conundrum—a lot of buttons may indicate a lot of features, but quantity does not necessarily lead to intuitive operation. Conversely, many consumers wanted one feature per button on larger appliances (Chapter 4); at least that was true until recently. They really did not want nested levels of menus. Computers, cell phones, and PDAs, however, are changing the marketplace for everyone. Better displays and ways of operating appliances are slowly emerging.

The next chapter has some observations on intuitive operations.

17.2   Product Teardown Summaries

My observations will derive from several sources. I will tear down and comment on several types of appliances. Several books can help with explaining product design. Bert Haskell's book is a particularly good reference for teardown summaries on cell phones, computers, and PDAs [1]. He provides four different perspectives on product design: power profiles, weight breakdowns of components and modules, component real estate, and cost breakdowns.

The cost drivers in most cell phones, laptop computers, and PDAs are the batteries and electronic components and assemblies—particularly displays. The housing and material hardware can be a significant portion of the cost for laptops.

Two differentiating factors between these portable devices and many appliances are circuit and computational densities. Most appliances do not need exotic circuit boards, such as flex circuits, or large processors capable of many computations.

17.3   Coffeemaker Teardown

17.3.1   Description

This coffeemaker grinds coffee beans and then drip-percolates hot water through the grounds to produce fresh coffee (Figure 17.1). It has several buttons to control the grinding of the beans and to produce the coffee. These buttons also program the coffeemaker for automatic operation. A small liquid crystal display (LCD) provides the time to help you set the automatic operation.

Figure 17.2 shows the sequence of operations to brew coffee. The grinder shatters the beans and then whips the grounds into a shoot that directs them down into the filter. Water is drawn from the reservoir, heated, and dripped from the top of the filter basket onto the grounds. The water percolates through the grounds to the bottom of the filter basket and into the glass carafe.

The coffeemaker represents a fairly prototypical embedded system because it has sensors, buttons, knobs, a display, and actuators. It can provide an interesting case study because it is a food-handling appliance that must adhere to commercial standards; it includes a user interface, a motor, and a heating element; it has manufacturing issues, which must contain cost.

FIGURE 17.1
A coffeemaker that grinds coffee beans to brew a fresh cup of coffee. (Photograph © 2006 by Kim Fowler, used with permission. All rights reserved.)

17.3.2 Architecture and Features

Appliances, like this coffeemaker, will typically last 5–10 years of occasional use. It is not a commercial unit; it will not withstand heavy daily use in a restaurant or cafe.

It has a centralized configuration where the sensors, buttons, display, and actuators all connect to the processor. There is no distributed processing. Features and functions combine for economy of production; the buttons, for instance, both direct immediate operations and program for later, automatic operation.

The coffeemaker has several features and interlocks to make it safer and more reliable. The design goes to some lengths to prevent wrong operation. It also avoids water damage in a number of ways.

17.3.3 Hardware

There are two circuit boards: one is the control panel near the top of the appliance behind the display; the other is a power regulator board buried deep in the base of the coffeemaker (Figures 17.3 and 17.4). The processor receives direct input from the buttons and sensors, drives the LCD display, and controls the motor and heating element. The power regulator board

FIGURE 17.2
Sequence of operations (a through g) to prepare the coffeemaker and brew the coffee. (a) Filter basket open with filter in place. (b) Adding coffee beans to grinder. (c) Covering beans in grinder. (d) Putting cover over filter basket. (e) Closing filter door. (f) Adding water to reservoir. (g) Turning on coffeemaker. (Photograph © 2006 by Kim Fowler, used with permission. All rights reserved.)

FIGURE 17.2
Continued

conditions and converts the AC power to DC and has two solid-state relays for controlling the motor and heating coil.

The processor in the control panel is either a microcontroller or an application-specific integrated circuit (ASIC); the lack of markings on the chip

(a)

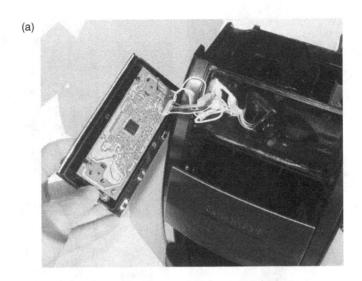

(b)

FIGURE 17.3
Central processing board with its (a) microcontroller, and (b) buttons, switches, peripheral components, and display. Note that some wires and the buttons and knobs are soldered directly to the circuit board; other wires attach to the board with a connector. (Photograph © 2006 by Kim Fowler, used with permission. All rights reserved.)

protects its identity. I would guess a microcontroller because it would be cost competitive for the market, and it would also allow modifications between models. Table 17.1 outlines one possible scenario for how an appliance manufacturer might determine the type of processing element that would be the most appropriate.

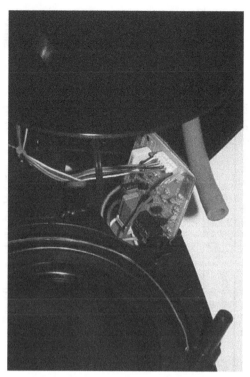

FIGURE 17.4
Power regulator board at the base of the coffeemaker. (Photograph © 2006 by Kim Fowler, used with permission. All rights reserved.)

- Purchasing an 8-bit microcontroller to be the processor in a coffeemaker can cost between US$0.50 and US$1 per chip in large quantities. The non-recurring engineering (NRE) is the cost of designing the circuitry, writing the software, and testing the design.

- Another possibility would be to build a fully-custom ASIC controller for the coffeemaker. The NRE to design a chip can easily be as high as US$1,000,000. Fabrication can approach US$3 per chip or more. This totals to US$4 per chip in final delivered form. Besides the expense, fully-custom ASICs cannot be easily revised, upgraded, or fixed.

- The circuit board in the control panel for this particular coffeemaker has the following components: four light-emitting diodes (LEDs), a connector, one LCD display, an annunciator to signal people that the coffee is brewed, four push-button switches, two rotary switches, and 36 discrete components (Figure 17.3).

- The circuit board in the power regulator has the following components: four transistors, two solid-state relays, an inductor, a connector, and 27 discrete components (Figure 17.4).

TABLE 17.1

One of Many Possible Scenarios for Cost (NRE + COGS) of the Electronics and Circuit Boards within a Coffeemaker.

	Microcontroller-Based				ASIC-Based			
	Qty.	Metric	Unit Cost (in $)	Extension (in $)	Qty.	Metric	Unit Cost (in $)	Extension (in $)
NRE								
Project management	2000	Hours	80.00	160,000	2000	Hours	80.00	160,000
Preparing specifications	800	Hours	60.00	48,000	800	Hours	60.00	48,000
Simulations and feasiblity	400	Hours	60.00	24,000	1200	Hours	60.00	72,000
Chip design								1,000,000
Software code (2 LOC/h)	2000	LOC	30.00	60,000	400	LOC	30.00	12,000
Hardware design	4000	Hours	60.00	240,000	2000	Hours	60.00	120,000
Development and test	4000	Hours	60.00	240,000	4000	Hours	60.00	240,000
Documentation	800	Hours	60.00	48,000	400	Hours	60.00	24,000
Field tests	4000	Hours	60.00	240,000	4000	Hours	60.00	240,000
UL certification	800	Hours	60.00	48,000	800	Hours	60.00	48,000
Manufacturing plans	4000	Hours	60.00	240,000	4000	Hours	60.00	240,000
Manufacturing test equipment	2000	Hours	60.00	120,000	2000	Hours	60.00	120,000
Customer meetings, product launch	4000	Hours	60.00	240,000	4000	Hours	60.00	240,000
Overhead—travel, meetings, etc.				200,000				200,000
			NRE subtotal =	$1,908,000			NRE subtotal =	$2,764,000
				NRE/unit				**NRE/unit**
	Quantity produced =	50,000		$38.16				$55.28
	Quantity produced =	100,000		$19.08				$27.64
	Quantity produced =	200,000		$9.54				$13.82

Cost to produce electronic circuit boards in 200,000 coffeemakers

Cost (NRE + COGS)
($40/h Burdened Labor Rate)

Processor NRE cost	1	NRE/unit	9.54		1	NRE/unit	13.82
Processor unit cost	1	Part	0.50		1	part	3.00
Peripheral component costs	1	Set	7.84		1	Set	7.84
Control processor PCB fabrication	1	Board	3.00		1	Board	3.00
Control processor PCB assembly	1	Board	10.00		1	Board	10.00
Power regulator component costs	1	Set	6.50		1	Set	6.50
Power regulator PCB fabrication	1	Board	1.00		1	Board	1.00
Power regulator PCB assembly	1	Board	4.00		1	Board	4.00
Final assembly and test	2	Board	1.33		2	Board	2.67
			2.67				1.33

Cost per unit = $45.05   Cost per unit = $51.83

Total cost for 200,000 sets of electronic boards   Total = $9,009,333   Total = $10,365,333

Cost (NRE + COGS)
($4/h Burdened Labor Rate)

Processor NRE cost	1	NRE/unit	9.54		1	NRE/unit	13.82
Processor unit cost	1	Part	0.50		1	Part	3.00
Peripheral component costs	1	Set	7.84		1	Set	7.84
Control processor PCB fabrication	1	Board	3.00		1	Board	3.00
Control processor PCB assembly	1	Board	1.00		1	Board	1.00
Power regulator component costs	1	Set	6.50		1	Set	6.50
Power regulator PCB fabrication	1	Board	1.00		1	Board	1.00
Power regulator PCB assembly	1	Board	0.40		1	Board	0.40
Final assembly and test	2	Board	0.13		2	Board	0.27
			0.27				0.13

Cost per unit = $30.05   Cost per unit = $36.83

Total cost for 200,000 sets of electronic boards   Total = $6,009,333   Total = $7,365,333

(Continued)

TABLE 17.1
Continued

	Microcontroller-Based					ASIC-Based			
	Qty.	Metric	Unit Cost (in $)	Extension (in $)		Qty.	Metric	Unit Cost (in $)	Extension (in $)
Cost (NRE+COGS) (Automated Assembly)									
Processor NRE cost	1	NRE/unit	9.54	9.54		1	NRE/unit	13.82	13.82
Processor unit cost	1	Part	0.50	0.50		1	Part	3.00	3.00
Peripheral component costs	1	Set	7.84	7.84		1	Set	7.84	7.84
Control processor PCB fabrication	1	Board	3.00	3.00		1	Board	3.00	3.00
Control processor PCB assembly	1	Board	0.20	0.20		1	Board	0.20	0.20
Power regulator component costs	1	Set	6.50	6.50		1	Set	6.50	6.50
Power regulator PCB fabrication	1	Board	1.00	1.00		1	Board	1.00	1.00
Power regulator PCB assembly	1	Board	0.20	0.20		1	Board	0.20	0.20
Final assembly and test	2	Board	0.20	0.40		2	Board	0.20	0.40
			Cost per unit =	$29.18				Cost per unit =	$35.96
Total cost for 200,000 sets of electronic boards			Total =	$5,836,000				Total =	$7,192,000

This Scenario does not Include the Cost of the Molding, Plastics, Sheet Metal, or Final Assembly (© 2006 by Kim Fowler, used with permission. All rights reserved.)

Table 17.1 is one scenario that compares these two approaches. Assuming a production run of 200,000 coffeemakers, the cost (NRE + cost of goods sold [COGS]) is between US$30 and US$40 per unit for the electronic boards alone. Assuming low labor rates outside of the United States or automated assembly, the highest expense is designing and programming and testing the central processor. Please note that you have many possible scenarios for how you might design the control for this coffeemaker.

17.3.4 Power

A separate circuit board resides in the base of the coffeemaker (Figure 17.4). It appears to be a power regulator with two solid-state relays and discrete transistors and components. The 120 VAC cord extends upwards from the base to the grinder motor in the upper structure of the appliance. It probably is a universal-type brush motor operating directly on 120 VAC. The processor controls a relay to turn the grinder motor on and off. The processor controls the other relay to turn on and off the heating coil. Figure 17.5 shows the heater coil and tubes to deliver the water for heating and dispensing.

17.3.5 Packaging

The design of the coffeemaker has simple features that help insure safe and robust operation. There are interlocks to prevent harmful operation.

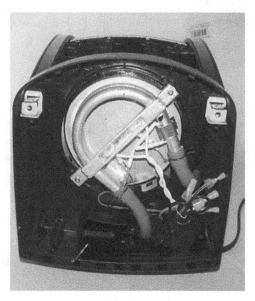

FIGURE 17.5
Power cord, heater coil, and tubes to deliver the water for heating and dispensing within the coffeemaker. (Photograph © 2006 by Kim Fowler, used with permission. All rights reserved.)

The design reduces potential damage to the circuitry from water or moisture.

Thermostat interlock: The heating coil has a thermostat that opens should temperature rise above a set point. This is important when the water empties out of the coil. It prevents runaway heating by the resistance coil that would melt down the coffeemaker and possibly cause a fire.

Grinder interlock: The grinder has a combination mechanical/electrical interlock to protect the user. It is a cover that functions both to shield fingers from the whirling blades and to contain the grounds so that beans do not spew everywhere. When the cover is in place on the grinder and the lid is closed over it, then a switch is closed to allow the grinder's motor to run. If either the cover is missing or the lid is not closed, then the switch remains open and the motor will not run. The grinder cover has a tube with a plastic plunger inside. The clear plastic tube on the grinder cover fits over a black plastic hollow post sticking up from the coffeemaker; that post has an opposing, spring-loaded plunger in it to activate the motor switch interlock (Figures 17.6 and 17.7).

Filter basket interlock: The filter basket has a mechanical interlock that closes if the coffee carafe is removed from the appliance. This is a spring-loaded valve that shuts and prevents coffee from dripping on to the hot

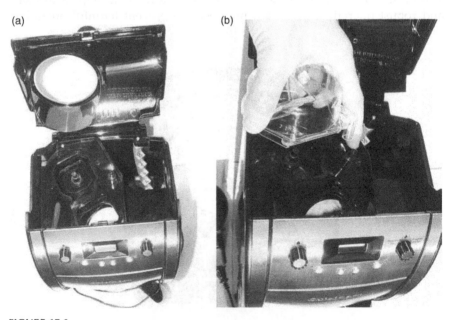

(a)   (b)

FIGURE 17.6
(a) Mechanical/electrical interlock on the coffee bean grinder; (b) the tip of my index finger touches the post with the switch contact for the interlock. (Photographs © 2006 by Kim Fowler, used with permission. All rights reserved.)

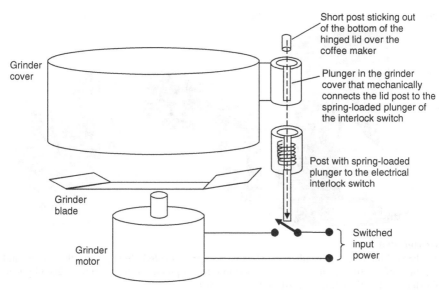

FIGURE 17.7
Action of the mechanical/electrical interlock on the coffee bean grinder. Only when the top lid closes and all three mechanical posts and plungers line up and compress together can the interlock switch closed to allow the grinder motor to turn on. (Drawing © 2006 by Kim Fowler, used with permission. All rights reserved.)

base of the appliance. The coffee carafe has a cover with a knob on it that pushes the valve plunger up; this action allows coffee to drip around the knob and through the cover into the carafe (Figure 17.8).

Avoiding water damage: The overflow hole near the top of the coffeemaker prevents overfilling the water reservoir (Figure 17.9). If that overflow hole were not there, spilled water could damage the grinder motor or the power regulator board.

Another interesting feature is the drip shield above the entry hole for the power cord, which is at the base of the coffeemaker. Should water spill out of the overflow hole, then the drip shield will reduce the chance that the water will enter the hole in the base with the power cord (Figure 17.9).

Avoiding condensation: Conformal coatings on circuit boards are polymer-based substances, such as paralene or polyurethane, that seal boards and components from moisture. The coffeemaker generates condensates that can infiltrate circuit boards and components and can degrade and destroy them. Not immediately obvious on the circuit board for the control panel is a broad swipe of conformal coating over the processor chip and two-thirds of the circuit board. It looks like the assembler took a 3-inch-wide brush and slapped conformal coating across the board. A conformal coating covers the entire power regulator board and all of its components in the base of the unit.

(a)   (b)

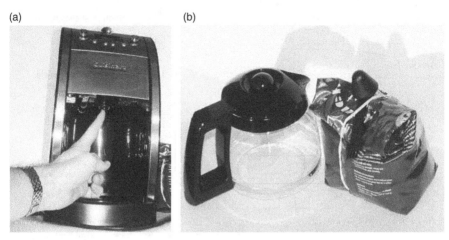

FIGURE 17.8
Mechanical interlock on the filter drip: (a) this is a spring-loaded plunger valve that closes until a coffee carafe (b) with its cover is pushed into the appliance. (Photographs © 2006 by Kim Fowler, used with permission. All rights reserved.)

FIGURE 17.9
Overflow hole for the water reservoir high up, just below the top cover cap, and the drip shield over the entry for the power cord. (Photograph © 2006 by Kim Fowler, used with permission. All rights reserved.)

The circuitry for the front panel resides within a plastic enclosure. Clear silicone caulking seals the edges of the plastic box enclosure. White silicone caulking surrounds the wires that run from the control panel to the power control circuit in the base of the coffeemaker. This seals the front panel electronics in a plastic box.

17.3.6 Manufacturing

The coffeemaker is a mass-produced appliance. It has been designed for assembly (DFA) to help reduce costs. One way to DFA is to solder the switches directly to the front of the circuit board while the processor is soldered to the back. This arrangement allows for automated assembly and reduces wiring. The front panel secures the push buttons and knobs, which align with the switches on the circuit board. The circuit board then snaps into place and is held by two screws.

The circuit board for the control panel has two layers while the circuit board for the power regulator board is single layer; these are simple and easy-to-fabricate circuit boards. All these design features contain cost. An insulating sheet covers the back side of the power regulator board to prevent chafing of the wires and premature failure—this is cheaper to do than staking the wires into another, more complex route.

The wires and cables attach to the enclosure in only two places with plastic cinch ties. The control wires have a six-row header connector on each end. Both of these features help reduce the cost of assembly.

The case snaps together and then is held in place by screws. While there are lots of screws (eight in the base alone, six attaching the front panel, and two holding each circuit board), they make for a stiff and sturdy construction for the coffeemaker.

17.3.7 Maintenance and Logistics

This appliance, like most appliances, is not designed for repair. The company has inventory to replace components like the carafe, should it break, or the grinder cover, should someone lose it. This is the extent of logistics for most consumer appliances.

Major components for larger appliances, such as motors, heating elements, and control boards, are stocked by the manufacturer in centralized warehouses and occasionally by specialized repair shops, but the components are not generally or widely available. I doubt that any of these parts are available for repairing this coffeemaker.

17.4   Remote Control Teardown

17.4.1   Market and Description

I hate these things. I hate the bad design most of them foist on consumers. More buttons do not mean better or easier operations. How many of you use all the buttons on a remote control? (See the next chapter to see the ridiculous extent to which remotes have grown.)

Remote controls control a wide variety of devices: televisions, VCRs, DVD players, CD players, stereo radios, and so forth. Amazingly, there are companies that do nothing but design and build remote controls. The market is very specialized and has unique design challenges.

17.4.2   Architecture and Features

Remote controls have a wide variety of buttons. Some new models have touch-sensitive visual displays. I will focus on a middle-of-the-road model that only has buttons (Figure 17.10).

This particular remote control can control televisions, DVD players, VCRs, and satellite or cable receivers. It accommodates nearly 100 brand names and hundreds of channels. It stores preferences for up to three users.

FIGURE 17.10
The remote control for a television. (Photograph © 2006 by Kim Fowler, used with permission. All rights reserved.)

17.4.3   Hardware

Many remote controls use a single ASIC to control the device. This particular device is no exception—it has the ASIC controller chip. The only peripheral devices are a transmitter driver chip and an LED for sending infrared energy (Figure 17.11).

The membrane buttons contact the single-layer circuit board. The ASIC controller reads these contacts directly.

17.4.4   Power

Two AA-sized batteries supply the power. Any regulation is done within the components. Reduced parts count helps hold the line on costs.

17.4.5   Manufacturing

Everything about this remote control is designed for cost containment. The circuit board clamps between the two shells of the enclosure, which snap together (Figure 17.12). The circuit board is a single layer. The battery contacts are spring steel soldered directly to the circuit board.

Table 17.2 gives one scenario for the cost—it is one among many possible scenarios. The cost per unit of a particular remote control is somewhere between US$5 and US$9. The most expensive component would be either the plastic enclosure or the circuit board. Labor could be the single greatest cost if not carefully managed.

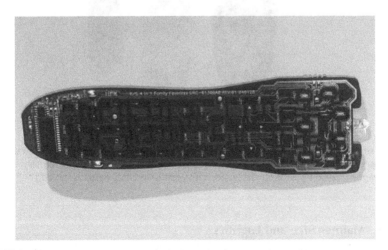

FIGURE 17.11
The circuit board for the remote control, it only has two integrated circuits—an ASIC controller and a driver for the infrared LED. The switch contacts for the buttons are etched directly onto the circuit board. (Photograph © 2006 by Kim Fowler, used with permission. All rights reserved.)

(a)

(b)

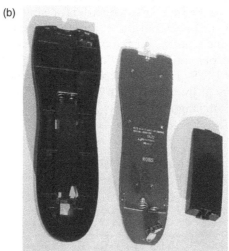

FIGURE 17.12
The remote control is designed for inexpensive manufacturing; it snaps together and the circuit board is single layer; it also functions as part of the battery holder. (a) Front of the circuit board with buttons, ASIC, and LED. (b) Back of the circuit board with clips for battery contacts. (Photographs © 2006 by Kim Fowler, used with permission. All rights reserved.)

17.4.6   Maintenance and Logistics

These devices are thrown away if they fail. They are not repaired and they have absolutely no diagnostics.

TABLE 17.2

One of Many Possible Scenarios for the Cost (NRE + COGS) of a Remote Control

NRE	Qty.	Metric	Unit Cost (in $)	Extension (in $)
Project management	100	Hours	80.00	8,000
Preparing specifications	50	Hours	60.00	3,000
Simulations and feasiblity	50	Hours	60.00	3,000
Hardware design	400	Hours	60.00	24,000
Development and test	400	Hours	60.00	24,000
Documentation	100	Hours	60.00	6,000
Field tests	200	Hours	60.00	12,000
Manufacturing plans	100	Hours	60.00	6,000
Manufacturing test equipment	50	Hours	60.00	3,000
Customer meetings, product launch	200	Hours	60.00	12,000
Overhead—travel, meetings, etc				5,000
			NRE subtotal =	**$106,000**
				NRE/unit
	Quantity produced =		50000	2.12
	Quantity produced =		100000	1.06
	Quantity produced =		200000	0.53

Cost to produce 200,000 remote controls
($40/h Burdened Labor Rate)

NRE cost	1	NRE/unit	0.53	0.53
ASIC unit cost	1	Part	0.50	0.50
Peripheral component costs	1	Set	0.70	0.70
Enclosure injection molding	1	Set	1.50	1.50
Button membrane molding	1	Set	0.40	0.40
springclips for batteries	1	Set	0.20	0.20
PCB fabrication	1	Board	1.50	1.50
PCB assembly	1	Board	2.67	2.67
Final assembly and test	2	Shells	0.67	1.33

Cost (NRE+COGS) per unit = $9.33

Total cost for 200,000 remote controls $ 1,866,000

($4/h Burdened Labor Rate)

NRE cost	1	NRE/unit	0.53	0.53
ASIC unit cost	1	Part	0.50	0.50
Peripheral component costs	1	Set	0.70	0.70
Enclosure injection molding	1	Set	1.50	1.50
Button membrane molding	1	Set	0.40	0.40
springclips for batteries	1	Set	0.20	0.20
PCB fabrication	1	Board	1.50	1.50

(Continued)

TABLE 17.2

Continued

NRE	Qty.	Metric	Unit Cost (in $)	Extension (in $)
PCB assembly	1	Board	0.27	0.27
Final assembly and test	2	shells	0.07	0.13
			Cost (NRE+COGS) per unit = $5.73	
Total cost for 200,000 remote controls				$1,146,000
Automated assembly				
NRE cost	1	NRE/unit	0.53	0.53
ASIC unit cost	1	Part	0.50	0.50
Peripheral component costs	1	Set	0.70	0.70
Enclosure injection molding	1	Set	1.50	1.50
Button membrane molding	1	Set	0.40	0.40
springclips for batteries	1	Set	0.20	0.20
PCB fabrication	1	Board	1.50	1.50
PCB assembly	1	Board	0.05	0.05
Final assembly and test	2	Shells	0.05	0.10
			Cost (NRE+COGS) per unit = $5.48	
Total cost for 200,000 remote controls				$1,096,000

17.5   Hobbies, Arts, and Crafts

17.5.1   Economics

Companies that produce supplies and electronics for hobbies, arts, and crafts range from part-time, single-person "garage" shops to large corporations. They can sell anywhere from tens to hundreds to thousands of units every year. Two distinguishing factors for hobbies, arts, and crafts are cost and longevity; everything is cost-sensitive and must last between 10 and 30 years.

Often there are "magical" cost points: a mechanical component such as a geared motor in a housing that mechanizes something usually needs to be less than US$25. Electronic control, such as individual model train controls usually need to be less than US$50 or US$70 per unit. More intricate model kits need to be less than US$80 or US$100 per unit. A thorough knowledge of the particular hobby or craft is necessary to understand acceptable price points.

17.5.2   Liabilities

One good thing about the hobby market is that liability is so low that for most products it is essentially nonexistent. If your product does not require line power but runs from batteries or low-voltage DC, then the shock hazard is eliminated. Most hobby equipment does not have a mechanical hazard either because it is small, light, and low power.

17.5.3   Considerations for Hobby, Arts, and Crafts

By their very nature, hobbies, arts, and crafts are "do-it-yourself" activities. Reliability and robustness usually are not a concern for the consumer—they are going to work around problems and repair or fix things themselves. The funny thing is this—if the customer is not an electronics hobbyist, then any sophisticated circuitry must have considerable thought put into the interface and into helping them to understand its operation. Requiring a customer to program in assembly or even some high-level language usually is bad form.

You may want to take some cues from large appliances when dealing in hobbies, arts, and crafts. Avoid complex keystrokes to program functions, unless you have a really well-written users' manual to walk customers through every step. Use the "one button, one function" approach if you can. Otherwise, simple pull-down menus on a display might work.

17.5.4   Hardware

If you need a processor, a simple 8-bit microcontroller will satisfy most applications. Unless you are building millions of units, an ASIC controller just does not make economic sense. Reduce the number of peripheral components to reduce cost. Use single-layer or two-layer circuit boards for cost containment as long as the frequency of operation stays below 20 MHz and is localized to the processor.

Try to reduce the wires and cables. They are always a weak point in any design. Flexing will eventually break the wires. Connectors add cost and can be another source of failure, too. One way to reduce the wiring is to put any buttons or displays right on the circuit board, just as the coffeemaker design did. Otherwise, consider ways to stake down the wiring to stop flexing and chafing.

17.5.5   Power

Batteries or low-voltage DC power many hobbies, arts, and crafts. You can avoid both shock hazards and Underwriters Laboratory (UL) certification costs by using low-voltage, low-current power sources. This also removes the need to filter the input power, though you may need to regulate it to a standard value such as 5 V or 3.3 V.

17.5.6   Manufacturing

Manufacturing should be simple using screws, glue, and snap-fits. You may want to seriously consider putting a conformal coat on the circuit boards and their edges to seal out moisture and ensure years of trouble-free operation.

17.5.7 Test, Maintenance, and Logistics

Most hobbies, arts, and crafts have no diagnostics and no repair (it is part of the fun of a hobby—fixing things). Logistics may include inventory of major components, such as motors or gearboxes, should they wear out.

17.6 Common Appliance Problems

Appliances can fail for any number of reasons. That considered, there are still some typical problems that occur regularly. The next two sections have some things to consider and avoid when designing appliances.

17.6.1 Frequent Problems

Table 17.3 summarizes six prevalent problems found in appliances. You can find this information at http://www.repairfaq.org [2].

TABLE 17.3

Six Primary Type of Failures Often Found in Appliances [2]

Failure	Result(s)	Cause(s)
1 Broken wiring inside cordset	Appliance stops working	Flexing, pulling, or long-term abuse
2 Bad internal connections— broken wires, corroded or loosened terminals	Either appliance stops working or its operation becomes intermittent or erratic	Vibration, corrosion, poor manufacturing, or thermal fatigue
3 Short circuits	Either appliance stops working, a shock hazard, or sparking and fire	Shoddy manufacturing with sharp edges—slice through wires due to vibrations or thermal cycles
4 Worn, dirty, or broken switches or thermostat contacts	Erratic or no action	Simple mechanical wear-out or abuse
5 Gummed-up lubrication, or worn or dry bearings	Sluggish or noisy operation or overheating, possibly blown fuse or burned out motor	Environmental conditions (dust, dirt, humidity) or poor quality control during manufacture (they forgot the oil)
6 Broken or worn drive belts or gears or parts	Appliance stops working	Normal wear and tear, improper use, accidents, or shoddy manufacturing

Broken wiring and bad internal connections are almost always from abuse or poor manufacturing quality. I repaired a television set once for a friend; the power plug needed replacing because she pulled the plug by its cord from the wall outlet after watching television every day. No cord is designed for that kind of abuse.

Short circuits often are the result of poor engineering that does not design for manufacture (DFM) or design for assembly (DFA). As the reference says, "... final assembly ... must sometimes be done blind—the wires get stuffed in and covers fastened—which may end up nicking or pinching wires between sharp metal parts. The appliance passes the final inspection and tests but fails down the road" [2].

17.6.2 Appliance Recalls

Appliance recalls can be instructive. They can give you a flavor of the type of failures that can happen. They also can give you an idea of the magnitude of problem a recall can present a company—millions of dollars or euros or your favorite currency lost—in fixing the problem.

Here are some examples of product recalls found during a recent internet search [3]:

- 45,000 heaters for defective thermostats that were improperly positioned, which could lead to the units overheating
- 3.1 million dishwashers for a slide switch (the lever that selects between heat drying and energy saving) that can melt and ignite over time, posing a fire hazard
- 5,500 toy flashlights because the batteries may overheat or leak and children could suffer burns from the leaking battery acids
- Upright vacuum cleaners because the power cord may break inside the handle, posing electrical shock and burn injury hazards

Table 17.4 gives more examples of recalls from the U.S. Consumer Product Safety Commission [4].

17.7 Summary

Cost is the primary concern for designing consumer appliances. DFM and DFA are important considerations for reducing the cost of manufacturing—examples include simple fasteners that install easily, even better if modules snap together, better yet if assembly and test can be automated.

TABLE 17.4

Examples of Recalls from the U.S. Consumer Product Safety Commission Web Site (http://www.cpsc.gov/cgi-bin/prod.aspx) [4]

Type of Product	Approx. Qty	Hazard	Incidents/Injuries
8-Cup coffee brewer	73,000	The coffee brewer has defective electrical wiring that can result in overheating, smoking, burning and melting, posing a possible fire hazard	23 reports of melting in the plastic housing of the brewers. No injuries have been reported
Coffeemaker	420,000	The coffeemaker may not turn off as programmed, causing the unit to overheat and melt, and posing a risk of fire and burn injury	14 reports of the coffeemakers overheating. This resulted in one report of a minor burn, and 12 reports of minor property damage to kitchen cabinets, countertops and floors
Rechargeable, lithium ion batteries	340,000	These lithium ion batteries can overheat, posing a fire hazard to consumers (3,080,000 sold worldwide)	16 reports of notebook computer batteries overheating, causing minor property damage and two minor burns. All of these reported incidents and injuries have been associated with earlier recalls of notebook computer batteries ...
Flat panel monitor	15,000	A ground clip inside the back plastic panel of these monitors can be incorrectly installed, posing a risk of electrical shock to consumers	1 report of a consumer receiving an electrical shock from one of these monitors. The consumer was not injured
Built-in dishwasher	74,300	These dishwashers have a connector that can short-circuit and overheat during normal use, posing a fire hazard to consumers	29 reports of connectors overheating, including one report of a fire that spread outside the dishwasher and caused minor property damage. No injuries have been reported
Espresso machine	6,600	The electrical connectors in the espresso machine can erode, posing a fire hazard	None
Treadmill	700	The treadmill can unexpectedly accelerate or decelerate, possibly causing the user to lose control and fall	9 reports of speed control problems. No injuries have been reported

(Continued)

TABLE 17.4

Continued

Type of Product	Approx. Qty	Hazard	Incidents/Injuries
Cordless electric lawnmower	160,000	An electrical component in the lawnmowers can overheat, posing a fire hazard	10 additional reports of electrical components overheating, including one additional report of a fire extending beyond the mower. Note: The original recall involved 11 reports of electrical components overheating. One of these resulted in a minor hand burn, and nine resulted in reports of minor property damage extending beyond the mower
DLP projector	21	If the lamp drive circuit touches the shield case, the unit is not grounded, and a person were to contact metal terminals at the back of the unit, there is a potential for electric shock	2 reports of the lamp-drive circuit contacting the shield case. No injuries have been reported
Radio-controlled airplane	7,500	The rechargeable battery pack inside the toy airplane can overheat posing a burn hazard	15 reports of the toy airplane's rechargeable battery pack overheating, including two reports of minor skin burns
Upright freezers and refrigerators	112,000	The defrost heater coil can become exposed inside the units, which poses a potential shock hazard to consumers. In some cases the exposed heater wire can also melt, or burn the unit's interior plastic food liner	45 reports of incidents of the defrost heater coil becoming exposed. Nine of those incidents resulted in an electrical short. The others melted and burned the unit's interior plastic liner. No injuries have been reported
Three-door refrigerator	20,000	A faulty component in the condenser fan motor can short circuit. This could cause the condenser fan motor to overheat, posing a potential fire hazard to consumers	82 reports of incidents involving a condenser fan motor failure due to a failed capacitor arcing and smoking. There has been smoke damage in a few incidents. There have been no injuries

(Continued)

TABLE 17.4

Continued

Type of Product	Approx. Qty	Hazard	Incidents/Injuries
Electric scooter	74,811	Improper wiring can cause a short circuit, posing a fire hazard in the scooter. In addition, inadequate insulation may expose electrical wiring, which poses a shock hazard	2 reports of the scooters catching fire. There have been 13 reports of scooters starting and/or moving on their own. One person reported receiving scratches as a result. There have also been five reports of property damage, including two reports of the scooters causing house fires
Sewing and embroidery machine	55,000	Electrical arcing can occur in the machine's power supply, posing a risk of fire	3 incidents of these machines overheating and catching fire. One incident resulted in extensive smoke damage to a consumer's home, and the other two incidents resulted in minor property damage
Upright vacuum cleaner	636,000	The recalled vacuums have defective on-off switches that can overheat the handle and tool holder areas of the vacuum, resulting in a fire hazard	249 reports of vacuums overheating, which caused the handle area to smoke, melt or catch fire. One minor burn injury requiring no medical attention was reported

References

1. Haskell, B., *Portable Electronics Product Design and Development*, McGraw-Hill, 2004, pp. 222, 257–264, 296–301.
2. http://www.repairfaq.org
3. http://www.matthewslawfirm.com, 2004. While available in 2004, it appears no longer available on the web.
4. U.S. Government's Consumer Product Safety Commission's website: http://www.cpsc.gov/cgi-bin/prod.aspx

18

Some Observations about User Interfaces

18.1 Why Are User Interfaces so Important?

The user interface, or UI, represents the purpose, function, and utility of a device or piece of equipment. The UI is the most powerful aspect of a device in setting the perception of a user who operates the device or equipment. Perception is one of those intangible, subjective things we must learn to accommodate.

Some basic awareness of UI principles can vastly improve the utility and perception of any product that you design. Even a simple two-button operation with one or two light-emitting diodes (LEDs) provides a UI that significantly affects users, customers, and owners. This chapter has some very simple basic principles and examples that will help you design better products.

18.2 Basic Principles for User Interfaces

18.2.1 The Goal

The simple goal of a UI is to establish a good relationship between the device and the user. We have often heard, "Make it intuitive"—but what does that mean? Intuitive for whom? How?

What defines a good relationship between the device and the user? Some attributes of a good relationship follow. They may not all be intuitive, but they can be the basis for an intuitive interface and a good relationship.

Visibility means that the function is clear to a user. It helps the user understand the system by showing the user the state of the system and the available options.

Control means that users can do what they want to do. They have direct control over a device and can do useful things. (Beware—automating sequences of operations can sometimes take control away from users.)

Appropriateness means that the cognitive mapping (or mental image) correctly defines the relationship between the controls and their results for the user. People using the device "expect" a certain action when they perform a specific operation (e.g., pushing a button). They get a measured response when they initiate an operation. An appropriate or measured response is analogous to the correct gain in an amplifier—you get a proportional response [1].

Feedback is an important aspect of interface design that ties together visibility, control, and appropriateness. If you push a button, a noticeable action occurs—maybe an acoustic click or a tactile snap or a visual display lights up. Feedback informs the user about the action performed and the result accomplished. It helps set the level of expectation of the user [1].

Perception is the most difficult and often the most important. It is subjective and has many dimensions to it. The interface is the major factor in forming perception. "From the user's point of view, the interface is the system . . . " [2]. Perception is reality!

Audience is the last component. Education and culture play into how the user relates with the interface. You need to understand who uses the device, when they use it, and how they use it.

18.2.2   UI Guidelines

There are five components of user interfaces: cognition, ergonomics, utility, image, and ownership. While identified as distinct terms, they do overlap and you cannot really isolate one from the others [3].

Cognition involves the mental tasks and computations for operating a device and relates to the capabilities and expectations of the user. The cognitive aspects of design include learning, memory, organization, and consistency.

Ergonomics is the physical form and fit found in human factors research. Examples include:

- Provide comfortable seating and arrangements
- Avoid awkward positions
- Avoid overextended limb movement
- Minimize actions that produce fatigue
- Supply appropriate lighting, not too dim, not too bright
- Avoid annoying or distracting noises

Utility is the metric for the usefulness of the device. It defines how much the device accomplishes against how much it theoretically could accomplish.

Image describes the users' perceptions of both the device and its operation. It involves more than styling; cognitive aspects, ergonomics, and utility all combine to form the image.

Ownership is the level of commitment that an user exercises to use your product. Ownership can be a powerful determinant in ongoing use of your device and the eventual purchase of newer models.

Organize the UI to fit the tasks. First, develop a good mental model of the device and its operations for the user. A good mental model provides visibility, which relates to cognition, utility, and ergonomics. Provide clear mappings from expectations of the user into operational results. A good mapping can be intuitive because it will fit the user's expectations [4].

Making the UI predictable and consistent will improve the mental model [1]. Strive for actions that cause the least astonishment. Operations that verify important actions are part of being predictable and reliable [5]. Economy of expression can help a more experienced user to be more efficient, but remember to be appropriate and not terse [1].

Making operations and functions easy to learn and remember is an important part of utility and cognition. Forcing users to endure a long period of learning reduces the effectiveness of the UI and ultimately the device; it can stress the relationship between users and the device [5]. Reference [6] is a handbook of concepts used in UI.

Finally, realize that people do what they know—not necessarily what is easy or even correct. We are creatures of habit, which helps reduce our mental workloads. We maintain mental models of what we want to accomplish and trust them to get results. We may well continue to do what we think works, even if it does not. We will not change unless the device makes clear to us a better or more correct method.

18.3   Vending Machine faux pas

Vending machines dispense any variety of commodities. Most of us have had experience with retrieving a soft drink from a vending machine. Some machines have simple buttons with corresponding labels indicating the particular drinks; you do not actually see the drinks stored; the machine delivers the drink container down a chute to you.

Some vending machines have glass fronts so that you can view the items inside. They have a variety of delivery mechanisms. One such mechanism is a wire "corkscrew" that turns and pushes items forward until they fall down into a tray that forms a delivery chute. Another mechanism is a robotic tray that moves up to the selected item and then down to deliver it to a chute.

Regardless of the mechanism, the slot for each set of items is usually uniquely identified by a letter to represent the row and a number to represent the column. A good UI for a vending machine with a glass front clearly maps the rows and columns to a keypad for selecting the desired slot. The UI should also group the bill and coin slot together and place the display above the keypad. Figure 18.1 shows one such machine that clearly violates these principles. The row letters and column numbers did not map spatially to the soft drink slots; this is confusing to users. The bill slot and coin slot are separated.

Make UIs clear. Reduce confusion; take the path of least astonishment! Figure 18.2 illustrates a better arrangement for the keypad, along with coin slot and display.

FIGURE 18.1
This is an actual vending machine that I spotted in an international airport. It breaks a number of rules for a good UI: a nonintuitive mapping for the keypad to displayed items, the bill slot over the display, and the coin slot separated from the bill slot. (Photograph © 2005 by Kim Fowler, used with permission. All rights reserved.)

18.4 Appliance Display faux pas

Figure 18.3 shows the display on a new refrigerator. It is a control panel to set the temperatures in the freezer and the refrigerator. It also displays the buttons that dispense water and ice.

It is unnecessarily complex. Too much information is presented on one screen to be useful. It should have hidden some of the information or put it in pages or menus behind the main screen. For instance, the buttons do not map intuitively to the display, it is not obvious how to set the temperature (and I never did figure how to do it), and the center panel of "Silent, Fuzzy, and Deodo" is essentially meaningless.

FIGURE 18.2
This is another vending machine with a better UI: a standardized keypad and the coin slot is under the display. (Photograph © 2007 by Kim Fowler, used with permission. All rights reserved.)

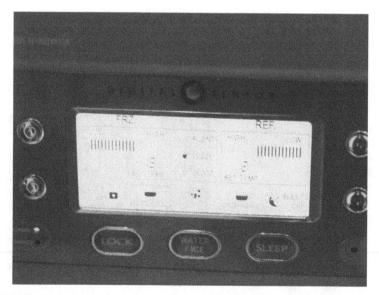

FIGURE 18.3
This is a display on a refrigerator. It is far too complex at a glance. There is a significant learning curve in using it. (Photograph © 2006 by Kim Fowler, used with permission. All rights reserved.)

18.5   Remote Control faux pas

Did I mention how much I hate remote controls? I hate them, in general. They tend to trade flexibility and quantity of functions with utility [7]. Do not most people just use the up/down arrows to channel surf and the +/− buttons to change volume? Yet remote controls continue to sprout more buttons year after year, as shown in Figure 18.4. Figure 18.5 shows one simple remote that most people can easily use—it does not have great flexibility with many functions, but it is easy to use.

Buttons on remote controls should be grouped for functionality and uniquely identified by shape and placement. The number pad for selecting an arbitrary channel should be a recognizable configuration, such as a telephone keypad. Figure 18.6 shows an old remote control from 1989 that controlled a television with a built-in VCR and a much later remote control for comparison. The old remote control had 64 identical "chiclet" style buttons. The numbered keypad had no kind of standard configuration.

New generations of remote controls contain liquid crystal display (LCD) screens with menus of selections. I can only hope that more thought has been put into these remote controls. The right thing would be to group similar functions on one page, such as controls for televisions or DVD players. Navigation between pages should be simple and easy.

FIGURE 18.4
Five different remote controls found in one beach vacation house. Six years after the photograph Figure 18.6 was taken and still they sport terrible UI—many buttons, confusing arrangements. (Photograph © 2005 by Kim Fowler, used with permission. All rights reserved.)

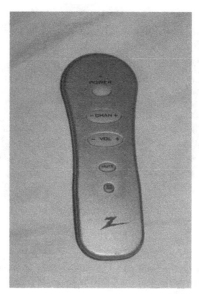

FIGURE 18.5
Where remote controls could be more useful and less confusing. This was found in a motel. (Photograph © 2006 by Kim Fowler, used with permission. All rights reserved.)

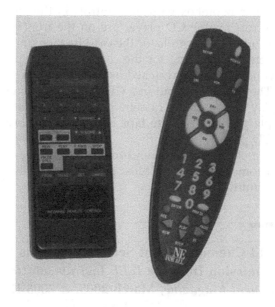

FIGURE 18.6
An old remote control (on the left) from 1989 and the one on the right from 1999. Designs began moving towards more usable and thoughtful configurations—standard keypad numbering, distinctly shaped buttons, and better fit to the hand. (Photograph © 2001 by Kim Fowler, used with permission. All rights reserved.)

FIGURE 18.7
Boombox with the controls in the wrong places—the cognitive mapping is counter intuitive. The controls for the CD are on the front panel just above the cassette tape unit and the controls for the cassette tape unit are on top near the CD. (Photograph © 2000 by Kim Fowler, used with permission. All rights reserved.)

18.6   Boombox faux pas

Figure 18.7 shows a boombox with the controls in the wrong places. The display and controls for the CD player are on the front panel just above the cassette tape unit. The controls for the cassette tape unit are on top near the CD. The cognitive mapping for the controls and display are counter-intuitive. Everyone that I have watched use this boombox for the first time gets it wrong. They invariably put in a CD in the top platen, close the lid, and push the play button on the cassette player next to the lid. Likewise, I have seen people put in a cassette tape and push the forward arrow (>) button for the CD play.

Clearly this is a cheap boombox and manufacturing ease was more important than customer satisfaction. Most people eventually get used to the inappropriate cognitive mapping.

18.7   Handheld Chemical Agent Sensors (Portions Reprinted, with Permission from the *IEEE Instrumentation & Measurement Magazine*, March 2005. © 2005 IEEE) [8]

18.7.1   Testing

I had the opportunity to test some equipment used to sniff out weapons of mass destruction. The equipment comprised handheld devices that were very sophisticated chemical sensors. It was fascinating to be seated in a laboratory containing high-flow fume hoods with nerve gases streaming through them and monitor the devices under test as they detected the agents.

Our goal was to evaluate the effectiveness of sensors in the presence of common household chemicals. We wanted to see if the sensors would detect these common chemicals and generate a false alarm. We also wanted to see if those same chemicals would mask the detection of truly dangerous agents and prevent a correct, positive alarm.

Each of the tested sensors used one of two different types of transducers to detect the various types of chemicals. One sensor used both types of transducers. They generally tended to be handheld devices with several buttons and a small LCD display. Further discussion of the exact technology is not necessary for this case study.

What really intrigued me was the software and graphical user displays for connecting the sensors to a computer and testing them in the laboratory. Each sensor approached the display very differently and none of them did it well. None of the software and displays adhered to good, basic principles for user interfaces; they did not seem to have a lot of thought behind them.

18.7.2 Basic Formats for Display

Each of the four devices was from a different manufacturer. Each approached the display of data differently. The common thread was that each used a histogram or graph of molecular ion peaks to display its current state of detection. The software in each device would update the histogram or trace the display every 5–15 sec. Three of the four also displayed a form of "waterfall" charts with a series of traces, that is, the graph of molecular ion peaks, overlaying each other. Every chemical compound has its own characteristic peaks in the display. Figure 18.8 gives an example of a waterfall chart. Furthermore, the devices would display similar histograms for positive (+) ion species and negative (−) ion species.

The software for each device would then display various other parameters, as well. In particular, each had its specific format for alarms if it detected a harmful or dangerous chemical. Some were more readily obvious than others.

Company A: The device from Company A had its screen split into four areas for displaying the traces and listing the chemicals. The largest areas displayed the traces. A horizontal bar divided the screen in half, with the trace of the current ion species at the top and the waterfall display of traces on the bottom. A vertical bar separated the left-hand column of chemical listings from the traces that dominated the majority of the display. It labeled the prominent peaks of the traces with the acronyms for the detected chemicals.

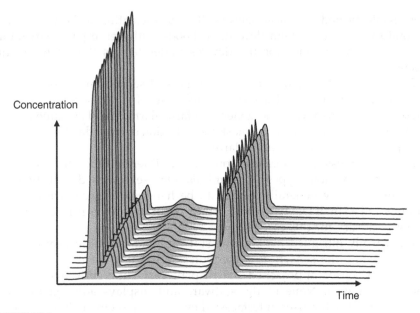

FIGURE 18.8
An example of what a "waterfall" chart of histograms might look like. (Reprinted with permission of the *IEEE Instrumentation & Measurement Magazine*, March 2005.)

When the device from Company A detected a harmful or dangerous chemical, it displayed an alarm by highlighting the name of the agent in red at the top of the column listing of chemical names. It also highlighted the acronym for the agent, which produced the alarm, above the prevalent peaks on the trace.

Here is the bad part, the software flashed the display screen between the positive ion trace, a gray screen, the negative ion trace, and another gray screen in a rapid, unending sequence. For each screen display of the ion traces it provided a dwell time of 3 sec on the screen and then 4 sec in the gray screen. This meant that the software:

- Simultaneously displayed the positive ion trace and its waterfall chart for 3 sec.
- Then displayed a blank, gray screen for 4 sec.
- Then simultaneously displayed the negative ion trace and its waterfall chart for 3 sec.
- And then displayed another blank, gray screen for 4 sec.

This made recognizing and assimilating the data nearly impossible. The software did manage to store the data on disk for later review, but immediate observation was extremely difficult and we regularly missed alarms.

Company B: The device from Company B had a fairly decent screen for displaying parameters. It had two simple histograms, one was a three-bar histogram for one type of transducer and the other was a 16-bar histogram for the other type of transducer. These remained on continuously; the histogram bars updated every 7 sec or so.

The device from Company B also had LED style alarms for four different general classes of toxic agents. Finally, it had some individual spaces for displaying numerical values of minor parameters.

The LED alarms were always present on the screen. The difference in intensity between an "off" LED and one that was "on" was not as great as I would have liked. I would have preferred it to flash on when detection of a toxic agent occurred. We occasionally would miss an alarm because of the minimal distinction between off and on.

The 16-bar histogram showed the bars moving upward from a midpoint to indicate positive ions. It showed the bars moving downward from the same midpoint to indicate negative ions. I liked this and found it fairly intuitive.

The histogram bars were each in a different color. I found this somewhat distracting. It gave more weight to certain bars, such as the red or green ones, over other colors, though this was not intentional or significant to the detection of chemical agents.

Company C: The device from Company C had several sets of traces, a current ion trace and a waterfall chart. We had to press a button to toggle between positive and negative ions. It only had several minor readouts for alarm intensities but no real indication of alarm (it did store on the data on the disk for later review).

I found the lack of immediate alarm indication somewhat irritating. It made it difficult to time the onset of detection.

Company D: The device from Company D had several sets of traces, a current ion trace and a waterfall chart. We had to select a small button item in the top menu bar to toggle between positive and negative ions. The detection alarm resided in a separate, small side-window and we had to scroll through the window to find the location of the alarm. It was somewhat cryptic too. It indicated a detection of a toxic agent by changing from a small or zero value to "exceed a threshold value." It had no separate indication of an alarm.

18.7.3   Basic Formats for Control

Each of the four devices approached the control differently. They used a variety of methods: pull-down menus, buttons, textboxes, and file logging.

Company A: The device from Company A had simple control. Basically, all we had to do was name a file and set some basic parameters in a small control panel on the screen and then press a start button. We stopped it by pulling down a menu selection to stop the logging process.

Company B: The device from Company B had a control panel to set up the log file name and several buttons on the display screen. We started and stopped the logging of data by pressing the appropriate button that was clearly marked on the screen. We could indicate that agent gas was on or off by pressing the appropriate button and the software would immediately log it in the recorded text file.

The layout of the buttons was OK and usable. I found that keystrokes on comments in the control panel window would stick and cause a variable delay of 1–1.5 sec. This was a function of the software; we ran the other devices on the same machine and did not experience similar behavior.

Company C: The device from Company C had sticky software, as well. When I typed in notes, comments, or file names, or edited comments, the keystrokes seemed to stick and delay. This device also had a start button to begin logging operation, but you had to remember to stop it and type in a file name to record the log on disk. Otherwise, data would be lost at the end of a trial. This was a minor annoyance.

Company D: The device from Company D had several quirks in the operation of its software. First, the cursor moved in a nonintuitive way to scroll through the window to find the alarm indication. You had to place the cursor in the window and move it up to drag the window upwards. It did not use a standard columnar scroll bar to move down the page (i.e., see the page move upwards). Second, it had these small, very hard-to-read pictograph buttons in the menu bar at the top of the screen to control the device. The schematics or diagrams on the pictograph buttons were odd, too, and you had to learn and remember their meaning.

Similar to the device from Company C, this device had a start button in the pictograph buttons in the menu bar to begin logging operation. Again, you had to remember to stop it by selecting and pressing a button and then type in a file name to record the log on disk. Otherwise, data would be lost at the end of a trial. As with Company C's device, this was an annoyance.

18.7.4 Considerations for Improvements

Each of these devices could have benefited by more consideration for the human interface. After spending several months running them, I have specific thoughts about how I would have designed the interface differently.

First, the setup and control should be clear and straightforward. It probably should have several separate screens that can be accessed intuitively—pull-down menus and a logical progression that pops up the necessary screens in order.

The main screen (Figure 18.9) would have a "start log" button that would pop up a screen (Figure 18.10) to select a file name and location if it has not

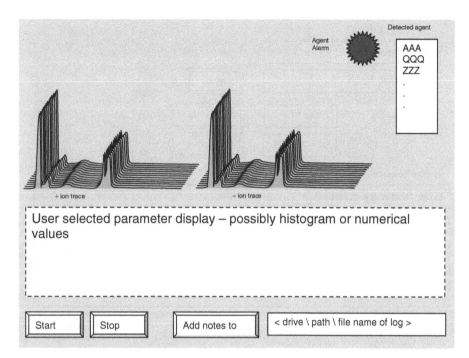

FIGURE 18.9
An example of what a main screen for the software GUI might look like in a sensor that detects toxic and deadly chemicals. (Reprinted with permission of the *IEEE Instrumentation & Measurement Magazine*, March 2005.)

been already set. That same pop up screen would have a text box for the inclusion of comments.

The main screen would have a "stop log" button and a display of the current file name for the log. It would also have a "add notes to log" button that would call up the pop up screen with its text box so that comments could be inserted at any time during the trial run.

The main screen would have the alarm indication, a big, flashing LED, near the top in an obvious location. Right next to the flashing LED would be a textbox to display the detected agent names.

The main screen would display both the positive and negative ion traces in a waterfall chart. It would also have a place for a histogram (for additional transducers) or a numerical display of selected parameters.

I would also consider having a separate control that allows users to select what specific parameters that they might want to see on the main screen. This allows users to focus on only two or three parameters, which is intuitive and typical of our human nature. This would mean that the device would do what they expect and want; it would add a measure of ownership to the interface because users could customize it to their applications.

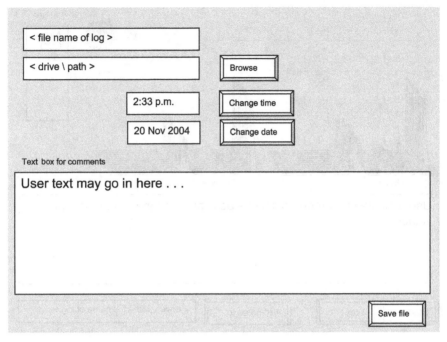

FIGURE 18.10
An example of what a log file screen might look like. (Reprinted with permission of the *IEEE Instrumentation & Measurement Magazine*, March 2005.)

18.7.5 Final Thoughts on UI for Chemical Sensors

Before I finish a design of the human interface, I would want to do two very important activities:

- Interview specific customers for needs and expectations at the beginning of the design.
- Hold focus groups, maybe Web meetings, to assess the functionality of the proposed interface.

My comments here do not reflect on the sensitivity, functionality, utility, or capability of the four sensors that my team and I tested. The various implementations gave me a platform for urging the engineering community for more considered effort in designing software.

18.8 Summary

The user interface, or UI, is important to any device design. It needs to be clear, obvious, and usable. I would recommend reading the following References 4, 5, and 6 to understand the basic principles of UI.

References

1. Marcus, A. and Van Dam, A., User-interface developments for the nineties. *Computer*, Vol. 24, No. 9, 1991, pp. 49–57.
2. Norman, D.A., Cognitive Engineering. In *User Centered System Design: New Perspectives on Human–Computer Interaction*, Erlbaum, Hillsdale, NJ, 1986, Chapter 3, p. 61.
3. Fowler, K., *Electronic Instrument Design, Architecting for the Life Cycle*, Oxford University Press, 1996, pp. 104–120.
4. Norman, D.A., *The Design of Everyday Things*, Doubleday, New York, 1988.
5. Constantine, L.L. and Lockwood, L.A.D., *Software for Use: A Practical Guide to the Models and Methods of Usage-Centered Design*, Addison–Wesley, Boston, MA, 1999, pp. 7, 23.
6. Lidwell, W., Holden, K., and Butler, J., *Universal Principles of Design*, Rockport Publishers, Gloucester, MA, 2003.
7. Lidwell, W., Holden, K., and Butler, J., *Universal Principles of Design*, Rockport Publishers, Gloucester, MA, 2003, pp. 86–87.
8. Fowler, K., Tried and True column, "Human Interface: A Case Study," *IEEE Instrumentation & Measurement Magazine*, Vol. 8, No. 1, March 2005, pp. 46–49.

Index